Applied Functional Analysis
and Its Applications

Applied Functional Analysis and Its Applications

Editors

Jen-Chih Yao
Shahram Rezapour

MDPI • Basel • Beijing • Wuhan • Barcelona • Belgrade • Manchester • Tokyo • Cluj • Tianjin

Editors
Jen-Chih Yao
China Medical University Hospital
China
China Medical University
Taiwan

Shahram Rezapour
Azerbaijan Shahid Madani
University Iran
China Medical University
Taiwan

Editorial Office
MDPI
St. Alban-Anlage 66
4052 Basel, Switzerland

This is a reprint of articles from the Special Issue published online in the open access journal *Mathematics* (ISSN 2227-7390) (available at: https://www.mdpi.com/journal/mathematics/special_issues/Applied_Functional_Analysis_Its_Applications).

For citation purposes, cite each article independently as indicated on the article page online and as indicated below:

LastName, A.A.; LastName, B.B.; LastName, C.C. Article Title. *Journal Name* **Year**, *Article Number*, Page Range.

ISBN 978-3-03936-776-4 (Hbk)
ISBN 978-3-03936-777-1 (PDF)

© 2020 by the authors. Articles in this book are Open Access and distributed under the Creative Commons Attribution (CC BY) license, which allows users to download, copy and build upon published articles, as long as the author and publisher are properly credited, which ensures maximum dissemination and a wider impact of our publications.

The book as a whole is distributed by MDPI under the terms and conditions of the Creative Commons license CC BY-NC-ND.

Contents

About the Editors vii

Preface to "Applied Functional Analysis and Its Applications" ix

Hsien-Chung Wu
Informal Norm in Hyperspace and Its Topological Structure
Reprinted from: *Mathematics* **2019**, *7*, 945, doi:10.3390/math7100945 1

Bing Tan, Shanshan Xu and Songxiao Li
Modified Inertial Hybrid and Shrinking Projection Algorithms for Solving Fixed Point Problems
Reprinted from: *Mathematics* **2020**, *8*, 236, doi:10.3390/math8020236 21

Jong Soo Jung
Convergence Theorems for Modified Implicit Iterative Methods with Perturbation for Pseudocontractive Mappings
Reprinted from: *Mathematics* **2020**, *8*, 72, doi:10.3390/math8010072 33

Jin Liang and Yunyi Mu
Properties for ψ-Fractional Integrals Involving a General Function ψ and Applications
Reprinted from: *Mathematics* **2019**, *7*, 517, doi:10.3390/math7060517 49

Badr Alqahtani, Hassen Aydi, Erdal Karapınar and Vladimir Rakočević
A Solution for Volterra Fractional Integral Equations by Hybrid Contractions
Reprinted from: *Mathematics* **2019**, *7*, 694, doi:10.3390/math7080694 63

Yinglin Luo, Bing Tan and Meijuan Shang
A General Inertial Viscosity Type Method for Nonexpansive Mappings and Its Applications in Signal Processing
Reprinted from: *Mathematics* **2020**, *8*, 288, doi:10.3390/math8020288 73

Lu-Chuan Ceng, Adrian Petruşel , Ching-Feng Wen and Jen-Chih Yao
Inertial-Like Subgradient Extragradient Methods for Variational Inequalities and Fixed Points of Asymptotically Nonexpansive and Strictly Pseudocontractive Mappings
Reprinted from: *Mathematics* **2019**, *7*, 860, doi:10.3390/math7090860 91

Lu-Chuan Ceng, Adrian Petruşel and Jen-Chih Yao
On Mann Viscosity Subgradient Extragradient Algorithms for Fixed Point Problems of Finitely Many Strict Pseudocontractions and Variational Inequalities
Reprinted from: *Mathematics* **2019**, *7*, 925, doi:10.3390/math7100925 111

Syed Shakaib Irfan, Mijanur Rahaman, Iqbal Ahmad, Rais Ahmadand Saddam Husain
Generalized Nonsmooth Exponential-Type Vector Variational-Like Inequalities and Nonsmooth Vector Optimization Problems in Asplund Spaces
Reprinted from: *Mathematics* **2019**, *7*, 345, doi:10.3390/math7040345 125

Liu He, Qi-Lin Wang, Ching-Feng Wen, Xiao-Yan Zhang and Xiao-Bing Li
A Kind of New Higher-Order Mond-Weir Type Duality for Set-Valued Optimization Problems
Reprinted from: *Mathematics* **2019**, *7*, 372, doi:10.3390/math7040372 137

Elisabeth Köbis, Markus A. Köbis and Xiaolong Qin
An Inequality Approach to Approximate Solutions of Set Optimization Problems in
Real Linear Spaces
Reprinted from: *Mathematics* **2020**, *8*, 143, doi:10.3390/math8010143 . . . . **155**

About the Editors

Jen-Chih Yao (Professor) is Chair Professor of the Center for General Education at China University (Taiwan) and Director of the Research Center for Interneural Computing, China Medical University Hospital, China Medical University. He has published numerous papers in different fields (e.g., variational inequalities, complementarity problems, fixed point theorems, variational analysis, optimization, vector optimization problems, etc.). He is the editor-in-chief and a member of the editorial board of several journals. He was also the chairman of the organizing and scientific committees of several international conferences and workshops on nonlinear analysis and optimization. His name has been included on lists of highly cited researchers in 2014, 2015, 2016, 2017, 2018, and 2019.

Shahram Rezapour (Professor) is a professor of mathematics at Azarbaijan Shahid Madani University (Iran). He simultaneously holds a visiting professor position at China Medical University (Taiwan). He has published numerous papers on different fields (e.g., approximation theory, fixed point theory, fractional integro-differential equations and inclusions, finite differences, numerical and approximate solutions of singular fractional differential equations, stability, and optimization). He is a member of the editorial boards of several journals. He was also the chairman of the organizing and scientific committees of several international conferences. His name has been included on lists of highly cited researchers in 2016 and 2017.

Preface to "Applied Functional Analysis and Its Applications"

It is well known that applied functional analysis is very important in most applied research fields and its influence has grown in recent decades. Many novel works have used techniques, ideas, notions, and methods of applied functional analysis. Furthermore, applied functional analysis includes linear and nonlinear problems.

The scope of this field is so wide that it cannot be expressed in a few books. This book covers a limited section of this field, namely, fixed point theory and applications, nonlinear methods and variational inequalities, and set-valued optimization problems.

The most important application of fixed point theory is proving the existence of solutions for fractional integro-differential equations and, therefore, increasing our ability to model different kinds of phenomena. In most everyday matters, we seek to use optimization. The importance of optimization has attracted many researchers to this field over the past few decades and provided new ideas, concepts, and techniques.

Jen-Chih Yao , Shahram Rezapour
Editors

Article
Informal Norm in Hyperspace and Its Topological Structure

Hsien-Chung Wu

Department of Mathematics, National Kaohsiung Normal University, Kaohsiung 802, Taiwan; hcwu@nknucc.nknu.edu.tw

Received: 4 September 2019; Accepted: 8 October 2019; Published: 11 October 2019

Abstract: The hyperspace consists of all subsets of a vector space. Owing to a lack of additive inverse elements, the hyperspace cannot form a vector space. In this paper, we shall consider a so-called informal norm to the hyperspace in which the axioms regarding the informal norm are almost the same as the axioms of the conventional norm. Under this consideration, we shall propose two different concepts of open balls. Based on the open balls, we shall also propose the different types of open sets. In this case, the topologies generated by these different concepts of open sets are investigated.

Keywords: hyperspace; informal open sets; informal norms; null set; open balls

1. Introduction

The topic of set-valued analysis (or multivalued analysis) has been studied for an extensive period. A detailed discussion can refer to Aubin and Frankowska [1], and Hu and Papageorgiou [2,3]. Applications in nonlinear analysis can refer to Agarwal and O'Regan [4], Burachik and Iusem [5], and Tarafdar and Chowdhury [6]. More specific applications in differential inclusion can also refer to Aubin and Cellina [7]. On the other hand, the fixed point theory for set-valued mappings can refer to Górniewicz [8], and set-valued optimization can refer to Chen et al. [9], Khan et al. [10] and Hamel et al. [11]. Also, the set optimization that is different from the set-valued optimization can refer to Wu [12] and the references therein.

Let $\mathcal{P}(X)$ be the collection of all subsets of a vector space X. The set-valued analysis usually studies the mathematical structure in $\mathcal{P}(X)$ in which each element in $\mathcal{P}(X)$ is treated as a subset of X. In this paper, we shall treat each element of $\mathcal{P}(X)$ as a "point". In other words, each subset of X is compressed as a point, and the family $\mathcal{P}(X)$ is treated as a universal set. In this case, the original vector space X plays no role in the settings. Therefore, we want to endow a vector structure to $\mathcal{P}(X)$. Although we can define the vector addition and scalar multiplication in $\mathcal{P}(X)$ in the usual way, owing to lacking an additive inverse element, the family $\mathcal{P}(X)$ cannot form a vector space. In this paper, we shall endow a so-called informal norm to $\mathcal{P}(X)$ even though $\mathcal{P}(X)$ is not a vector space. Then, the conventional techniques of functional analysis and topological vector space based on the vector space can be used by referring to the monographs [13–23]. The main purpose of this paper is to study the topological structures of informally normed space $\mathcal{P}(X)$. Based on these topological structures, the potential applications in nonlinear analysis, differential inclusion and set-valued optimization (or set optimization) are possible after suitable formulation.

Given a (conventional) vector space X, we denote by $\mathcal{P}(X)$ the collection of all subsets of X. For any $A, B \in \mathcal{P}(X)$, the set addition is defined by

$$A \oplus B = \{a + b : a \in A \text{ and } b \in B\}.$$

Given a scalar λ in $\mathbb{R}$, the scalar multiplication in $\mathcal{P}(X)$ is defined by

$$\lambda A = \{\lambda a : a \in A\}.$$

The substraction between A and B is denoted and defined by

$$A \ominus B \equiv A \oplus (-B) = \{a - b : a \in A \text{ and } b \in B\}.$$

We denote by θ_X the zero element of X. Let $\theta_{\mathcal{P}(X)} = \{\theta_X\}$ be a singleton set. We see that

$$A \oplus \theta_{\mathcal{P}(X)} = A \oplus \{\theta_X\} = A,$$

which says that $\{\theta_X\}$ is the zero element of $\mathcal{P}(X)$. It is clear to see that $A \ominus A \neq \{\theta_X\}$, which says that $A \ominus A$ cannot be the zero element of $\mathcal{P}(X)$. That is to say, the additive inverse element of A in $\mathcal{P}(X)$ does not exist. Therefore, the hyperspace $\mathcal{P}(X)$ cannot form a vector space under the above set of addition and scalar multiplication. Since $A \ominus A$ is not the zero element, we consider the null set of $\mathcal{P}(X)$ defined by

$$\Omega = \{A \ominus A : A \in \mathcal{P}(X)\}, \tag{1}$$

which may be treated as a kind of "zero element" of $\mathcal{P}(X)$. It is clear to see that the null set is closed under the addition.

In this paper, we shall consider the so-called informal norm in $\mathcal{P}(X)$. The axioms of informal norm will be almost the same as the axioms of conventional norm. The only difference is that the null set will be involved in the axioms of informal norm. In order to study the topological structure of $(\mathcal{P}(X), \|\cdot\|)$, we need to consider the open balls. Let us recall that if $(X, \|\cdot\|)$ is a (conventional) normed hyperspace, then we see that

$$\{y : \|x - y\| < \epsilon\} = \{x + z : \|z\| < \epsilon\}$$

by taking $y = x + z$. However, for the space $(\mathcal{P}(X), \|\cdot\|)$ and $A, B, C \in \mathcal{P}(X)$, the following equality

$$\{B : \|A \ominus B\| < \epsilon\} = \{A \oplus C : \|C\| < \epsilon\}$$

does not hold. The reason is that, by taking $B = A \oplus C$, we can just have

$$\|A \ominus B\| = \|A \ominus (A \oplus C)\| = \|\omega \ominus C\| \neq \|C\|,$$

where $\omega = A \ominus A \in \Omega$. In this case, two types of open balls will be considered in $(\mathcal{P}(X), \|\cdot\|)$. Therefore, many types of open sets will also be considered. Based on the different types of openness, we shall study the topological structure of the normed hyperspace $(\mathcal{P}(X), \|\cdot\|)$.

In Section 2, many interesting properties in $\mathcal{P}(X)$ are presented in order to study the the topology generated by the so-called informal norm. In Section 3, we introduce the concept of informal norms and provide many useful properties for further investigation. In Section 4, we provide the non-intuitive properties for the open balls. In Section 5, we propose many types of informal open sets based on the different types of open balls. Finally, in Section 6, we investigate the topologies generated by these different types of open sets.

2. Hyperspaces

Since the null set Ω defined in (1) can be treated as a kind of "zero element", we propose the almost identical concept for elements in $\mathcal{P}(X)$ as follows.

Definition 1. For any $A, B \in \mathcal{P}(X)$, the elements A and B are said to be almost identical if there exist $\omega_1, \omega_2 \in \Omega$ satisfying $A \oplus \omega_1 = B \oplus \omega_2$. In this case, we write $A \stackrel{\Omega}{=} B$.

For $A \ominus B = C$, we cannot have $A = B \oplus C$. However, we can obtain $A \stackrel{\Omega}{=} B \oplus C$. Let $B \ominus B \equiv \omega \in \Omega$. Since $A \ominus B = C$, by adding B on both sides, we have $A \oplus \omega = B \oplus C$, which says that $A \stackrel{\Omega}{=} B \oplus C$.

Proposition 1. *Given any $A, B \in \mathcal{P}(X)$, we have the following properties.*

(i) *Suppose that $A \ominus B \in \Omega$. Then $A \stackrel{\Omega}{=} B$.*
(ii) *Suppose that $A \stackrel{\Omega}{=} B$. Then there exists $\omega \in \Omega$ satisfying $A \ominus B \oplus \omega \in \Omega$.*

Proof. To prove part (i), we first note that there exists $\omega_1 \in \Omega$ such that
$$A \ominus B = A \oplus (-B) = \omega_1.$$
By adding B on both sides, we obtain $A \oplus (-B) \oplus B = \omega_1 \oplus B$. Therefore, we have $A \oplus \omega_2 = \omega_1 \oplus B$, where $\omega_2 = B \ominus B \in \Omega$.

To prove part (ii), since $A \stackrel{\Omega}{=} B$, there exist $\omega_1, \omega_2 \in \Omega$ such that $A \oplus \omega_2 = \omega_1 \oplus B$. By adding $-B$ on both sides, we obtain $A \ominus B \oplus \omega_2 = \omega_1 \oplus \omega_3 \in \Omega$, where $\omega_3 = B \ominus B \in \Omega$. This completes the proof. □

Proposition 2. *The following statements hold true.*

(i) *Given any subset $\mathcal{A}$ of $\mathcal{P}(X)$, we have $\mathcal{A} \subseteq \mathcal{A} \oplus \Omega$.*
(ii) *We have $\Omega \oplus \Omega = \Omega$. Given any subset $\mathcal{A}$ of $\mathcal{P}(X)$, let $\bar{\mathcal{A}} = \mathcal{A} \oplus \Omega$. Then $\bar{\mathcal{A}} \oplus \Omega = \bar{\mathcal{A}}$.*
(iii) *Given any $\omega = B \ominus B \in \Omega$ for some $B \subseteq X$, we have $\omega = \omega_1 \oplus \omega_2$ for some $\omega_1, \omega_2 \in \Omega$. If $B \neq \{\theta_X\}$ then we can take $\omega_1 \neq \{\theta_X\}$ and $\omega_2 \neq \{\theta_X\}$.*

Proof. To prove part (i), since $\theta_{\mathcal{P}(X)} \equiv \{\theta_X\} \in \Omega$, given any $A \in \mathcal{A}$, we have
$$A = A \oplus \{\theta_X\} = A \oplus \theta_{\mathcal{P}(X)} \in \mathcal{A} \oplus \Omega.$$
To prove part (ii), given any $\omega_1, \omega_2 \in \Omega$, we have $\omega_1 = A \ominus A$ and $\omega_2 = B \ominus B$ for some $A, B \in \mathcal{P}(X)$. Therefore we obtain
$$\omega_1 \oplus \omega_2 = A \ominus A \oplus B \ominus B = (A \oplus B) \ominus (A \oplus B) \in \Omega,$$
which says that $\Omega \oplus \Omega \subseteq \Omega$. Now, for any $\omega \in \Omega$, since $\theta_{\mathcal{P}(X)} \equiv \{\theta_X\} \in \Omega$, we have
$$\omega = \omega \oplus \{\theta_X\} = \omega \oplus \theta_{\mathcal{P}(X)} \in \Omega \oplus \Omega,$$
which says that $\Omega \subseteq \Omega \oplus \Omega$. Therefore we obtain $\Omega \oplus \Omega = \Omega$. On the other hand, we have
$$\bar{\mathcal{A}} \oplus \Omega = \mathcal{A} \oplus \Omega \oplus \Omega = \mathcal{A} \oplus \Omega = \bar{\mathcal{A}}.$$
To prove part (iii), given any $B \subseteq X$, we have $B = B_1 \oplus B_2$ for some subsets B_1 and B_2 of X. For example, we can take $B_1 = \{b\}$ and $B_2 = B \ominus \{b\}$ for some $b \in B$. Therefore we have
$$\omega = B \ominus B = (B_1 \oplus B_2) \ominus (B_1 \oplus B_2) = (B_1 \ominus B_1) \oplus (B_2 \ominus B_2) \equiv \omega_1 \oplus \omega_2.$$
This completes the proof. □

The following interesting results will be used for discussing the topological structure of informal normed hyperspace.

Proposition 3. *Let $\mathcal{A}_1$ and $\mathcal{A}_2$ be subsets of $\mathcal{P}(X)$. Then the following inclusion is satisfied:*

$$(\mathcal{A}_1 \cap \mathcal{A}_2) \oplus \Omega \subseteq [(\mathcal{A}_1 \oplus \Omega) \cap (\mathcal{A}_2 \oplus \Omega)].$$

If we further assume that $\mathcal{A}_1 \oplus \Omega \subseteq \mathcal{A}_1$ and $\mathcal{A}_2 \oplus \Omega \subseteq \mathcal{A}_2$, then the following equality is satisfied:

$$[(\mathcal{A}_1 \oplus \Omega) \cap (\mathcal{A}_2 \oplus \Omega)] = (\mathcal{A}_1 \cap \mathcal{A}_2) \oplus \Omega.$$

Proof. For $B \in (\mathcal{A}_1 \cap \mathcal{A}_2) \oplus \Omega$, we have $B = A \oplus \omega$ with $A \in \mathcal{A}_i$ for $i = 1, 2$ and $\omega \in \Omega$, which also says that $B \in [(\mathcal{A}_1 \oplus \Omega) \cap (\mathcal{A}_2 \oplus \Omega)]$, i.e., $(\mathcal{A}_1 \cap \mathcal{A}_2) \oplus \Omega \subseteq [(\mathcal{A}_1 \oplus \Omega) \cap (\mathcal{A}_2 \oplus \Omega)]$. Under the assumption, using part (i) of Proposition 2, we have

$$[(\mathcal{A}_1 \oplus \Omega) \cap (\mathcal{A}_2 \oplus \Omega)] \subseteq \mathcal{A}_1 \cap \mathcal{A}_2 \subseteq (\mathcal{A}_1 \cap \mathcal{A}_2) \oplus \Omega.$$

This completes the proof. □

3. Informal Norms

Many kinds of informal norms on $\mathcal{P}(X)$ are proposed below.

Definition 2. *Consider the nonnegative real-valued function $\|\cdot\|: \mathcal{P}(X) \to \mathbb{R}_+$ and the following conditions:*

(i) $\|\lambda A\| = |\lambda| \|A\|$ *for any $A \in \mathcal{P}(X)$ and $\lambda \in \mathbb{R}$.*
(i') $\|\lambda A\| = |\lambda| \|A\|$ *for any $A \in \mathcal{P}(X)$ and $\lambda \in \mathbb{R}$ with $\lambda \neq 0$.*
(ii) $\|A \oplus B\| \leq \|A\| + \|B\|$ *for any $A, B \in \mathcal{P}(X)$.*
(iii) $\|A\| = 0$ *implies $A \in \Omega$.*

The informal norm $\|\cdot\|$ is said to satisfy the null condition when condition (iii) is replaced by $\|A\| = 0$ if and only if $A \in \Omega$.

Different kinds of informal normed hyperspaces are defined below.

- *The ordered pair $(\mathcal{P}(X), \|\cdot\|)$ is said to be an informal pseudo-seminormed hyperspace when conditions (i') and (ii) are satisfied.*
- *The ordered pair $(\mathcal{P}(X), \|\cdot\|)$ is said to be an informal seminormed hyperspace when conditions (i) and (ii) are satisfied.*
- *The ordered pair $(\mathcal{P}(X), \|\cdot\|)$ is said to be an informal pseudo-normed hyperspace when conditions (i'), (ii) and (iii) are satisfied.*
- *The ordered pair $(\mathcal{P}(X), \|\cdot\|)$ is said to be an informal normed hyperspace when conditions (i), (ii) and (iii) are satisfied.*

We further consider the following conditions:

- *The informal norm $\|\cdot\|$ is said to satisfy the null super-inequality when $\|A \oplus \omega\| \geq \|A\|$ for any $A \in \mathcal{P}(X)$ and $\omega \in \Omega$.*
- *The informal norm $\|\cdot\|$ is said to satisfy the null sub-inequality when $\|A \oplus \omega\| \leq \|A\|$ for any $A \in \mathcal{P}(X)$ and $\omega \in \Omega$.*
- *The informal norm $\|\cdot\|$ is said to satisfy the null equality when $\|A \oplus \omega\| = \|A\|$ for any $A \in \mathcal{P}(X)$ and $\omega \in \Omega$.*

Example 1. *Let $(X, \|\cdot\|_X)$ be a (conventional) normed space. Given any element $A \in \mathcal{P}(X)$, we define*

$$\|A\| = \sup_{a \in A} \|a\|_X.$$

We are going to claim that $(\mathcal{P}(X), \|\cdot\|)$ is an informal normed hyperspace.

- If $A = \{\theta\}$, then we have $\|A\| = 0$. If $\|A\| = 0$, then also we have $\|a\|_X = 0$ for all $a \in A$, i.e., $A = \{\theta\}$. Therefore, we obtain that $\|A\| = 0$ if and only if $A = \{\theta\} \in \Omega$.
- We have
$$\|\lambda A\| = \sup_{a \in \lambda A} \|a\|_X = \sup_{b \in A} \|\lambda b\|_X = |\lambda| \sup_{b \in A} \|b\|_X = |\lambda| \|A\|.$$
- We want to prove the triangle inequality $\|A \oplus B\| \leq \|A\| + \|B\|$. Let
$$\zeta_1 = \sup_{\{(a,b): a \in A, b \in B\}} \|a\|_X \text{ and } \zeta_2 = \sup_{\{(a,b): a \in A, b \in B\}} \|b\|_X.$$

It is clear to see that $\|a\| + \|b\| \leq \zeta_1 + \zeta_2$ for all $a \in A$ and $b \in B$, which implies
$$\sup_{\{(a,b): a \in A, b \in B\}} (\|a\|_X + \|b\|_X) \leq \zeta_1 + \zeta_2 = \sup_{\{(a,b): a \in A, b \in B\}} \|a\|_X + \sup_{\{(a,b): a \in A, b \in B\}} \|b\|_X.$$

Then, we obtain
$$\|A \oplus B\| = \sup_{c \in A \oplus B} \|c\|_X = \sup_{\{(a,b): a \in A, b \in B\}} \|a+b\|_X$$
$$\leq \sup_{\{(a,b): a \in A, b \in B\}} (\|a\|_X + \|b\|_X)$$
$$\leq \sup_{\{(a,b): a \in A, b \in B\}} \|a\|_X + \sup_{\{(a,b): a \in A, b \in B\}} \|b\|_X$$
$$= \sup_{a \in A} \|a\|_X + \sup_{b \in B} \|b\|_X = \|A\| + \|B\|.$$

Therefore, we conclude that $(\mathcal{P}(X), \|\cdot\|)$ is indeed an informal normed hyperspace. Given any $\omega \in \Omega$, there exists $B \in \mathcal{P}(X)$ satisfying $\omega = B \ominus B$. Therefore, we obtain
$$\|\omega\| = \|B \ominus B\| = \sup_{\{(b_1, b_2): b_1, b_2 \in B\}} \|b_1 - b_2\|_X.$$

Since $\|\omega\|$ is not equal to zero in general, it means that the null condition is not satisfied.

Proposition 4. *Let $(\mathcal{P}(X), \|\cdot\|)$ be an informal pseudo-seminormed hyperspace. Suppose that the informal norm $\|\cdot\|$ satisfies the null super-inequality. For any $A, C, B_1, \cdots, B_m \in \mathcal{P}(X)$, we have*
$$\|A \ominus C\| \leq \|A \ominus B_1\| + \|B_1 \ominus B_2\| + \cdots + \|B_j \ominus B_{j+1}\| + \cdots + \|B_m \ominus C\|.$$

Proof. We have
$$\|A \ominus C\| \leq \|A \oplus (-C) \oplus B_1 \oplus \cdots \oplus B_m \oplus (-B_1) \oplus \cdots \oplus (-B_m)\|$$
$$\text{(using the null super-inequality for } m \text{ times)}$$
$$= \|[A \oplus (-B_1)] \oplus [B_1 \oplus (-B_2)] \oplus \cdots + [B_j \oplus (-B_{j+1})] + \cdots + [B_m \oplus (-C)]\|$$
$$\leq \|A \ominus B_1\| + \|B_1 \ominus B_2\| + \cdots + \|B_j \ominus B_{j+1}\| + \cdots + \|B_m \ominus C\|$$
(using the triangle inequality).

This completes the proof. □

4. Open Balls

If $(X, \|\cdot\|)$ is a (conventional) seminormed space, then we see that

$$\{y : \| x - y \| < \epsilon\} = \{x + z : \| z \| < \epsilon\}$$

by taking $y = x + z$. Let $(\mathcal{P}(X), \|\cdot\|)$ be an informal seminormed hyperspace. Then the following equality

$$\{B : \| A \ominus B \| < \epsilon\} = \{A \oplus C : \| C \| < \epsilon\}$$

does not hold. The reason is that, by taking $B = A \oplus C$, we can just have

$$\| A \ominus B \| = \| A \ominus (A \oplus C) \| = \| -C \oplus \omega \| \neq \| C \|,$$

where $\omega = A \ominus A \in \Omega$. Therefore we can define two types of open ball.

Definition 3. *Let $(\mathcal{P}(X), \|\cdot\|)$ be an informal pseudo-seminormed hyperspace. Two types of open balls with radius ϵ are defined by*

$$\mathcal{B}^\circ(A; \epsilon) = \{A \oplus C : \| C \| < \epsilon\}$$

and

$$\mathcal{B}(A; \epsilon) = \{B : \| A \ominus B \| = \| B \ominus A \| < \epsilon\}.$$

Example 2. *Continued from Example 1, for any $A \in \mathcal{P}(X)$, we define*

$$\| A \| = \sup_{a \in A} \| a \|_X .$$

The open balls $\mathcal{B}(A; \epsilon)$ and $\mathcal{B}^\circ(A; \epsilon)$ with radius ϵ are given by

$$\mathcal{B}(A; \epsilon) = \{B \in \mathcal{P}(X) : \| A \ominus B \| < \epsilon\} = \left\{ B \in \mathcal{P}(X) : \sup_{a \in A \ominus B} \| a \|_X < \epsilon \right\}$$

and

$$\mathcal{B}^\circ(A; \epsilon) = \{A \oplus C \in \mathcal{P}(X) : \| C \| < \epsilon\} = \left\{ A \oplus C \in \mathcal{P}(X) : \sup_{c \in C} \| c \|_X < \epsilon \right\}.$$

Remark 1. *Let $(\mathcal{P}(X), \|\cdot\|)$ be an informal pseudo-seminormed hyperspace. Then we have the following observations.*

- *For any $A \in \mathcal{P}(X)$, the equality $\| A \ominus A \| = 0$ does not necessarily hold true, unless $\|\cdot\|$ satisfies the null condition. In other words, the properties $A \in \mathcal{B}(A; \epsilon)$ can only hold true when $\|\cdot\|$ satisfies the null condition.*
- *Suppose that $\| \theta_{\mathcal{P}(X)} \| = \| \{\theta_X\} \| = 0$. Then $A \in \mathcal{B}^\circ(A; \epsilon)$, since $A = A \oplus \theta_{\mathcal{P}(X)}$.*

Proposition 5. *Let $(\mathcal{P}(X), \|\cdot\|)$ be an informal pseudo-seminormed hyperspace.*
- (i) *For $A \in \mathcal{P}(X)$ with $\omega_A = A \ominus A \in \Omega$, we have $\mathcal{B}(A; \epsilon) \oplus \omega_A \subseteq \mathcal{B}^\circ(A; \epsilon)$.*
- (ii) *If $\|\cdot\|$ satisfies the null sub-inequality, then $\mathcal{B}^\circ(A; \epsilon) \subseteq \mathcal{B}(A; \epsilon)$.*
- (iii) *If $\|\cdot\|$ satisfies the null sub-inequality, for any $A \in \mathcal{P}(X)$ with $\omega_A = A \ominus A \in \Omega$, then $\mathcal{B}(A; \epsilon) \oplus \omega_A \subseteq \mathcal{B}(A; \epsilon)$ and $\mathcal{B}^\circ(A; \epsilon) \oplus \omega_A \subseteq \mathcal{B}^\circ(A; \epsilon)$.*

Proof. To prove part (i), for any $B \in \mathcal{B}(A; \epsilon)$, i.e., $\| B \ominus A \| < \epsilon$, if we take $C = B \ominus A$, then $\| C \| < \epsilon$ and $B \oplus \omega_A = A \oplus C$. This shows the inclusion

$$\mathcal{B}(A; \epsilon) \oplus \omega_A \subseteq \{A \oplus C : \| C \| < \epsilon\} = \mathcal{B}^\circ(A; \epsilon).$$

To prove part (ii), for $C \in \mathcal{P}(X)$ with $\| C \| < \epsilon$, since $\| \cdot \|$ satisfies the null sub-inequality, it follows that
$$\| (A \oplus C) \ominus A \| = \| \omega_A \oplus C \| \leq \| C \| < \epsilon,$$
which says that $A \oplus C \in \mathcal{B}(A; \epsilon)$ and shows the inclusion
$$\mathcal{B}^{\circ}(A; \epsilon) = \{A \oplus C : \| C \| < \epsilon\} \subseteq \mathcal{B}(A; \epsilon).$$

Part (iii) follows from parts (i) and (ii) immediately. This completes the proof. □

Proposition 6. *Let $(\mathcal{P}(X), \| \cdot \|)$ be an informal pseudo-seminormed hyperspace.*

(i) *If $\| \cdot \|$ satisfies the null super-inequality, then $\mathcal{B}(A \oplus \omega; \epsilon) \subseteq \mathcal{B}(A; \epsilon)$ for any $\omega \in \Omega$.*

(ii) *If $\| \cdot \|$ satisfies the null sub-inequality, then we have the following inclusions:*

- $\mathcal{B}(A; \epsilon) \subseteq \mathcal{B}(A \oplus \omega; \epsilon)$ *for any $\omega \in \Omega$.*
- $\mathcal{B}^{\circ}(A \oplus \omega; \epsilon) \subseteq \mathcal{B}^{\circ}(A; \epsilon)$ *for any $\omega \in \Omega$.*

(iii) *If $\| \cdot \|$ satisfies the null equality, then $\mathcal{B}(A \oplus \omega; \epsilon) = \mathcal{B}(A; \epsilon)$ for any $\omega \in \Omega$.*

Proof. To prove part (i), the inclusion $\mathcal{B}(A \oplus \omega; \epsilon) \subseteq \mathcal{B}(A; \epsilon)$ follows from the following expression
$$\epsilon > \| (A \oplus \omega) \ominus B \| = \| (A \ominus B) \oplus \omega \| \geq \| A \ominus B \|,$$
and the inclusion $\mathcal{B}(A \oplus \omega; \epsilon) \subseteq \mathcal{B}(A; \epsilon)$ follows from the following expression
$$\epsilon > \| B \ominus (A \oplus \omega) \| = \| (B \ominus A) \oplus \omega \| \geq \| B \ominus A \|.$$

To prove the first case of part (ii), the inclusion $\mathcal{B}(A; \epsilon) \subseteq \mathcal{B}(A \oplus \omega; \epsilon)$ follows from the following expression
$$\epsilon > \| A \ominus B \| \geq \| (A \ominus B) \oplus \omega \| = \| (A \oplus \omega) \ominus B \|.$$

To prove the second case of part (ii), for $B = A \oplus \omega \oplus C \in \mathcal{B}^{\circ}(A \oplus \omega; \epsilon)$ with $\| C \| < \epsilon$, let $\tilde{C} = \omega \oplus C$. Then, using the null sub-inequality, we have
$$\| \tilde{C} \| = \| \omega \oplus C \| \leq \| C \| < \epsilon, \qquad (2)$$
which says that $B = A \oplus \tilde{C} \in \mathcal{B}^{\circ}(A; \epsilon)$. Therefore we obtain the inclusion $\mathcal{B}^{\circ}(A \oplus \omega; \epsilon) \subseteq \mathcal{B}^{\circ}(A; \epsilon)$. Part (iii) follows from parts (i) and (ii) immediately. This completes the proof. □

In the (conventional) normed hyperspace $(X, \| \cdot \|)$, we have the equality
$$\mathcal{B}(x; \epsilon) \oplus \hat{x} = \mathcal{B}(x \oplus \hat{x}; \epsilon). \qquad (3)$$

However, in the informal normed hyperspace $(\mathcal{P}(X), \| \cdot \|)$, the intuitive observation (3) will not hold true in general. The following proposition presents the exact relationship.

Proposition 7. *Let $(\mathcal{P}(X), \| \cdot \|)$ be an informal pseudo-seminormed hyperspace.*

(i) *We have the equality $\mathcal{B}^{\circ}(A; \epsilon) \oplus \hat{A} = \mathcal{B}^{\circ}(A \oplus \hat{A}; \epsilon)$. In particular, for any $\omega \in \Omega$, we also have $\mathcal{B}^{\circ}(A; \epsilon) \oplus \omega = \mathcal{B}^{\circ}(A \oplus \omega; \epsilon)$.*

(ii) *Suppose that $\| \cdot \|$ satisfies the null sub-inequality. Then we have the inclusion $\mathcal{B}(A; \epsilon) \oplus \hat{A} \subseteq \mathcal{B}(A \oplus \hat{A}; \epsilon)$. We further assume that $\| \cdot \|$ satisfies the null equality. Then, for any $\omega \in \Omega$, we also have the inclusions $\mathcal{B}(A; \epsilon) \oplus \omega \subseteq \mathcal{B}(A; \epsilon)$ and $\mathcal{B}(\omega; \epsilon) \oplus \hat{A} \subseteq \mathcal{B}(\hat{A}; \epsilon)$.*

(iii) *Suppose that $\| \cdot \|$ satisfies the null sub-inequality. For any $A \in \mathcal{P}(X)$ with $\omega_A = A \ominus A$, we have the inclusion $\mathcal{B}(A; \epsilon) \oplus \omega_A \subseteq A \oplus \mathcal{B}(\omega_A; \epsilon)$.*

(iv) For any $\widehat{A} \in \mathcal{P}(X)$ with $\omega_{\widehat{A}} = \widehat{A} \ominus \widehat{A}$, we have the inclusion

$$\mathcal{B}(A \oplus \widehat{A}; \epsilon) \oplus \omega_{\widehat{A}} \subseteq \mathcal{B}(A; \epsilon) \oplus \widehat{A}.$$

Proof. Part (i) follows from the following equality

$$(A \oplus C) \oplus \widehat{A} = (A \oplus \widehat{A}) \oplus C \text{ for } \| C \| < \epsilon.$$

To prove part (ii), for $B \in \mathcal{B}(A; \epsilon) \oplus \widehat{A}$, we have $B = \widehat{B} \oplus \widehat{A}$ with $\| A \ominus \widehat{B} \| < \epsilon$. Then, by the null sub-inequality, we can obtain

$$\| (A \oplus \widehat{A}) \ominus B \| = \| (A \oplus \widehat{A}) \ominus (\widehat{B} \oplus \widehat{A}) \| = \| (A \ominus \widehat{B}) \oplus (\widehat{A} \ominus \widehat{A}) \| \leq \| A \ominus \widehat{B} \| < \epsilon,$$

which says that $B \in \mathcal{B}(A \oplus \widehat{A}; \epsilon)$. Therefore we obtain the inclusion $\mathcal{B}(A; \epsilon) \oplus \widehat{A} \subseteq \mathcal{B}(A \oplus \widehat{A}; \epsilon)$. Now we take $\widehat{A} = \omega$. By part (iii) of Proposition 6, we have

$$\mathcal{B}(A; \epsilon) \oplus \omega \subseteq \mathcal{B}(A \oplus \omega; \epsilon) = \mathcal{B}(A; \epsilon).$$

Similarly, if we take $A = \omega$, then we have

$$\mathcal{B}(\omega; \epsilon) \oplus \widehat{A} \subseteq \mathcal{B}(\omega \oplus \widehat{A}; \epsilon) = \mathcal{B}(\widehat{A}; \epsilon).$$

To prove part (iii), for $\widehat{A} \in \mathcal{B}(A; \epsilon)$, we have $\widehat{A} \oplus \omega_A = A \oplus (\widehat{A} \ominus A)$. The null sub-inequality gives

$$\| \omega_A \ominus (\widehat{A} \ominus A) \| \leq \| \widehat{A} \ominus A \| < \epsilon,$$

which says that $\widehat{A} \ominus A \in \mathcal{B}(\omega; \epsilon)$, i.e.,

$$\widehat{A} \oplus \omega_A = A \oplus (\widehat{A} \ominus A) \in A \oplus \mathcal{B}(\omega_A; \epsilon).$$

To prove part (iv), for $B \in \mathcal{B}(A \oplus \widehat{A}; \epsilon)$, we have $\| B \ominus (A \oplus \widehat{A}) \| < \epsilon$. We also have

$$\epsilon > \| B \ominus (A \oplus \widehat{A}) \| = \| (B \ominus \widehat{A}) \ominus A \|.$$

This shows that $B \ominus \widehat{A} \in \mathcal{B}(A; \epsilon)$. Let $\omega_{\widehat{A}} = \widehat{A} \ominus \widehat{A} \in \Omega$. Since $B \oplus \omega_{\widehat{A}} = (B \ominus \widehat{A}) \oplus \widehat{A}$, it says that $B \oplus \omega_{\widehat{A}} \in \mathcal{B}(A; \epsilon) \oplus \widehat{A}$. In other words, we have the inclusion

$$\mathcal{B}(A \oplus \widehat{A}; \epsilon) \oplus \omega_{\widehat{A}} \subseteq \mathcal{B}(A; \epsilon) \oplus \widehat{A}.$$

This completes the proof. □

Proposition 8. *Let $(\mathcal{P}(X), \| \cdot \|)$ be an informal pseudo-seminormed hyperspace.*

(i) *The following statements hold true:*

- *Suppose that $\| \cdot \|$ satisfies the null super-inequality. For any $\omega \in \Omega$, if $A \oplus \omega \in \mathcal{B}(A_0; \epsilon)$, then $A \in \mathcal{B}(A_0; \epsilon)$.*
- *Suppose that $\| \cdot \|$ satisfies the null sub-inequality. For any $\omega \in \Omega$, if $A \in \mathcal{B}(A_0; \epsilon)$, then $A \oplus \omega \in \mathcal{B}(A_0; \epsilon)$, and if $A \in \mathcal{B}^\circ(A_0; \epsilon)$, then $A \oplus \omega \in \mathcal{B}^\circ(A_0; \epsilon)$.*
- *Suppose that $\| \cdot \|$ satisfies the null equality. Then, for any $\omega \in \Omega$, $A \oplus \omega \in \mathcal{B}(A_0; \epsilon)$ if and only if $A \in \mathcal{B}(A_0; \epsilon)$.*

(ii) *We have the inclusions*

$$\mathcal{B}(A; \epsilon) \subseteq \mathcal{B}(A; \epsilon) \oplus \Omega \text{ and } \mathcal{B}^\circ(A; \epsilon) \subseteq \mathcal{B}^\circ(A; \epsilon) \oplus \Omega.$$

If we further assume that $\|\cdot\|$ satisfies the null sub-inequality, then

$$\mathcal{B}(A;\epsilon)\oplus\Omega = \mathcal{B}(A;\epsilon) \text{ and } \mathcal{B}^\circ(A;\epsilon)\oplus\Omega = \mathcal{B}^\circ(A;\epsilon).$$

(iii) Suppose that $\|\cdot\|$ satisfies the null condition. Given a fixed $\omega \in \Omega$, we have

$$\Omega\oplus\omega \subseteq \mathcal{B}^\circ(\omega;\epsilon) \text{ and } \Omega \subseteq \mathcal{B}(\omega;\epsilon).$$

(iv) Suppose that $\|\cdot\|$ satisfies the null equality. Given any fixed $\omega \in \Omega$ and $\alpha \neq 0$, we have $\alpha\mathcal{B}(\omega;\epsilon) \subseteq \mathcal{B}(\omega;|\alpha|\epsilon)$.

(v) Given any fixed $\omega \in \Omega$ and $\alpha \neq 0$, we have

$$\alpha\mathcal{B}^\circ(\omega;\epsilon) \subseteq \mathcal{B}^\circ(\alpha\omega;|\alpha|\epsilon) \text{ and } \mathcal{B}^\circ(\alpha\omega;|\alpha|\epsilon) \subseteq \alpha\mathcal{B}^\circ(\omega;\epsilon).$$

Proof. The first case of part (i) follows from the following expression

$$\|A\ominus A_0\| \leq \|(A\oplus\omega)\ominus A_0\| < \epsilon.$$

The second case of part (i) regarding the open ball $\mathcal{B}(A_0;\epsilon)$ follows from the following expression

$$\|(A\oplus\omega)\ominus A_0\| \leq \|A\ominus A_0\| < \epsilon. \tag{4}$$

For the open ball $\mathcal{B}^\circ(A_0;\epsilon)$, if $A \in \mathcal{B}^\circ(A_0;\epsilon)$, then $A = A_0 \oplus C$ with $\|C\| < \epsilon$. Given an $\omega \in \Omega$, let $\tilde{C} = C \oplus \omega$. Therefore we have $A \oplus \omega = A_0 \oplus \tilde{C}$, where

$$\|\tilde{C}\| = \|C\oplus\omega\| \leq \|C\| < \epsilon, \tag{5}$$

which says that $A \oplus \omega \in \mathcal{B}^\circ(A_0;\epsilon)$. The third case of part (i) follows from the previous two cases.

To prove part (ii), since $\theta_{\mathcal{P}(X)} \in \Omega$ is the zero element of $\mathcal{P}(X)$, it follows that $B = B \oplus \theta_{\mathcal{P}(X)}$. Therefore we have $\mathcal{B}(A;\epsilon) \subseteq \mathcal{B}(A;\epsilon) \oplus \Omega$ and $\mathcal{B}^\circ(A;\epsilon) \subseteq \mathcal{B}^\circ(A;\epsilon) \oplus \Omega$. On the other hand, for $A \in \mathcal{B}(A_0;\epsilon)$ and $\omega \in \Omega$, from (4), we see that $A \oplus \omega \in \mathcal{B}(A_0;\epsilon)$, which shows the inclusion $\mathcal{B}(A_0;\epsilon) \oplus \Omega \subseteq \mathcal{B}(A_0;\epsilon)$. Also, for $B = A \oplus C \in \mathcal{B}^\circ(A;\epsilon)$ with $\|C\| < \epsilon$, let $\tilde{C} = \omega \oplus C$. By (5), we have $B \oplus \omega = A \oplus \tilde{C} \in \mathcal{B}^\circ(A;\epsilon)$, which shows the inclusion $\mathcal{B}^\circ(A;\epsilon) \oplus \Omega \subseteq \mathcal{B}^\circ(A;\epsilon)$. This proves part (ii).

To prove part (iii), for any $\omega' \in \Omega$, we have $\|\omega'\| = 0$, which says that $\omega \oplus \omega' \in \mathcal{B}^\circ(\omega;\epsilon)$. Therefore we obtain the inclusion $\Omega \oplus \omega \subseteq \mathcal{B}^\circ(\omega;\epsilon)$. On the other hand, we also have

$$\|\omega'\ominus\omega\| = \|\omega'\oplus(-\omega)\| \leq \|\omega'\| + \|-\omega\| = \|\omega'\| + \|\omega\| = 0,$$

which shows that $\omega' \in \mathcal{B}(\omega;\epsilon)$, i.e., $\Omega \subseteq \mathcal{B}(\omega;\epsilon)$.

To prove part (iv), for $A \in \mathcal{B}(\omega;\epsilon)$, since $\alpha\omega \in \Omega$, we have

$$\|\omega\ominus\alpha A\| = \|(\omega\oplus\alpha\omega)\ominus\alpha A\| = \|\alpha\omega\ominus\alpha A\| = \|\alpha(A\ominus\omega)\| = |\alpha|\|A\ominus\omega\| < |\alpha|\epsilon,$$

i.e., $\alpha A \in \mathcal{B}(\omega;|\alpha|\epsilon)$. This shows the inclusion $\alpha\mathcal{B}(\omega;\epsilon) \subseteq \mathcal{B}(\omega;|\alpha|\epsilon)$.

To prove the first inclusion of part (v), for $A \in \mathcal{B}(\omega;\epsilon)$, we have $A = \omega \oplus C$ with $\|C\| < \epsilon$. It follows that $\alpha A = \alpha\omega \oplus \alpha C$. Let $\tilde{C} = \alpha C$. Then $\|\tilde{C}\| < |\alpha|\epsilon$, which shows the inclusion $\alpha\mathcal{B}^\circ(\omega;\epsilon) \subseteq \mathcal{B}^\circ(\alpha\omega;|\alpha|\epsilon)$. To prove the second inclusion of part (v), for $A \in \mathcal{B}^\circ(\alpha\omega;|\alpha|\epsilon)$, we have $A = \alpha\omega \oplus C$ with $\|C\| < |\alpha|\epsilon$. Let $\hat{C} = C/\alpha$. Then

$$A = \alpha\omega\oplus C = \alpha\omega\oplus\alpha(C/\alpha) = \alpha\omega\oplus\alpha\hat{C} = \alpha(\omega\oplus\hat{C}) \text{ with } \|\hat{C}\| < \epsilon,$$

which says that $A \in \alpha\mathcal{B}^\circ(\omega;\epsilon)$. This completes the proof. □

5. Informal Open Sets

Let $(\mathcal{P}(X), \|\cdot\|)$ be an informal pseudo-seminormed hyperspace. We are going to consider the open subsets of $\mathcal{P}(X)$.

Definition 4. *Let $(\mathcal{P}(X), \|\cdot\|)$ be an informal pseudo-seminormed hyperspace, and let $\mathcal{A}$ be a nonempty subset of $\mathcal{P}(X)$.*

- *A point $A_0 \in \mathcal{A}$ is said to be an informal interior point of $\mathcal{A}$ if there exists $\epsilon > 0$ such that $\mathcal{B}(A_0; \epsilon) \subseteq \mathcal{A}$. The collection of all informal interior points of $\mathcal{A}$ is called the informal interior of $\mathcal{A}$ and is denoted by $int(\mathcal{A})$.*
- *A point $A_0 \in \mathcal{A}$ is said to be an informal type-I-interior point of $\mathcal{A}$ if there exists $\epsilon > 0$ such that $\mathcal{B}(A_0; \epsilon) \oplus \Omega \subseteq \mathcal{A}$. The collection of all informal type-I-interior points of $\mathcal{A}$ is called the informal type-I-interior of $\mathcal{A}$ and is denoted by $int^{(I)}(\mathcal{A})$.*
- *A point $A_0 \in \mathcal{A}$ is said to be an informal type-II-interior point of $\mathcal{A}$ if there exists $\epsilon > 0$ such that $\mathcal{B}(A_0; \epsilon) \subseteq \mathcal{A} \oplus \Omega$. The collection of all informal type-II-interior points of $\mathcal{A}$ is called the informal type-II-interior of $\mathcal{A}$ and is denoted by $int^{(II)}(\mathcal{A})$.*
- *A point $A_0 \in \mathcal{A}$ is said to be an informal type-III-interior point of $\mathcal{A}$ if there exists $\epsilon > 0$ such that $\mathcal{B}(A_0; \epsilon) \oplus \Omega \subseteq \mathcal{A} \oplus \Omega$. The collection of all informal type-III-interior points of $\mathcal{A}$ is called the informal type-III-interior of $\mathcal{A}$ and is denoted by $int^{(III)}(\mathcal{A})$.*

The different types of informal $\diamond$-interior points based on the open ball $\mathcal{B}^\diamond(A_0; \epsilon)$ can be similarly defined. For example, $int^{(\diamond III)}(\mathcal{A})$ denotes the informal $\diamond$-type-III-interior of $\mathcal{A}$.

Remark 2. *Recall that we cannot have the property $A \in \mathcal{B}(A; \epsilon)$ in general by Remark 1, unless $\|\cdot\|$ satisfies the null condition. Given any $A \in \mathcal{I}$ with $\| A \ominus A \| \neq 0$, it follows that $A \notin \mathcal{B}(A; \epsilon^*)$ for $\epsilon^* < \| A \ominus A \|$. Now, given $\epsilon < \epsilon^*$, it is clear that $\mathcal{B}(A; \epsilon) \subseteq \mathcal{B}(A; \epsilon^*)$. Let us take $\mathcal{A} = \mathcal{B}(A; \epsilon^*)$. It means that the open ball $\mathcal{B}(A; \epsilon)$ is contained in $\mathcal{A}$ even though the center A is not in $\mathcal{A}$.*

Remark 3. *From Remark 2, it can happen that there exists an open ball such that $\mathcal{B}(A; \epsilon)$ is contained in $\mathcal{A}$ even though the center A is not in $\mathcal{A}$. In this situation, we will not say that A is an informal interior point, since A is not in $\mathcal{A}$. Also, the sets $\mathcal{B}(A; \epsilon) \oplus \Omega$ and $\mathcal{B}^\diamond(A; \epsilon) \oplus \Omega$ will not necessarily contain the center A. In other words, it can happen that there exists an open ball such that $\mathcal{B}(A; \epsilon) \oplus \Omega$ is contained in $\mathcal{A}$ even though the center A is not in $\mathcal{A}$. In this situation, we will not say that A is an informal type-I-interior point, since A is not in $\mathcal{A}$. We also have the following observations.*

- *Suppose that $\|\cdot\|$ satisfies the null condition. Then $A \in \mathcal{B}(A; \epsilon)$. Since $A = A \oplus \theta_{\mathcal{P}(X)}$, we also have $A \in \mathcal{B}(A; \epsilon) \oplus \Omega$.*
- *Suppose that $\| \theta_{\mathcal{P}(X)} \| = 0$. The second observation of Remark 1 says that $A \in \mathcal{B}^\diamond(A; \epsilon)$. Since $A = A \oplus \theta_{\mathcal{P}(X)}$, it follows that $A \in \mathcal{B}^\diamond(A; \epsilon) \oplus \Omega$.*

According to Remark 3, we can define the different concepts of informal pseudo-interior point.

Definition 5. *Let $(\mathcal{P}(X), \|\cdot\|)$ be an informal pseudo-seminormed hyperspace, and let $\mathcal{A}$ be a nonempty subset of $\mathcal{P}(X)$.*

- *A point $A_0 \in \mathcal{P}(X)$ is said to be an informal pseudo-interior point of $\mathcal{A}$ if there exists $\epsilon > 0$ such that $\mathcal{B}(A_0; \epsilon) \subseteq \mathcal{A}$. The collection of all informal pseudo-interior points of $\mathcal{A}$ is called the informal pseudo-interior of $\mathcal{A}$ and is denoted by $pint(\mathcal{A})$.*
- *A point $A_0 \in \mathcal{P}(X)$ is said to be an informal type-I-pseudo-interior point of $\mathcal{A}$ if, and only if, there exists $\epsilon > 0$ such that $\mathcal{B}(A_0; \epsilon) \oplus \Omega \subseteq \mathcal{A}$. The collection of all informal type-I-pseudo-interior points of $\mathcal{A}$ is called the informal type-I-pseudo-interior of $\mathcal{A}$ and is denoted by $pint^{(I)}(\mathcal{A})$.*

- A point $A_0 \in \mathcal{P}(X)$ is said to be an informal type-II-pseudo-interior point of $\mathcal{A}$ if there exists $\epsilon > 0$ such that $\mathcal{B}(A_0; \epsilon) \subseteq \mathcal{A} \oplus \Omega$. The collection of all informal type-II-pseudo-interior points of $\mathcal{A}$ is called the informal type-II-pseudo-interior of $\mathcal{A}$ and is denoted by $pint^{(II)}(\mathcal{A})$.
- A point $A_0 \in \mathcal{P}(X)$ is said to be an informal type-III-pseudo-interior point of $\mathcal{A}$ if there exists $\epsilon > 0$ such that $\mathcal{B}(A_0; \epsilon) \oplus \Omega \subseteq \mathcal{A} \oplus \Omega$. The collection of all informal type-III-pseudo-interior points of $\mathcal{A}$ is called the informal type-III-pseudo-interior of $\mathcal{A}$ and is denoted by $pint^{(III)}(\mathcal{A})$.

The different types of informal $\diamond$-pseudo-interior point based on the open ball $\mathcal{B}^\diamond(A_0; \epsilon)$ can be similarly defined.

Remark 4. *We have to remark that the difference between Definitions 4 and 5 is that we consider $A_0 \in \mathcal{A}$ in Definition 4, and consider $A_0 \in \mathcal{P}(X)$ in Definition 5. From Remark 2, if $\epsilon^* < \| A \ominus A \|$, then A is a pseudo-interior point of $\mathcal{B}(A; \epsilon^*)$. We also have the following observations.*

- *It is clear that $int(\mathcal{A}) \subseteq pint(\mathcal{A})$, $int^{(I)}(\mathcal{A}) \subseteq pint^{(I)}(\mathcal{A})$, $int^{(II)}(\mathcal{A}) \subseteq pint^{(II)}(\mathcal{A})$ and $int^{(III)}(\mathcal{A}) \subseteq pint^{(III)}(\mathcal{A})$. The same inclusions can also apply to the different types of informal $\diamond$-interior and $\diamond$-pseudo-interior.*
- *It is clear that $int(\mathcal{A}) \subseteq \mathcal{A}$, $int^{(I)}(\mathcal{A}) \subseteq \mathcal{A}$, $int^{(II)}(\mathcal{A}) \subseteq \mathcal{A}$ and $int^{(III)}(\mathcal{A}) \subseteq \mathcal{A}$. However, the above kinds of inclusions cannot hold true for the informal pseudo-interior.*
- *From Remark 1, we have the following observations.*
    - *Suppose that $\| \cdot \|$ satisfies the null condition. Then these concepts of informal interior point and informal pseudo-interior point are equivalent, since A_0 is in the open ball $\mathcal{B}(A_0; \epsilon)$.*
    - *Suppose that $\| \theta \| = 0$. Then these concepts of informal $\diamond$-type of interior point and informal $\diamond$-type of pseudo-interior point are equivalent, since A_0 is in the open ball $\mathcal{B}^\diamond(A_0; \epsilon)$.*

Remark 5. *From part (ii) of Proposition 8, if $\| \cdot \|$ satisfies the null sub-inequality, then these concepts of informal interior point and informal type-I-interior point are equivalent, and these concepts of informal type-II-interior point and informal type-III-interior point are equivalent. The same situation also applies to the cases of informal pseudo-interior points. We also remark that if $\| \cdot \|$ satisfies the null condition, then $\| \cdot \|$ satisfies the null sub-inequality, since we have $\| A \oplus \omega \| \leq \| A \| + \| \omega \| = \| A \|$ for any $\omega \in \Omega$.*

Remark 6. *Suppose that $\| \cdot \|$ satisfies the null sub-inequality. From part (ii) of Proposition 5, we see that if A_0 is an informal interior (respectively type-I-interior, type-II-interior, type-III-interior) point then it is also an informal $\diamond$-interior (resp. $\diamond$-type-I-interior, $\diamond$-type-II-interior, $\diamond$-type-III-interior) point. In other words, from Remark 5, we have*

$$int(\mathcal{A}) = int^{(I)}(\mathcal{A}) \subseteq int^{(\diamond I)}(\mathcal{A}) = int^\diamond(\mathcal{A})$$

and

$$int^{(II)}(\mathcal{A}) = int^{(III)}(\mathcal{A}) \subseteq int^{(\diamond III)}(\mathcal{A}) = int^{(\diamond II)}(\mathcal{A}).$$

Regarding the different concepts of pseudo-interior point, we also have

$$pint(\mathcal{A}) = pint^{(I)}(\mathcal{A}) \subseteq pint^{(\diamond I)}(\mathcal{A}) = pint^\diamond(\mathcal{A})$$

and

$$pint^{(II)}(\mathcal{A}) = pint^{(III)}(\mathcal{A}) \subseteq pint^{(\diamond III)}(\mathcal{A}) = pint^{(\diamond II)}(\mathcal{A}).$$

Remark 7. *Let $(\mathcal{P}(X), \| \cdot \|)$ be an informal pseudo-seminormed hyperspace.*

- *Suppose that the center A_0 is in the open ball $\mathcal{B}(A_0; \epsilon)$. Then the concepts of informal interior point and informal pseudo-interior point are equivalent. It follows that $pint(\mathcal{A}) = int(\mathcal{A}) \subseteq \mathcal{A}$. Similarly, if the center A_0 is in the open ball $\mathcal{B}^\diamond(A_0; \epsilon)$, then $pint^\diamond(\mathcal{A}) = int^\diamond(\mathcal{A}) \subseteq \mathcal{A}$.*

- From part (ii) of Proposition 8, we have $\mathcal{B}(\mathcal{A};\epsilon) \subseteq \mathcal{B}(\mathcal{A};\epsilon) \oplus \Omega$ and $\mathcal{B}^\diamond(\mathcal{A};\epsilon) \subseteq \mathcal{B}^\diamond(\mathcal{A};\epsilon) \oplus \Omega$. Suppose that the center A_0 is in the open ball $\mathcal{B}(A_0;\epsilon)$. Let A_0 be an informal type-I-pseudo-interior point of $\mathcal{A}$. Since
$$A_0 \in \mathcal{B}(A_0;\epsilon) \subseteq \mathcal{B}(A_0;\epsilon) \oplus \Omega \subseteq \mathcal{A},$$
using Remark 4, we obtain
$$pint^{(I)}(\mathcal{A}) \subseteq int(\mathcal{A}) \subseteq \mathcal{A} \text{ and } pint^{(I)}(\mathcal{A}) \subseteq int^{(I)}(\mathcal{A}) \subseteq pint^{(I)}(\mathcal{A}),$$
which also implies $pint^{(I)}(\mathcal{A}) = int^{(I)}(\mathcal{A})$. Similarly, if the center A_0 is in the open ball $\mathcal{B}^\diamond(A_0;\epsilon)$, then $pint^{(\diamond I)}(\mathcal{A}) = int^{(\diamond I)}(\mathcal{A})$.

- Suppose that $\mathcal{A} \oplus \Omega \subseteq \mathcal{A}$. We have the following observations. Assume that the center A_0 is in the open ball $\mathcal{B}(A_0;\epsilon)$. Let A_0 be an informal type-II-pseudo-interior point of $\mathcal{A}$. Since
$$A_0 \in \mathcal{B}(A_0;\epsilon) \subseteq \mathcal{A} \oplus \Omega \subseteq \mathcal{A},$$
we obtain
$$pint^{(II)}(\mathcal{A}) \subseteq int(\mathcal{A}) \subseteq \mathcal{A} \text{ and } pint^{(II)}(\mathcal{A}) \subseteq int^{(II)}(\mathcal{A}) \subseteq pint^{(II)}(\mathcal{A}),$$
which also implies $pint^{(II)}(\mathcal{A}) = int^{(II)}(\mathcal{A})$. Similarly, if the center A_0 is in the open ball $\mathcal{B}^\diamond(A_0;\epsilon)$, then $pint^{(\diamond II)}(\mathcal{A}) = int^{(\diamond II)}(\mathcal{A})$.

- Suppose that $\mathcal{A} \oplus \Omega \subseteq \mathcal{A}$. We have the following observations. From part (ii) of Proposition 8, we have $\mathcal{B}(\mathcal{A};\epsilon) \subseteq \mathcal{B}(\mathcal{A};\epsilon) \oplus \Omega$ and $\mathcal{B}^\diamond(\mathcal{A};\epsilon) \subseteq \mathcal{B}^\diamond(\mathcal{A};\epsilon) \oplus \Omega$. Assume that the center A_0 is in the open ball $\mathcal{B}(A_0;\epsilon)$. Let A_0 be an informal type-III-pseudo-interior point of $\mathcal{A}$. Since
$$A_0 \in \mathcal{B}(A_0;\epsilon) \subseteq \mathcal{B}(A_0;\epsilon) \oplus \Omega \subseteq \mathcal{A} \oplus \Omega \subseteq \mathcal{A},$$
we obtain
$$pint^{(III)}(\mathcal{A}) \subseteq int(\mathcal{A}) \subseteq \mathcal{A} \text{ and } pint^{(III)}(\mathcal{A}) \subseteq int^{(III)}(\mathcal{A}) \subseteq pint^{(III)}(\mathcal{A}),$$
which also implies $pint^{(III)}(\mathcal{A}) = int^{(III)}(\mathcal{A})$. Similarly, if the center A_0 is in the open ball $\mathcal{B}^\diamond(A_0;\epsilon)$, then $pint^{(\diamond III)}(\mathcal{A}) = int^{(\diamond III)}(\mathcal{A})$.

Definition 6. *Let $(\mathcal{I}, \|\cdot\|)$ be an informal pseudo-seminormed hyperspace, and let $\mathcal{A}$ be a nonempty subset of $\mathcal{I}$. The set $\mathcal{A}$ is said to be informally open if $\mathcal{A} = int(\mathcal{A})$. The set $\mathcal{A}$ is said to be informally type-I-open if $\mathcal{A} = int^{(I)}(\mathcal{A})$. The set $\mathcal{A}$ is said to be informally type-II-open if $\mathcal{A} = int^{(II)}(\mathcal{A})$. The set $\mathcal{A}$ is said to be informally type-III-open if $\mathcal{A} = int^{(III)}(\mathcal{A})$. We can similarly define the informal $\diamond$-open set based on the informal $\diamond$-interior. Also, the informal pseudo-openness can be similarly defined.*

We adopt the convention $\emptyset \oplus \Omega = \emptyset$.

Remark 8. *Let $(\mathcal{P}(X), \|\cdot\|)$ be an informal pseudo-seminormed hyperspace, and let $\mathcal{A}$ be a nonempty subset of $\mathcal{P}(X)$. We consider the extreme cases of the empty set $\emptyset$ and whole set $\mathcal{P}(X)$.*

- *Since the empty set $\emptyset$ contains no elements, it means that $\emptyset$ is informally open and pseudo-open (we can regard the empty set as an open ball). It is clear that $\mathcal{P}(X)$ is also informally open and pseudo-open, since $A \in \mathcal{B} \subseteq X$ for any open ball $\mathcal{B}$, i.e., $\mathcal{P}(X) \subseteq int(\mathcal{P}(X))$ and $\mathcal{P}(X) \subseteq pint(\mathcal{P}(X))$.*
- *Since $\emptyset \oplus \Omega = \emptyset \subseteq \emptyset$, the emptyset $\emptyset$ is informally type-I-open and type-I-pseudo-open. It is clear that $\mathcal{P}(X)$ is also informally type-I-open and type-I-pseudo-open, since $A \in \mathcal{B} \oplus \Omega \subseteq X$ for any open ball $\mathcal{B}$, i.e., $\mathcal{P}(X) \subseteq int^{(I)}(\mathcal{P}(X))$ and $\mathcal{P}(X) \subseteq pint^{(I)}(\mathcal{P}(X))$.*
- *Since $\emptyset \subseteq \emptyset = \Omega \oplus \emptyset$, it means that $\emptyset$ is informally type-II-open and type-II-pseudo-open. We also see that $\mathcal{P}(X)$ is an informal type-II-open and type-II-pseudo-open set, since, for any $A \in \mathcal{P}(X)$ and any open*

ball $\mathcal{B}$, we have $A \in \mathcal{B} \subseteq \mathcal{P}(X) \subseteq \mathcal{P}(X) \oplus \Omega$ by part (i) of Proposition 2, i.e., $\mathcal{P}(X) \subseteq \text{int}^{(\text{II})}(\mathcal{P}(X))$ and $\mathcal{P}(X) \subseteq \text{pint}^{(\text{II})}(\mathcal{P}(X))$.

- Since $\emptyset \oplus \Omega \subseteq \Omega \oplus \emptyset$, it means that $\emptyset$ is informally type-III-open and type-III-pseudo-open. Now for any $A \in \mathcal{P}(X)$ and any open ball $\mathcal{B}$, we have $A \in \mathcal{B} \subseteq X$, which says that $\mathcal{B} \oplus \Omega \subseteq X \oplus \Omega$, i.e., $\mathcal{P}(X) \subseteq \text{int}^{(\text{III})}(\mathcal{P}(X))$ and $\mathcal{P}(X) \subseteq \text{pint}^{(\text{III})}(\mathcal{P}(X))$. This shows that $\mathcal{P}(X)$ is informally type-III-open and type-III-pseudo-open.

We have the above similar results for the different types of informal $\diamond$-open sets.

Proposition 9. *Let $(\mathcal{P}(X), \|\cdot\|)$ be an informal pseudo-seminormed hyperspace, and let $\mathcal{A}$ be a nonempty subset of $\mathcal{I}$.*

- *If $\mathcal{A}$ is informally pseudo-open, i.e., $\mathcal{A} = \text{pint}(\mathcal{A})$, then $\mathcal{A}$ is also informally open, i.e., $\mathcal{A} = \text{pint}(\mathcal{A}) = \text{int}(\mathcal{A})$. If $\mathcal{A} = \text{pint}^{\circ}(\mathcal{A})$, then $\mathcal{A} = \text{pint}^{\circ}(\mathcal{A}) = \text{int}^{\circ}(\mathcal{A})$.*
- *If $\mathcal{A} = \text{pint}^{(\text{I})}(\mathcal{A})$, then $\mathcal{A} = \text{pint}^{(\text{I})}(\mathcal{A}) = \text{int}^{(\text{I})}(\mathcal{A})$. If $\mathcal{A} = \text{pint}^{(\circ\text{I})}(\mathcal{A})$, then $\mathcal{A} = \text{pint}^{(\circ\text{I})}(\mathcal{A}) = \text{int}^{(\circ\text{I})}(\mathcal{A})$.*
- *If $\mathcal{A} = \text{pint}^{(\text{II})}(\mathcal{A})$, then $\mathcal{A} = \text{pint}^{(\text{II})}(\mathcal{A}) = \text{int}^{(\text{II})}(\mathcal{A})$. If $\mathcal{A} = \text{pint}^{(\circ\text{II})}(\mathcal{A})$, then $\mathcal{A} = \text{pint}^{(\circ\text{II})}(\mathcal{A}) = \text{int}^{(\circ\text{II})}(\mathcal{A})$.*
- *If $\mathcal{A} = \text{pint}^{(\text{III})}(\mathcal{A})$, then $\mathcal{A} = \text{pint}^{(\text{III})}(\mathcal{A}) = \text{int}^{(\text{III})}(\mathcal{A})$. If $\mathcal{A} = \text{pint}^{(\circ\text{III})}(\mathcal{A})$, then $\mathcal{A} = \text{pint}^{(\circ\text{III})}(\mathcal{A}) = \text{int}^{(\circ\text{III})}(\mathcal{A})$.*

Proof. If A is an informal pseudo-interior point, i.e., $A \in \text{pint}(\mathcal{A}) = \mathcal{A}$, then there exists $\epsilon > 0$ such that $\mathcal{B}(A_0; \epsilon) \subseteq \mathcal{A}$. Since $A \in \mathcal{A}$, it follows that A is also an informal interior point, i.e., $\text{pint}(\mathcal{A}) \subseteq \text{int}(\mathcal{A})$. From the first observation of Remark 4, we obtain the desired result. The remaining cases can be similarly realized, and the proof is complete. $\square$

Proposition 10. *Let $(\mathcal{P}(X), \|\cdot\|)$ be an informal pseudo-seminormed hyperspace.*

(i) *Suppose that $\|\cdot\|$ satisfies the null super-inequality.*

- *If $\mathcal{A}$ is any type of informally pseudo-open, then $A \in \mathcal{A}$ implies $A \oplus \omega \in \mathcal{A}$ for any $\omega \in \Omega$.*
- *If $\mathcal{A}$ is informally open, then $A \in \mathcal{A}$ implies $A \oplus \omega \in \text{pint}(\mathcal{A})$ for any $\omega \in \Omega$.*
- *If $\mathcal{A}$ is informally type-I-open, then $A \in \mathcal{A}$ implies $A \oplus \omega \in \text{pint}^{(\text{I})}(\mathcal{A})$ for any $\omega \in \Omega$.*
- *If $\mathcal{A}$ is informally type-II-open, then $A \in \mathcal{A}$ implies $A \oplus \omega \in \text{pint}^{(\text{II})}(\mathcal{A})$ for any $\omega \in \Omega$.*
- *If $\mathcal{A}$ is informally type-III-open, then $A \in \mathcal{A}$ implies $A \oplus \omega \in \text{pint}^{(\text{III})}(\mathcal{A})$ for any $\omega \in \Omega$.*

(ii) *Suppose that $\|\cdot\|$ satisfies the null sub-inequality, and that $\mathcal{A}$ is any type of informally pseudo-open. Then the following statements hold true.*

- $A \oplus \omega \in \mathcal{A}$ *implies* $A \in \mathcal{A}$ *for any* $\omega \in \Omega$.
- $A \oplus \omega \subseteq \mathcal{A}$ *for any* $\omega \in \Omega$ *and* $\mathcal{A} \oplus \Omega \subseteq \mathcal{A}$.
- $A \oplus \omega \in \mathcal{A} \oplus \omega$ *implies* $A \in \mathcal{A}$ *for any* $\omega \in \Omega$.
- *We have* $\mathcal{A} = \mathcal{A} \oplus \Omega$.

(iii) *Suppose that $\|\cdot\|$ satisfies the null sub-inequality, and that $\mathcal{A}$ is any type of informal $\diamond$-pseudo-open. Then $A \in \mathcal{A}$ implies $A \oplus \omega \in \mathcal{A}$ for any $\omega \in \Omega$.*

Proof. To prove part (i), suppose that $\mathcal{A}$ is informally type-III-pseudo-open. For $A \in \mathcal{A} = \text{pint}^{(\text{III})}(\mathcal{A})$, by definition, there exists $\epsilon > 0$ such that $\mathcal{B}(A; \epsilon) \oplus \Omega \subseteq \mathcal{A} \oplus \Omega$. From part (i) of Proposition 6, we also have $\mathcal{B}(A \oplus \omega; \epsilon) \oplus \Omega \subseteq \mathcal{A} \oplus \Omega$, which says that $A \oplus \omega \in \text{pint}^{(\text{III})}(\mathcal{A}) = \mathcal{A}$. Now we assume that $\mathcal{A}$ is informally type-III-open. Then $A \in \mathcal{A} = \text{int}^{(\text{III})}(\mathcal{A}) \subseteq \text{pint}^{(\text{III})}(\mathcal{A})$. We can also obtain $A \oplus \omega \in \text{pint}^{(\text{III})}(\mathcal{A})$. The other openness can be similarly obtained.

To prove the first case of part (ii), we consider the informal type-III-pseudo-open sets. If $A \oplus \omega \in \mathcal{A} = \text{pint}^{\text{(III)}}(\mathcal{A})$, there exists $\epsilon > 0$ such that $\mathcal{B}(A \oplus \omega; \epsilon) \oplus \Omega \subseteq \mathcal{A} \oplus \Omega$. From part (ii) of Proposition 6, we also have $\mathcal{B}(A; \epsilon) \oplus \Omega \subseteq \mathcal{A} \oplus \Omega$, which shows that $A \in \text{pint}^{\text{(III)}}(\mathcal{A}) = \mathcal{A}$.

To prove the second case of part (ii), we consider the informal type-III-pseudo-open sets. If $A \in \mathcal{A} \oplus \omega$, then $A = \hat{A} \oplus \omega$ for some $\hat{A} \in \mathcal{A} = \text{pint}^{\text{(III)}}(\mathcal{A})$. Therefore there exists $\epsilon > 0$ such that $\mathcal{B}(\hat{A}; \epsilon) \oplus \Omega \subseteq \mathcal{A} \oplus \Omega$. Since $\mathcal{B}(A; \epsilon) \subseteq \mathcal{B}(A \oplus \omega; \epsilon) = \mathcal{B}(\hat{A}; \epsilon)$ by part (ii) of Proposition 6, we see that $\mathcal{B}(A; \epsilon) \oplus \Omega \subseteq \mathcal{A} \oplus \Omega$, i.e., $A \in \text{pint}^{\text{(III)}}(\mathcal{A}) = \mathcal{A}$. Now, for $A \in \mathcal{A} \oplus \Omega$, we see that $A \in \mathcal{A} \oplus \omega$ for some $\omega \in \Omega$, which implies $A \in \mathcal{A}$. Therefore we obtain $\mathcal{A} \oplus \Omega \subseteq \mathcal{A}$.

To prove the third case of part (ii), using the second case of part (ii), we have

$$A \oplus \omega \in \mathcal{A} \oplus \omega \subseteq \mathcal{A} \oplus \Omega \subseteq \mathcal{A}.$$

Using the first case of part (ii), we obtain $A \in \mathcal{A}$.

To prove the fourth case of part (ii), since $\mathcal{A} = \mathcal{A} \oplus \{\theta_X\}$ and $\{\theta_X\} \in \Omega$, it follows that $\mathcal{A} \subseteq \mathcal{A} \oplus \Omega$. By the second case of part (ii), we obtain the desired result.

To prove part (iii), from part (ii) of Proposition 6, we have $\mathcal{B}^\circ(A \oplus \omega; \epsilon) \subseteq \mathcal{B}^\circ(A; \epsilon)$. Therefore, using the similar argument in the proof of part (i), we can obtain the desired results. This completes the proof. □

We remark that the results in Proposition 10 will not be true for any types of informal open sets. For example, in the proof of part (i), the inclusion $\mathcal{B}(A \oplus \omega; \epsilon) \oplus \Omega \subseteq \mathcal{A} \oplus \Omega$ can just say that $A \oplus \omega \in \text{pint}^{\text{(III)}}(\mathcal{A})$, since we do not know whether $A \oplus \omega$ is in $\mathcal{A}$ or not.

Proposition 11. *Let $(\mathcal{P}(X), \|\cdot\|)$ be an informal pseudo-seminormed hyperspace.*

(i) *Suppose that $\|\cdot\|$ satisfies the null condition.*

- *We have $\text{int}(\mathcal{A}) = \text{int}^{\text{(I)}}(\mathcal{A}) \oplus \Omega \subseteq \mathcal{A}$. In particular, if $\mathcal{A}$ is informally open or type-I-open, then $\mathcal{A} \oplus \Omega \subseteq \mathcal{A}$.*
- *We have $\text{int}^{\text{(II)}}(\mathcal{A}) = \text{int}^{\text{(III)}}(\mathcal{A}) \subseteq \mathcal{A} \oplus \Omega$.*

Moreover the concept of informal (resp. type-I, type-II, type-III) open set is equivalent to the concept of informal (resp. type-I, type-II, type-III) pseudo-open set.

(ii) *Suppose that $\|\cdot\|$ satisfies the null sub-inequality. Then*

$$(\text{pint}^{\text{(II)}}(\mathcal{A}))^c \oplus \Omega = (\text{pint}^{\text{(III)}}(\mathcal{A}))^c \oplus \Omega \subseteq (\text{pint}^{\text{(II)}}(\mathcal{A}))^c = (\text{pint}^{\text{(III)}}(\mathcal{A}))^c.$$

In particular, if $\mathcal{A}$ is informally type-II-pseudo-open or type-III-pseudo-open, then $\mathcal{A}^c \oplus \Omega \subseteq \mathcal{A}^c$.

Proof. To prove the first case of part (i), for any $A \in \text{int}^{\text{(I)}}(\mathcal{A})$, there exists an open ball $\mathcal{B}(A; \epsilon)$ such that $\mathcal{B}(A; \epsilon) \oplus \Omega \subseteq \mathcal{A}$. Since $A \in \mathcal{B}(A; \epsilon)$ by the first observation of Remark 1, we have $A \oplus \Omega \subseteq \mathcal{B}(A; \epsilon) \oplus \Omega \subseteq \mathcal{A}$. This shows $\text{int}^{\text{(I)}}(\mathcal{A}) \oplus \Omega \subseteq \mathcal{A}$. Using Remark 5, we obtain the desired results.

To prove the second case of part (i), for any $A \in \text{int}^{\text{(II)}}(\mathcal{A})$, there exists an open ball $\mathcal{B}(A; \epsilon)$ such that $\mathcal{B}(A; \epsilon) \subseteq \mathcal{A} \oplus \Omega$. Then we have $A \in \mathcal{A} \oplus \Omega$, since $A \in \mathcal{B}(A; \epsilon)$. This shows $\text{int}^{\text{(II)}}(\mathcal{A}) \subseteq \mathcal{A} \oplus \Omega$. Using Remark 5, we obtain the desired results. From Remark 4, we see that the concept of informal (resp. type-I, type-II, type-III) open set is equivalent to the concept of informal (resp. type-I, type-II, type-III) pseudo-open set.

To prove part (ii), for any $A \in (\text{pint}^{\text{(II)}}(\mathcal{A}))^c \oplus \Omega$, we have $A = \hat{A} \oplus \hat{\omega}$ for some $\hat{A} \in (\text{pint}^{\text{(II)}}(\mathcal{A}))^c$ and $\hat{\omega} \in \Omega$. By definition, we see that $\mathcal{B}(\hat{A}; \epsilon) \not\subseteq \mathcal{A} \oplus \Omega$ for every $\epsilon > 0$. By part (ii) of Proposition 6, we also have $\mathcal{B}(A; \epsilon) \not\subseteq \mathcal{A} \oplus \Omega$ for every $\epsilon > 0$. This says that A is not an informal type-II-pseudo-interior point of $\mathcal{A}$, i.e., $A \notin \text{pint}^{\text{(II)}}(\mathcal{A})$. This completes the proof. □

Proposition 12. *Let $(\mathcal{P}(X), \|\cdot\|)$ be an informal pseudo-seminormed hyperspace.*

(i) $\mathcal{B}^\circ(A_0;\epsilon)$ is informally $\diamond$-open, $\diamond$-type-II-open and $\diamond$-type-III-open. We also have the inclusions $\mathcal{B}^\circ(A_0;\epsilon) \subseteq \text{pint}(\mathcal{B}^\circ(A_0;\epsilon))$, $\mathcal{B}^\circ(A_0;\epsilon) \subseteq \text{pint}^{(\circ\text{II})}(\mathcal{B}^\circ(A_0;\epsilon))$ and $\mathcal{B}^\circ(A_0;\epsilon) \subseteq \text{pint}^{(\circ\text{III})}(\mathcal{B}^\circ(A_0;\epsilon))$.

(ii) $\mathcal{B}(A_0;\epsilon)$ is informally open, type-II-open and type-III-open. We also have the inclusions $\mathcal{B}(A_0;\epsilon) \subseteq \text{pint}(\mathcal{B}(A_0;\epsilon))$, $\mathcal{B}(A_0;\epsilon) \subseteq \text{pint}^{(\text{II})}(\mathcal{B}(A_0;\epsilon))$ and $\mathcal{B}(A_0;\epsilon) \subseteq \text{pint}^{(\text{III})}(\mathcal{B}(A_0;\epsilon))$.

(iii) Suppose that $\|\cdot\|$ satisfies the null sub-inequality. Then $\mathcal{B}^\circ(A_0;\epsilon)$ is informally $\diamond$-type-I-open, and $\mathcal{B}(A_0;\epsilon)$ is informally type-I-open. We also have the inclusions $\mathcal{B}^\circ(A_0;\epsilon) \subseteq \text{pint}^{(\circ\text{I})}(\mathcal{B}^\circ(A_0;\epsilon))$ and $\mathcal{B}(A_0;\epsilon) \subseteq \text{pint}^{(\text{I})}(\mathcal{B}(A_0;\epsilon))$.

Proof. To prove part (i), for any $A \in \mathcal{B}^\circ(A_0;\epsilon)$, we have $A = A_0 \oplus C$ with $\|C\| < \epsilon$. Let $\hat{\epsilon} = \epsilon - \|C\| > 0$. For any $\hat{A} \in \mathcal{B}^\circ(A;\hat{\epsilon})$, i.e., $\hat{A} = A \oplus D$ with $\|D\| < \hat{\epsilon}$, we obtain $\hat{A} = A_0 \oplus C \oplus D$ and

$$\|C \oplus D\| \leq \|C\| + \|D\| = \epsilon - \hat{\epsilon} + \|D\| < \epsilon - \hat{\epsilon} + \hat{\epsilon} = \epsilon,$$

which means that $\hat{A} \in \mathcal{B}^\circ(A_0;\epsilon)$, i.e.,

$$\mathcal{B}^\circ(A;\hat{\epsilon}) \subseteq \mathcal{B}^\circ(A_0;\epsilon). \tag{6}$$

This shows that $\mathcal{B}^\circ(A_0;\epsilon) \subseteq \text{int}(\mathcal{B}^\circ(A_0;\epsilon))$. Therefore we obtain $\mathcal{B}^\circ(A_0;\epsilon) = \text{int}(\mathcal{B}^\circ(A_0;\epsilon))$. We can similarly obtain the inclusion $\mathcal{B}^\circ(A_0;\epsilon) \subseteq \text{pint}(\mathcal{B}^\circ(A_0;\epsilon))$. However, we cannot have the equality $\mathcal{B}^\circ(A_0;\epsilon) = \text{pint}(\mathcal{B}^\circ(A_0;\epsilon))$, since $\text{pint}(\mathcal{B}^\circ(A_0;\epsilon))$ is not necessarily contained in $\mathcal{B}^\circ(A_0;\epsilon)$. From (6), we have $\mathcal{B}^\circ(x;\hat{\epsilon}) \oplus \Omega \subseteq \mathcal{B}^\circ(A_0;\epsilon) \oplus \Omega$. This says that $\mathcal{B}^\circ(A_0;\epsilon)$ is informally $\diamond$-type-III-open. On the other hand, from (6) and part (ii) of Proposition 8, we also have

$$\mathcal{B}^\circ(A;\hat{\epsilon}) \subseteq \mathcal{B}^\circ(A_0;\epsilon) \subseteq \mathcal{B}^\circ(A_0;\epsilon) \oplus \Omega.$$

This shows that $\mathcal{B}^\circ(A_0;\epsilon)$ is informally $\diamond$-type-II-open.

To prove part (ii), for any $A \in \mathcal{B}(A_0;\epsilon)$, we have $\|A \ominus A_0\| < \epsilon$. Let $\hat{\epsilon} = \|A \ominus A_0\|$. For any $\hat{A} \in \mathcal{B}(A;\epsilon - \hat{\epsilon})$, we have $\|\hat{A} \ominus A\| < \epsilon - \hat{\epsilon}$. Therefore, by Proposition 4, we obtain

$$\|\hat{A} \ominus A_0\| \leq \|\hat{A} \ominus A\| + \|A \ominus A_0\| = \hat{\epsilon} + \|\hat{A} \ominus A\| < \hat{\epsilon} + \epsilon - \hat{\epsilon} = \epsilon,$$

which means that $\hat{A} \in \mathcal{B}(A_0;\epsilon)$, i.e.,

$$\mathcal{B}(A;\epsilon - \hat{\epsilon}) \subseteq \mathcal{B}(A_0;\epsilon). \tag{7}$$

This shows that $\mathcal{B}(A_0;\epsilon) \subseteq \text{int}(\mathcal{B}(A_0;\epsilon))$.

Therefore we obtain $\mathcal{B}(A_0;\epsilon) = \text{int}(\mathcal{B}(A_0;\epsilon))$. We can similarly obtain the inclusion $\mathcal{B}(A_0;\epsilon) \subseteq \text{pint}(\mathcal{B}(A_0;\epsilon))$. From (7), we have $\mathcal{B}(A;\epsilon - \hat{\epsilon}) \oplus \Omega \subseteq \mathcal{B}(A_0;\epsilon) \oplus \Omega$. This says that $\mathcal{B}(A_0;\epsilon)$ is informally type-III-open. On the other hand, from (7) and part (ii) of Proposition 8, we also have

$$\mathcal{B}(A;\epsilon - \hat{\epsilon}) \subseteq \mathcal{B}(A_0;\epsilon) \subseteq \mathcal{B}(A_0;\epsilon) \oplus \Omega.$$

This shows that $\mathcal{B}(A_0;\epsilon)$ is informally type-II-open.

To prove part (iii), from (6), (7) and part (ii) of Proposition 8, we have

$$\mathcal{B}^\circ(A;\hat{\epsilon}) \oplus \Omega \subseteq \mathcal{B}^\circ(A_0;\epsilon) \oplus \Omega = \mathcal{B}^\circ(A_0;\epsilon)$$

and

$$\mathcal{B}(A;\epsilon - \hat{\epsilon}) \oplus \Omega \subseteq \mathcal{B}(A_0;\epsilon) \oplus \Omega = \mathcal{B}(A_0;\epsilon).$$

This shows that $\mathcal{B}^\circ(A_0;\epsilon)$ is informally $\diamond$-type-I-open, and that $\mathcal{B}(A_0;\epsilon)$ is informally type-I-open. We complete the proof. □

Proposition 13. *Let $(\mathcal{P}(X), \|\cdot\|)$ be an informal pseudo-seminormed hyperspace. Suppose that the center A_0 is in the open balls $\mathcal{B}(A_0; \epsilon)$ and $\mathcal{B}^\circ(A_0; \epsilon)$. The following statements hold true:*

(i) $\mathcal{B}(A_0; \epsilon)$ *is informally pseudo-open and $\diamond$-pseudo-open.*

(ii) *Suppose that $\|\cdot\|$ satisfies the null sub-inequality. Then $\mathcal{B}(A_0; \epsilon)$ is informally type-I-pseudo-open, type-II-pseudo-open and type-III-pseudo-open.*

(iii) *Suppose that $\|\cdot\|$ satisfies the null sub-inequality. Then $\mathcal{B}(A_0; \epsilon)$ is informally $\diamond$-type-I-pseudo-open, $\diamond$-type-II-pseudo-open and $\diamond$-type-III-pseudo-open.*

Proof. The results follow from Proposition 12, Remark 7 and part (ii) of Proposition 8 immediately. □

6. Topoloigcal Spaces

Now we are in a position to investigate the topological structure generated by the informal pseudo-seminormed hyperspace $(\mathcal{P}(X), \|\cdot\|)$ based on the different kinds of openness. We denote by τ_0 and $\tau_0^{(\diamond)}$ the set of all informal open and informal $\diamond$-open subsets of $\mathcal{P}(X)$, respectively, and by $p\tau_0$ and $p\tau_0^{(\diamond)}$ the set of all informal pseudo-open and informal $\diamond$-pseudo-open subsets of $\mathcal{P}(X)$, respectively. We denote by $\tau^{(I)}$ and $\tau^{(\diamond I)}$ the set of all informal type-I-open and informal $\diamond$-type-I-open subsets of $\mathcal{P}(X)$, respectively, and by $p\tau^{(I)}$ and $p\tau^{(\diamond I)}$ the set of all informal type-I-pseudo-open and informal $\diamond$-type-I-pseudo-open subsets of $\mathcal{P}(X)$, respectively. We can similarly define the families $\tau^{(II)}$, $\tau^{(III)}$, $\tau^{(\diamond II)}$, $\tau^{(\diamond III)}$, $p\tau^{(II)}$, $p\tau^{(III)}$, $p\tau^{(\diamond II)}$ and $p\tau^{(\diamond III)}$.

Proposition 14. *Let $(\mathcal{P}(X), \|\cdot\|)$ be an informal pseudo-seminormed hyperspace.*

(i) $(\mathcal{P}(X), \tau^{(I)})$ *and $(\mathcal{P}(X), \tau^{(\diamond I)})$ are topological spaces.*

(ii) *Suppose that each open ball $\mathcal{B}(A_0; \epsilon)$ contains the center A_0. Then $(\mathcal{P}(X), p\tau^{(I)}) = (\mathcal{P}(X), \tau^{(I)})$ is a topological space.*

(iii) *Suppose that each open ball $\mathcal{B}^\circ(A_0; \epsilon)$ contains the center A_0. Then $(\mathcal{P}(X), p\tau^{(\diamond I)}) = (\mathcal{P}(X), \tau^{(\diamond I)})$ is a topological space.*

Proof. To prove part (i), by the second observation of Remark 8, we see that $\emptyset \in \tau^{(I)}$ and $\mathcal{P}(X) \in \tau^{(I)}$. Let $\mathcal{A} = \bigcap_{i=1}^n \mathcal{A}_i$, where $\mathcal{A}_i$ are informal type-I-open sets for all $i = 1, \cdots, n$. For $A \in \mathcal{A}$, we have $A \in \mathcal{A}_i$ for all $i = 1, \cdots, n$. Then there exist ϵ_i such that $\mathcal{B}(A; \epsilon_i) \oplus \Omega \subseteq \mathcal{A}_i$ for all $i = 1, \cdots, n$. Let $\epsilon = \min\{\epsilon_1, \cdots, \epsilon_n\}$. Then $\mathcal{B}(A; \epsilon) \oplus \Omega \subseteq \mathcal{B}(A; \epsilon_i) \oplus \Omega \subseteq \mathcal{A}_i$ for all $i = 1, \cdots, n$, which says that $\mathcal{B}(A; \epsilon) \oplus \Omega \subseteq \bigcap_{i=1}^n \mathcal{A}_i = \mathcal{A}$, i.e., $\mathcal{A} \subseteq \text{int}^{(I)}(\mathcal{A})$. Therefore the intersection $\mathcal{A}$ is informally type-I-open by Remark 4. On the other hand, let $\mathcal{A} = \bigcup_\delta \mathcal{A}_\delta$. Then $A \in \mathcal{A}$ implies that $A \in \mathcal{A}_\delta$ for some δ. This indicates that $\mathcal{B}(A; \epsilon) \oplus \Omega \subseteq \mathcal{A}_\delta \subseteq \mathcal{A}$ for some $\epsilon > 0$, i.e., $\mathcal{A} \subseteq \text{int}^{(I)}(\mathcal{A})$. Therefore the union $\mathcal{A}$ is informally type-I-open. This shows that $(\mathcal{P}(X), \tau^{(I)})$ is a topological space. For the case of informal $\diamond$-type-I-open subsets of $\mathcal{P}(X)$, we can similarly obtain the desired result. Parts (ii) and (iii) follow from Remark 7 and part (i) immediately. This completes the proof. □

Remark 1 shows the sufficient conditions for the open ball $\mathcal{B}(A; \epsilon)$ containing the center A.

Proposition 15. *Let $(\mathcal{P}(X), \|\cdot\|)$ be an informal pseudo-seminormed hyperspace.*

(i) $(\mathcal{P}(X), \tau_0)$ *and $(\mathcal{P}(X), \tau_0^{(\diamond)})$ are topological spaces.*

(ii) *Suppose that each open ball $\mathcal{B}(A_0; \epsilon)$ contains the center A_0. Then $(\mathcal{P}(X), \tau_0) = (\mathcal{P}(X), p\tau_0)$ is a topological space.*

(iii) *Suppose that each open ball $\mathcal{B}^\circ(A_0; \epsilon)$ contains the center A_0. Then $(\mathcal{P}(X), \tau_0^{(\diamond)}) = (\mathcal{P}(X), p\tau_0^{(\diamond)})$ is a topological space.*

Proof. The empty set $\emptyset$ and $\mathcal{P}(X)$ are informal open by the first observation of Remark 8. The remaining proof follows from the similar argument of Proposition 14 without considering the null set Ω. □

Let $(\mathcal{P}(X), \|\cdot\|)$ be an informal pseudo-seminormed hyperspace. We consider the following families:
$$\widetilde{\tau}^{(\text{II})} = \{\mathcal{A} \in \tau^{(\text{II})} : \mathcal{A} \oplus \Omega \subseteq \mathcal{A}\}$$

and
$$\widetilde{\tau}^{(\text{III})} = \{\mathcal{A} \in \tau^{(\text{III})} : \mathcal{A} \oplus \Omega \subseteq \mathcal{A}\}.$$

We can similarly define $\widetilde{\tau}^{(\circ\text{II})}$ and $\widetilde{\tau}^{(\circ\text{III})}$. Then $\widetilde{\tau}^{(\text{II})} \subseteq \tau^{(\text{II})}$, $\widetilde{\tau}^{(\text{III})} \subseteq \tau^{(\text{III})}$, $\widetilde{\tau}^{(\circ\text{II})} \subseteq \tau^{(\circ\text{II})}$ and $\widetilde{\tau}^{(\circ\text{III})} \subseteq \tau^{(\circ\text{III})}$. We can also similarly define $\widetilde{p\tau}^{(\text{II})}$, $\widetilde{p\tau}^{(\text{III})}$, $\widetilde{p\tau}^{(\circ\text{II})}$ and $\widetilde{p\tau}^{(\circ\text{III})}$ regarding the informal pseudo-openness. Then $\widetilde{p\tau}^{(\text{II})} \subseteq p\tau^{(\text{II})}$, $\widetilde{p\tau}^{(\text{III})} \subseteq p\tau^{(\text{III})}$, $\widetilde{p\tau}^{(\circ\text{II})} \subseteq p\tau^{(\circ\text{II})}$ and $\widetilde{p\tau}^{(\circ\text{III})} \subseteq p\tau^{(\circ\text{III})}$.

Proposition 16. *Let $(\mathcal{P}(X), \|\cdot\|)$ be an informal pseudo-seminormed hyperspace. Suppose that $\|\cdot\|$ satisfies the null sub-inequality. Then*
$$\widetilde{p\tau}^{(\text{II})} = p\tau^{(\text{II})} = p\tau^{(\text{III})} = \widetilde{p\tau}^{(\text{III})} \text{ and } \widetilde{\tau}^{(\text{II})} = \tau^{(\text{II})} = \tau^{(\text{III})} = \widetilde{\tau}^{(\text{III})}.$$

Proof. The results follow from Remark 5 and part (ii) of Proposition 10 immediately. □

Proposition 17. *Let $(\mathcal{P}(X), \|\cdot\|)$ be an informal pseudo-seminormed hyperspace.*

(i) *$(\mathcal{P}(X), \widetilde{\tau}^{(\text{II})})$ and $(\mathcal{P}(X), \widetilde{\tau}^{(\circ\text{II})})$ are topological spaces.*
(ii) *The following statements hold true.*

- *Suppose that each open ball $\mathcal{B}(\mathcal{A}; \epsilon)$ contains the center $\mathcal{A}$. Then $(\mathcal{P}(X), \widetilde{p\tau}^{(\text{II})}) = (\mathcal{P}(X), \widetilde{\tau}^{(\text{II})})$ is a topological space.*
- *Suppose that each open ball $\mathcal{B}^\circ(\mathcal{A}; \epsilon)$ contains the center $\mathcal{A}$. Then $(\mathcal{P}(X), \widetilde{p\tau}^{(\circ\text{II})}) = (\mathcal{P}(X), \widetilde{\tau}^{(\circ\text{II})})$ is a topological space.*

Proof. To prove part (i), given $\mathcal{A}_1, \mathcal{A}_2 \in \widetilde{\tau}^{(\text{II})}$, let $\mathcal{A} = \mathcal{A}_1 \cap \mathcal{A}_2$. For $A \in \mathcal{A}$, we have $A \in \mathcal{A}_i$ for $i = 1, 2$. Then there exist ϵ_i such that $\mathcal{B}(A; \epsilon_i) \subseteq \mathcal{A}_i \oplus \Omega$ for all $i = 1, 2$. Let $\epsilon = \min\{\epsilon_1, \epsilon_2\}$. Then
$$\mathcal{B}(A; \epsilon) \subseteq \mathcal{B}(A; \epsilon_i) \subseteq \mathcal{A}_i \oplus \Omega$$
for all $i = 1, 2$, which says that
$$\mathcal{B}(A; \epsilon) \subseteq [(\mathcal{A}_1 \oplus \Omega) \cap (\mathcal{A}_2 \oplus \Omega)] = (\mathcal{A}_1 \cap \mathcal{A}_2) \oplus \Omega = \mathcal{A} \oplus \Omega$$
by Proposition 3. This shows that $\mathcal{A}$ is informally type-II-open. For $A \in \mathcal{A} \oplus \Omega$, we have $A = \bar{A} \oplus \omega$ for some $\bar{A} \in \mathcal{A}$ and $\omega \in \Omega$. Since $\bar{A} \in \mathcal{A}_1 \cap \mathcal{A}_2$, it follows that $A \in \mathcal{A}_1 \oplus \Omega \subseteq \mathcal{A}_1$ and $A \in \mathcal{A}_2 \oplus \Omega \subseteq \mathcal{A}_2$, which says that $A \in \mathcal{A}_1 \cap \mathcal{A}_2 = \mathcal{A}$, i.e., $\mathcal{A} \oplus \Omega \subseteq \mathcal{A}$. This shows that $\mathcal{A}$ is indeed in $\widetilde{\tau}^{(\text{II})}$. Therefore, the intersection of finitely many members of $\widetilde{\tau}^{(\text{II})}$ is a member of $\widetilde{\tau}^{(\text{II})}$.

Now, given a family $\{\mathcal{A}_\delta\}_{\delta \in \Lambda} \subset \widetilde{\tau}^{(\text{II})}$, let $\mathcal{A} = \bigcup_{\delta \in \Lambda} \mathcal{A}_\delta$. Then $A \in \mathcal{A}$ implies that $A \in \mathcal{A}_\delta$ for some $\delta \in \Lambda$. This says that
$$\mathcal{B}(A; \epsilon) \subseteq \mathcal{A}_\delta \oplus \Omega \subseteq \mathcal{A} \oplus \Omega$$
for some $\epsilon > 0$. Therefore, the union $\mathcal{A}$ is informally type-II-open. For $A \in \mathcal{A} \oplus \Omega$, we have $A = \bar{A} \oplus \omega$, where $\bar{A} \in \mathcal{A}$, i.e., $\bar{A} \in \mathcal{A}_\delta$ for some $\delta \in \Lambda$. It also says that $A \in \mathcal{A}_\delta \oplus \Omega \subseteq \mathcal{A}_\delta \subseteq \mathcal{A}$, i.e., $\mathcal{A} \oplus \Omega \subseteq \mathcal{A}$. This shows that $\mathcal{A}$ is indeed in $\widetilde{\tau}^{(\text{II})}$. By the third observation of Remark 8, we see that $\emptyset$ and $\mathcal{P}(X)$ are also informal type-II-open. It is not hard to see that $\emptyset \oplus \Omega = \emptyset$ and $\mathcal{P}(X) \oplus \Omega \subseteq \mathcal{P}(X)$, which shows that $\emptyset, X \in \widetilde{\tau}^{(\text{II})}$. Therefore, $(\mathcal{P}(X), \widetilde{\tau}^{(\text{II})})$ is indeed a topological space. The above arguments are also valid for $\widetilde{\tau}^{(\circ\text{II})}$.

Part (ii) follows immediately from the third observation of Remark 7 and part (i). This completes the proof. □

Proposition 18. *Let $(\mathcal{P}(X), \|\cdot\|)$ be an informal pseudo-seminormed hyperspace.*

(i) *$(\mathcal{P}(X), \widetilde{\tau}^{(\text{III})})$ and $(\mathcal{P}(X), \widetilde{\tau}^{(\circ\text{III})})$ are topological spaces.*

(ii) *The following statements hold true.*

- *Suppose that each open ball $\mathcal{B}(A; \epsilon)$ contains the center A. Then $(\mathcal{P}(X), \widetilde{p\tau}^{(\text{III})}) = (\mathcal{P}(X), \widetilde{\tau}^{(\text{III})})$ is a topological space.*
- *Suppose that each open ball $\mathcal{B}^\circ(A; \epsilon)$ contains the center A. Then $(\mathcal{P}(X), \widetilde{p\tau}^{(\circ\text{III})}) = (\mathcal{P}(X), \widetilde{\tau}^{(\circ\text{III})})$ is a topological space.*

Proof. To prove part (i), by the fourth observation of Remark 8, it is clear to see that $\emptyset, \mathcal{P}(X) \in \widetilde{\tau}^{(\text{III})}$. Since $\emptyset \oplus \Omega = \emptyset$ and $\mathcal{P}(X) \oplus \Omega \subseteq \mathcal{P}(X)$, it follows that $\emptyset, \mathcal{P}(X) \in \widetilde{\tau}^{(\text{III})}$. Given $\mathcal{A}_1, \mathcal{A}_2 \in \widetilde{\tau}^{(\text{III})}$, let $\mathcal{A} = \mathcal{A}_1 \cap \mathcal{A}_2$. For $A \in \mathcal{A}$, there exist ϵ_i such that $\mathcal{B}(A; \epsilon_i) \oplus \Omega \subseteq \mathcal{A}_i \oplus \Omega$ for all $i = 1, 2$. Let $\epsilon = \min\{\epsilon_1, \epsilon_2\}$. Then

$$\mathcal{B}(A; \epsilon) \oplus \Omega \subseteq \mathcal{B}(A; \epsilon_i) \oplus \Omega \subseteq \mathcal{A}_i \oplus \Omega$$

for all $i = 1, 2$, which says that

$$\mathcal{B}(A; \epsilon) \oplus \Omega \subseteq [(\mathcal{A}_1 \oplus \Omega) \cap (\mathcal{A}_2 \oplus \Omega)] = (\mathcal{A}_1 \cap \mathcal{A}_2) \oplus \Omega = \mathcal{A} \oplus \Omega$$

by Proposition 3. This shows that $\mathcal{A}$ is informally type-III-open. From the proof of Proposition 17, we also see that $\mathcal{A} \oplus \Omega \subseteq \mathcal{A}$. Therefore, the intersection of finitely many members of $\widetilde{\tau}^{(\text{III})}$ is a member of $\widetilde{\tau}^{(\text{III})}$.

Now, given a family $\{\mathcal{A}_\delta\}_{\delta \in \Lambda} \subset \widetilde{\tau}^{(\text{III})}$, let $\mathcal{A} = \cup_{\delta \in \Lambda} \mathcal{A}_\delta$. Then $A \in \mathcal{A}$ implies that $A \in \mathcal{A}_\delta$ for some $\delta \in \Lambda$. This says that

$$\mathcal{B}(A; \epsilon) \oplus \Omega \subseteq \mathcal{A}_\delta \oplus \Omega \subseteq \mathcal{A} \oplus \Omega$$

for some $\epsilon > 0$. Therefore, the union $\mathcal{A}$ is informally type-III-open. From the proof of Proposition 17, we also see that $\mathcal{A} \oplus \Omega \subseteq \mathcal{A}$, i.e., $\mathcal{A} \in \widetilde{\tau}^{(\text{III})}$. This shows that $(\mathcal{P}(X), \widetilde{\tau}^{(\text{III})})$ is indeed a topological space. The above arguments are also valid for $\widetilde{\tau}^{(\circ\text{III})}$.

Part (ii) follows immediately from the fourth observation of Remark 7 and part (i). This completes the proof. □

Proposition 19. *Let $(\mathcal{P}(X), \|\cdot\|)$ be an informal pseudo-seminormed hyperspace. Suppose that $\|\cdot\|$ satisfies the null sub-inequality. If each open ball $\mathcal{B}(A; \epsilon)$ contains the center A, then $(\mathcal{P}(X), p\tau^{(\text{III})}) = (\mathcal{P}(X), p\tau^{(\text{III})})$ is a topological space.*

Proof. By the third observation of Remark 8, we see that $\emptyset, \mathcal{P}(X) \in p\tau^{(\text{III})}$. Given $\mathcal{A}_1, \mathcal{A}_2 \in p\tau^{(\text{III})}$, let $\mathcal{A} = \mathcal{A}_1 \cap \mathcal{A}_2$. We want to show $\mathcal{A} = \text{pint}^{(\text{III})}(\mathcal{A})$. For $A \in \mathcal{A}$, we have $A \in \mathcal{A}_i$ for $i = 1, 2$. There exist ϵ_i such that $\mathcal{B}(A; \epsilon_i) \subseteq \mathcal{A}_i \oplus \Omega$ for all $i = 1, 2$. Let $\epsilon = \min\{\epsilon_1, \epsilon_2\}$. Then $\mathcal{B}(A; \epsilon) \subseteq \mathcal{B}(A; \epsilon_i) \subseteq \mathcal{A}_i \oplus \Omega$ for $i = 1, 2$, which says that, using part (ii) of Proposition 10,

$$\mathcal{B}(A; \epsilon) \subseteq [(\mathcal{A}_1 \oplus \Omega) \cap (\mathcal{A}_2 \oplus \Omega)] = \mathcal{A}_1 \cap \mathcal{A}_2 = (\mathcal{A}_1 \cap \mathcal{A}_2) \oplus \Omega = \mathcal{A} \oplus \Omega.$$

This shows that $A \in \text{int}^{(\text{III})}(\mathcal{A})$, i.e., $\mathcal{A} \subseteq \text{int}^{(\text{III})}(\mathcal{A}) \subseteq \text{pint}^{(\text{III})}(\mathcal{A})$ by Remark 4. On the other hand, for $A \in \text{pint}^{(\text{III})}(\mathcal{A})$, using part (ii) of Proposition 10, we have

$$A \in \mathcal{B}(A; \epsilon) \subseteq \mathcal{A} \oplus \Omega = (\mathcal{A}_1 \cap \mathcal{A}_2) \oplus \Omega \subseteq \mathcal{A}_1 \oplus \Omega = \mathcal{A}_1.$$

We can similarly obtain $A \in \mathcal{A}_2$, i.e., $A \in \mathcal{A}_1 \cap \mathcal{A}_2 = \mathcal{A}$. This shows that $\text{pint}^{(\text{III})}(\mathcal{A}) \subseteq \mathcal{A}$. Therefore, we conclude that the intersection of finitely many members of $p\tau^{(\text{III})}$ is a member of $p\tau^{(\text{III})}$.

Now, given a family $\{\mathcal{A}_\delta\}_{\delta \in \Lambda} \subset \mathfrak{p}\tau^{(\text{II})}$, let $\mathcal{A} = \bigcup_{\delta \in \Lambda} \mathcal{A}_\delta$. Then $A \in \mathcal{A}$ implies that $A \in \mathcal{A}_\delta$ for some $\delta \in \Lambda$. This says that
$$\mathcal{B}(A; \epsilon) \subseteq \mathcal{A}_\delta \oplus \Omega \subseteq \mathcal{A} \oplus \Omega$$
for some $\epsilon > 0$. Therefore we obtain $\mathcal{A} \subseteq \text{int}^{(\text{II})}(\mathcal{A}) \subseteq \text{pint}^{(\text{II})}(\mathcal{A})$. On the other hand, for $A \in \text{pint}^{(\text{II})}(\mathcal{A})$, we have
$$A \in \mathcal{B}(A; \epsilon) \subseteq \mathcal{A} \oplus \Omega = \mathcal{A}$$
by part (ii) of Proposition 10. This shows that $\text{pint}^{(\text{II})}(\mathcal{A}) \subseteq \mathcal{A}$, i.e., $\mathcal{A} = \text{pint}^{(\text{II})}(\mathcal{A})$. Therefore, by Remark 5, we conclude that $(\mathcal{P}(X), \mathfrak{p}\tau^{(\text{II})}) = (\mathcal{P}(X), \mathfrak{p}\tau^{(\text{III})})$ is a topological space. This completes the proof. $\square$

7. Conclusions

The hyperspace denoted by $\mathcal{P}(X)$ is the collection of all subsets of a vector space X. Under the set addition
$$A \oplus B = \{a + b : a \in A \text{ and } b \in B\}$$
and the scalar multiplication
$$\lambda A = \{\lambda a : a \in A\},$$
the hyperspace $\mathcal{P}(X)$ cannot form a vector space. The reason is that each $A \in \mathcal{P}(X)$ cannot have the additive inverse element. In this paper, the null set defined by
$$\Omega = \{A \ominus A : A \in \mathcal{P}(X)\}$$
can be treated as a kind of "zero element" of $\mathcal{P}(X)$. Although $\mathcal{P}(X)$ is not a vector space, a so-called informal norm is introduced to $\mathcal{P}(X)$, which will mimic the conventional norm. Using this informal norm, two different concepts of open balls are proposed, which are used to define many types of open sets. Therefore, we can generate many types of topologies based on these different concepts of open sets.

As we mentioned before, the theory of set-valued analysis has been applied to nonlinear analysis, differential inclusion, fixed point theory and set-valued optimization, which treats each element in $\mathcal{P}(X)$ as a subset of X. In this paper, each element of $\mathcal{P}(X)$ is treated as a "point", and the family $\mathcal{P}(X)$ is treated as a universal set. The topological structures studied in this paper may provide the potential applications in nonlinear analysis, differential inclusion, fixed point theory and set-valued optimization (or set optimization) based on the different point of view regarding the elements of $\mathcal{P}(X)$, which will be for future research.

Funding: This research received no external funding.

Conflicts of Interest: The author declares no conflict of interest.

References

1. Aubin, J.-P.; Frankowska, H. *Set-Valued Analysis*; Springer: Berlin, Germany, 1990.
2. Hu, S.; Papageorgiou, N.S. *Handbook of Multivalued Analysis I: Theory*; Kluwer Academic Publishers: New York, NY, USA, 1997.
3. Hu, S.; Papageorgiou, N.S. *Handbook of Multivalued Analysis II: Applications*; Kluwer Academic Publishers: New York, NY, USA, 2000.
4. Agarwal, R.P.; O'Regan, D. *Set-Valued Mappings with Applications in Nonlinear Analysis*; Taylor and Francis: London, UK, 2002.
5. Burachik, R.S.; Iusem, A.N. *Set-Valued Mappings and Enlargements of Monotone Operators*; Springer: Berlin, Germany, 2008.
6. Tarafdar, E.U.; Chowdhury, M.S.R. *Topological Methods for Set-Valued Nonlinear Analysis*; World Scientific Publishing Company: Singapore, 2008.

7. Aubin, J.-P.; Cellina, A. *Differential Inclusion: Set-Valued Maps and Viability Theory*; Springer: Berlin, Germany, 1984.
8. Górniewicz, L. *Topological Fixed Point Theory of Multivalued Mappings*; Springer: Berlin, Germany, 2006.
9. Chen, G.-Y.; Huang, X.; Yang, X. *Vector Optimization: Set-Valued and Variational Analysis*; Springer: Berlin, Germany, 2005.
10. Khan, A.A.; Tammer, C.; Zălinescu, C. *Set-Valued Optimization*; Springer: Berlin, Germany, 2015
11. Hamel, A.H.; Heyde F.; Lohne, A.; Rudloff, B.; Schrage, C. *Set Optimization and Applications*; Springer: Berlin, Germany, 2015
12. Wu, H.-C. Solving Set Optimization Problems Based on the Concept of Null Set. *J. Math. Anal. Appl.* **2019**, *472*, 1741–1761. [CrossRef]
13. Khaleelulla, S.M. *Counterexamples in Topological Vector Spaces*; Springer: Berlin, Germany, 1982.
14. Aubin, J.-P. *Applied Functional Analysis*, 2nd ed.; John Wiley and Sons: Hoboken, NJ, USA, 2000.
15. Adasch, N.; Ernst, B.; Keim, D. *Topological Vector Spaces: The Theory without Convexity Conditions*; Springer: Berlin, Germany, 1978.
16. Conway, J.B. *A Course in Functional Analysis*, 2nd ed.; Springer: Berlin, Germany, 1990.
17. Dunford, N.; Schwartz, J.T. *Linear Operators, Part I: General Theory*; John Wiley & Sons: Hoboken, NJ, USA, 1957.
18. Holmes, R.B. *Geometric Functional Analysis and Its Applications*; Springer: Berlin, Germany, 1975.
19. Peressini, A.L. *Ordered Topological Vector Spaces*; Harper and Row: New York, NY, USA, 1967.
20. Schaefer, H.H. *Topological Vector Spaces*; Springer: Berlin, Germany, 1966.
21. Rudin, W. *Functional Analysis*; McGraw-Hill: New York, NY, USA, 1973.
22. Wong, Y.-C.; Ng, K.-F. *Partially Ordered Topological Vector Spaces*; Oxford University Press: Oxford, UK, 1973.
23. Yosida, K., *Functional Analysis*, 6th ed.; Springer: New York, NY, USA, 1980.

 © 2019 by the author. Licensee MDPI, Basel, Switzerland. This article is an open access article distributed under the terms and conditions of the Creative Commons Attribution (CC BY) license (http://creativecommons.org/licenses/by/4.0/).

Article

Modified Inertial Hybrid and Shrinking Projection Algorithms for Solving Fixed Point Problems

Bing Tan [1], Shanshan Xu [2] and Songxiao Li [1,*]

[1] Institute of Fundamental and Frontier Sciences, University of Electronic Science and Technology of China, Chengdu 611731, China; bingtan72@std.uestc.edu.cn
[2] School of Mathematical Sciences, University of Electronic Science and Technology of China, Chengdu 611731, China; xss0702@std.uestc.edu.cn
* Correspondence: jyulsx@163.com

Received: 22 January 2020; Accepted: 10 February 2020; Published: 12 February 2020

Abstract: In this paper, we introduce two modified inertial hybrid and shrinking projection algorithms for solving fixed point problems by combining the modified inertial Mann algorithm with the projection algorithm. We establish strong convergence theorems under certain suitable conditions. Finally, our algorithms are applied to convex feasibility problem, variational inequality problem, and location theory. The algorithms and results presented in this paper can summarize and unify corresponding results previously known in this field.

Keywords: conjugate gradient method; steepest descent method; hybrid projection; shrinking projection; inertial Mann; strongly convergence; nonexpansive mapping

MSC: 49J40; 47H05; 90C52

---

1. Introduction

Throughout this paper, let C denote a nonempty closed convex subset of real Hilbert spaces H with standard inner products $\langle \cdot, \cdot \rangle$ and induced norms $\|\cdot\|$. For all $x, y \in C$, there is $\|Tx - Ty\| \leq \|x - y\|$, and the mapping $T : C \to C$ is said to be nonexpansive. We use $\text{Fix}(T) := \{x \in C : Tx = x\}$ to represent the set of fixed points of a mapping $T : C \to C$. The main purpose of this paper is to consider the following fixed point problem: Find $x^* \in C$, such that $T(x^*) = x^*$, where $T : C \to C$ is nonexpansive with $\text{Fix}(T) \neq \emptyset$.

There are various specific applications for approximating fixed point problems with nonexpansive mappings, such as monotone variational inequalities, convex optimization problems, convex feasibility problems, and image restoration problems; see, e.g., [1–6]. It is well known that the Picard iteration method may not converge, and an effective way to overcome this difficulty is to use Mann iterative method, which generates sequences $\{x_n\}$ recursively:

$$x_{n+1} = \alpha_n x_n + (1 - \alpha_n) T x_n, \quad n \geq 0, \tag{1}$$

the iterative sequence $\{x_n\}$ defined by (1) weakly converges to a fixed point of T when the condition $\sum_{n=1}^{\infty} \alpha_n (1 - \alpha_n) = +\infty$ is satisfied, where $\{\alpha_n\} \subset (0, 1)$.

Many practical applications, for instance, quantum physics and image reconstruction, are in infinite dimensional spaces. To investigate these problems, norm convergence is usually preferable to weak convergence. Therefore, modifying the Mann iteration method to obtain strong convergence is an important research topic. For recent works, see [7–12] and the references therein. On the other hand, the Ishikawa iterative method can strongly converge to the fixed point of nonlinear mappings.

For more discussion, see [13–16]. In 2003, Nakajo and Takahashi [17] established strong convergence of the Mann iteration with the aid of projections. Indeed, they considered the following algorithm:

$$\begin{cases} y_n = \alpha_n x_n + (1 - \alpha_n) T x_n, \\ C_n = \{z \in C : \|y_n - z\| \leq \|x_n - z\|\}, \\ Q_n = \{z \in C : \langle x_n - z, x_n - x_0 \rangle \leq 0\}, \\ x_{n+1} = P_{C_n \cap Q_n} x_0, \quad n \in N, \end{cases} \quad (2)$$

where $\{\alpha_n\} \subset [0,1)$, T is a nonexpansive mapping on C and $P_{C_n \cap Q_n}$ is the metric projection from C onto $C_n \cap Q_n$. This method is now referred to as the hybrid projection method. Inspired by Nakajo and Takahashi [17], Takahashi, Takeuchi, and Kubota [18] also proposed a projection-based method and obtained strong convergence results, which is now called the shrinking projection method. In recent years, many authors gained new algorithms based on projection method; see [10,18–23].

Generally, the Mann algorithm has a slow convergence rate. In recent years, there has been tremendous interest in developing the fast convergence of algorithms, especially for the inertial type extrapolation method, which was first proposed by Polyak in [24]. Recently, some researchers have constructed different fast iterative algorithms by means of inertial extrapolation techniques, for example, inertial Mann algorithm [25], inertial forward–backward splitting algorithm [26,27], inertial extragradient algorithm [28,29], inertial projection algorithm [30,31], and fast iterative shrinkage–thresholding algorithm (FISTA) [32]. The results of these algorithms and other related ones not only theoretically analyze the convergence properties of inertial type extrapolation algorithms, but also numerically demonstrate their computational performance on some data analysis and image processing problems.

In 2008, Mainge [25] proposed the following inertial Mann algorithm based on the idea of the Mann algorithm and inertial extrapolation:

$$\begin{cases} w_n = x_n + \delta_n (x_n - x_{n-1}), \\ x_{n+1} = (1 - \eta_n) w_n + \eta_n T w_n, \quad n \geq 1. \end{cases} \quad (3)$$

It should be pointed out that the iteration sequence $\{x_n\}$ defined by (3) only obtains weak convergence results under the following assumptions:

(C1) $\delta_n \in [0,1)$ and $0 < \inf_{n \geq 1} \eta_n \leq \sup_{n \geq 1} \eta_n < 1$;
(C2) $\sum_{n=1}^{\infty} \delta_n \|x_n - x_{n-1}\|^2 < +\infty$.

It should be noted that the condition (C2) is very strong, which prohibits execution of related algorithms. Recently, Bot and Csetnek [33] got rid of the condition (C2); for more details, see Theorem 5 in [33].

In 2014, Sakurai and Iiduka [34] introduced an algorithm to accelerate the Halpern fixed point algorithm in Hilbert spaces by means of conjugate gradient methods that can accelerate the convergence rate of the steepest descent method. Very recently, inspired by the work of Sakurai and Iiduka [34], Dong et al. [35] proposed a modified inertial Mann algorithm by combining the inertial method, the Picard algorithm and the conjugate gradient method. Their numerical results showed that the proposed algorithm has some advantages over other algorithms. Indeed, they obtained the following result:

Theorem 1. Let $T : C \to C$ be a nonexpansive mapping with $\text{Fix}(T) \neq \emptyset$. Set $\mu \in (0,1], \eta > 0$ and $x_0, x_1 \in H$ arbitrarily and set $d_0 := (Tx_0 - x_0)/\eta$. Define a sequence $\{x_n\}$ by the following algorithm:

$$\begin{cases} w_n = x_n + \delta_n(x_n - x_{n-1}), \\ d_{n+1} = \frac{1}{\eta}(Tw_n - w_n) + \psi_n d_n, \\ y_n = w_n + \eta d_{n+1}, \\ x_{n+1} = \mu v_n w_n + (1 - \mu v_n) y_n, n \geq 1. \end{cases} \quad (4)$$

The iterative sequence $\{x_n\}$ defined by (4) converges weakly to a point in $\text{Fix}(T)$ under the following conditions:

(D1) $\{\delta_n\} \subset [0,\delta]$ is nondecreasing with $\delta_1 = 0$ and $0 \leq \delta < 1$, $\sum_{n=1}^{\infty} \psi_n < \infty$;

(D2) Exists $v, \sigma, \varphi > 0$ such that $\varphi > \frac{\delta^2(1+\delta)+\delta\sigma}{1-\delta^2}$ and $0 < 1 - \mu v \leq 1 - \mu v_n \leq \frac{\varphi - \delta[\delta(1+\delta)+\delta\varphi+\sigma]}{\varphi[1+\delta(1+\delta)+\delta\varphi+\sigma]}$;

(D3) $\{w_n\}$ defined in (4) assume that $\{Tw_n - w_n\}$ is bounded and $\{Tw_n - y\}$ is bounded for any $y \in \text{Fix}(T)$.

Inspired and motivated by the above works, in this paper, based on the modified inertial Mann algorithm (4) and the projection algorithm (2), we propose two new modified inertial hybrid and shrinking projection algorithms, respectively. We obtain strong convergence results under some mild conditions. Finally, our algorithms are applied to a convex feasibility problem, a variational inequality problem, and location theory.

The structure of the paper is the following. Section 2 gives the mathematical preliminaries. Section 3 present modified inertial hybrid and shrinking projection algorithms for nonexpansive mappings in Hilbert spaces and analyzes their convergence. Section 4 gives some numerical experiments to compare the convergence behavior of our proposed algorithms with previously known algorithms. Section 5 concludes the paper with a brief summary.

2. Preliminaries

We use the notation $x_n \to x$ and $x_n \rightharpoonup x$ to denote the strong and weak convergence of a sequence $\{x_n\}$ to a point $x \in H$, respectively. Let $\omega_w \{x_n\} := \{x : \exists x_{n_j} \rightharpoonup x\}$ denote the weak w-limit set of $\{x_n\}$. For any $x, y \in H$ and $t \in R$, we have $\|tx + (1-t)y\|^2 = t\|x\|^2 + (1-t)\|y\|^2 - t(1-t)\|x-y\|^2$.

For any $x \in H$, there is a unique nearest point $P_C x$ in C, such that $P_C(x) := \text{argmin}_{y \in C} \|x - y\|$. P_C is called the metric projection of H onto C. $P_C x$ has the following characteristics:

$$P_C x \in C \quad \text{and} \quad \langle P_C x - x, P_C x - y \rangle \leq 0, \quad \forall y \in C. \quad (5)$$

From this characterization, the following inequality can be obtained

$$\|x - P_C x\|^2 + \|y - P_C x\|^2 \leq \|x - y\|^2, \quad \forall x \in H, \forall y \in C. \quad (6)$$

We give some special cases with simple analytical solutions:

(1) The Euclidean projection of x_0 onto an Euclidean ball $B[c, r] = \{x : \|x - c\| \leq r\}$ is given by

$$P_{B[c,r]}(x) = c + \frac{r}{\max\{\|x-c\|, r\}}(x - c).$$

(2) The Euclidean projection of x_0 onto a box $\text{Box}[\ell, u] = \{x : \ell \leq x \leq u\}$ is given by

$$P_{\text{Box}[\ell,u]}(x)_i = \min\{\max\{x_i, \ell_i\}, u_i\}.$$

(3) The Euclidean projection of x_0 onto a halfspace $H_{a,b}^- = \{x : \langle a, x \rangle \leq b\}$ is given by

$$P_{H_{a,b}^-}(x) = x - \frac{[\langle a, x \rangle - b]_+}{\|a\|^2} a.$$

Next we give some results that will be used in our main proof.

Lemma 1. *[36] Let C be a nonempty closed convex subset of real Hilbert spaces H and let $T : C \to H$ be a nonexpansive mapping with $\mathrm{Fix}(T) \neq \emptyset$. Assume that $\{x_n\}$ be a sequence in C and $x \in H$ such that $x_n \rightharpoonup x$ and $Tx_n - x_n \to 0$ as $n \to \infty$, then $x \in \mathrm{Fix}(T)$.*

Lemma 2. *[37] Let C be a nonempty closed convex subset of real Hilbert spaces H. For any $x, y, z \in H$ and $a \in R$. $\{v \in C : \|y - v\|^2 \leq \|x - v\|^2 + \langle z, v \rangle + a\}$ is convex and closed.*

Lemma 3. *[38] Let C be a nonempty closed convex subset of real Hilbert spaces H. Let $\{x_n\} \subset H$, $u \in H$ and $m = P_C u$. If $\omega_w \{x_n\} \subset C$ and satisfies the condition $\|x_n - u\| \leq \|u - m\|$, $\forall n \in N$. Then $x_n \to m$.*

3. Modified Inertial Hybrid and Shrinking Projection Algorithms

In this section, we introduce two modified inertial hybrid and shrinking projection algorithms for nonexpansive mappings in Hilbert spaces using the ideas of the inertial method, the Picard algorithm, the conjugate gradient method, and the projection method. We prove the strong convergence of the algorithms under suitable conditions.

Theorem 2. *Let C be a bounded closed convex subset of real Hilbert spaces H and let $T : C \to C$ be a nonexpansive mapping with $\mathrm{Fix}(T) \neq \emptyset$. Assume that the following conditions are satisfied:*

$$\eta > 0, \delta_n \subset [\delta_1, \delta_2], \delta_1 \in (-\infty, 0], \delta_2 \in [0, \infty), \psi_n \subset [0, \infty), \lim_{n \to \infty} \psi_n = 0, \nu_n \subset (0, \nu], 0 < \nu < 1.$$

Set $x_{-1}, x_0 \in H$ arbitrarily and set $d_0 := (Tx_0 - x_0)/\eta$. Define a sequence $\{x_n\}$ by the following:

$$\begin{cases} w_n = x_n + \delta_n(x_n - x_{n-1}), \\ d_{n+1} = \frac{1}{\eta}(Tw_n - w_n) + \psi_n d_n, \\ y_n = w_n + \eta d_{n+1}, \\ z_n = \nu_n w_n + (1 - \nu_n) y_n, \\ C_n = \left\{ z \in H : \|z_n - z\|^2 \leq \|w_n - z\|^2 - \nu_n(1 - \nu_n)\|w_n - y_n\|^2 + \xi_n \right\}, \\ Q_n = \{z \in H : \langle x_n - z, x_n - x_0 \rangle \leq 0\}, \\ x_{n+1} = P_{C_n \cap Q_n} x_0, \quad n \geq 0, \end{cases} \quad (7)$$

where the sequence $\{\xi_n\}$ is defined by $\xi_n := \eta \psi_n M_2 [\eta \psi_n M_2 + 2 M_1]$, $M_1 := \mathrm{diam}\, C = \sup_{x,y \in C} \|x - y\|$ and $M_2 := \max\left\{\max_{1 \leq k \leq n_0} \|d_k\|, \frac{2}{\eta} M_1\right\}$, where n_0 satisfies $\psi_n \leq \frac{1}{2}$ for all $n \geq n_0$. Then the iterative sequence $\{x_n\}$ defined by (7) converges to $P_{\mathrm{Fix}(T)} x_0$ in norm.

Proof. We divided our proof in three steps.
Step 1. To begin with, we need to show that $\mathrm{Fix}(T) \subset C_n \cap Q_n$. It is easy to check that C_n is convex by Lemma 2. Next we prove $\mathrm{Fix}(T) \subset C_n$ for all $n \geq 0$. Assume that $\|d_n\| \leq M_1$ for some $n \geq n_0$. The triangle inequality ensures that

$$\|d_{n+1}\| = \left\|\frac{1}{\eta}(Tw_n - w_n) + \psi_n d_n\right\| \leq \frac{1}{\eta}\|Tw_n - w_n\| + \psi_n \|d_n\| \leq M_2,$$

which implies that $\|d_n\| \leq M_2$ for all $n \geq 0$, that is, $\{d_n\}$ is bounded. Due to $w_n \in C$, we get that $\|w_n - p\| \leq M_1$ for all $u \in \text{Fix}(T)$. From the definition of $\{y_n\}$ and nonexpansivity of T we obtain

$$\|y_n - u\| = \left\|w_n + \eta\left(\frac{1}{\eta}(Tw_n - w_n) + \psi_n d_n\right) - u\right\| = \|Tw_n + \eta\psi_n d_n - u\|$$
$$\leq \|w_n - u\| + \eta\psi_n M_2.$$

Therefore,

$$\|z_n - u\|^2 = \|v_n(w_n - u) + (1 - v_n)(y_n - u)\|^2$$
$$= v_n\|w_n - u\|^2 + (1 - v_n)\|y_n - u\|^2 - v_n(1 - v_n)\|w_n - y_n\|^2$$
$$\leq \|w_n - u\|^2 + 2\eta\psi_n M_2\|w_n - u\| + (\eta\psi_n M_2)^2 - v_n(1 - v_n)\|w_n - y_n\|^2$$
$$\leq \|w_n - u\|^2 - v_n(1 - v_n)\|w_n - y_n\|^2 + \xi_n,$$

where $\xi_n = \eta\psi_n M_2 [\eta\psi_n M_2 + 2M_1]$. Thus, we have $u \in C_n$ for all $n \geq 0$ and hence $\text{Fix}(T) \subset C_n$ for all $n \geq 0$. On the other hand, it is easy to see that $\text{Fix}(T) \subset C = Q_0$ when $n = 0$. Suppose that $\text{Fix}(T) \subset Q_{n-1}$, by combining the fact that $x_n = P_{C_{n-1} \cap Q_{n-1}} x_0$ and (5) we obtain $\langle x_n - z, x_n - x_0 \rangle \leq 0$ for any $z \in C_{n-1} \cap Q_{n-1}$. According to the induction assumption we have $\text{Fix}(T) \subset C_{n-1} \cap Q_{n-1}$, and it follows from the definition of Q_n that $\text{Fix}(T) \subset Q_n$. Therefore, we get $\text{Fix}(T) \subset C_n \cap Q_n$ for all $n \geq 0$.

Step 2. We prove that $\|x_{n+1} - x_n\| \to 0$ as $n \to \infty$. Combining the definition of Q_n and $\text{Fix}(T) \subset Q_n$, we obtain

$$\|x_n - x_0\| \leq \|u - x_0\|, \quad \text{for all } u \in \text{Fix}(T).$$

We note that $\{x_n\}$ is bounded and

$$\|x_n - x_0\| \leq \|x^* - x_0\|, \quad \text{where } x^* = P_{\text{Fix}(T)} x_0. \tag{8}$$

The fact $x_{n+1} \in Q_n$, we have $\|x_n - x_0\| \leq \|x_{n+1} - x_0\|$, which means that $\lim_{n \to \infty} \|x_n - x_0\|$ exists. Using (6), one sees that

$$\|x_n - x_{n+1}\|^2 \leq \|x_{n+1} - x_0\|^2 - \|x_n - x_0\|^2,$$

which implies that $\|x_{n+1} - x_n\| \to 0$ as $n \to \infty$. Next, by the definition of w_n, we have

$$\|w_n - x_n\| = |\delta_n|\|x_n - x_{n-1}\| \leq \delta_2\|x_n - x_{n-1}\| \to 0 \ (n \to \infty),$$

which further yields that

$$\|w_n - x_{n+1}\| \leq \|w_n - x_n\| + \|x_n - x_{n+1}\| \to 0 \ (n \to \infty).$$

Step 3. It remains to show that $x_n \to x^*$, where $x^* = P_{\text{Fix}(T)} x_0$. From $x_{n+1} \in C_n$ we get

$$\|z_n - x_{n+1}\|^2 \leq \|w_n - x_{n+1}\|^2 - v_n(1 - v_n)\|w_n - y_n\|^2 + \xi_n.$$

Therefore,

$$\|z_n - x_{n+1}\| \leq \|w_n - x_{n+1}\| + \sqrt{\xi_n}.$$

On the other hand, since $z_n = \nu_n w_n + (1-\nu_n) T w_n + (1-\nu_n) \eta \psi_n d_n$ and $\nu_n \leq \nu$, we obtain

$$\|Tw_n - w_n\| = \frac{1}{1-\nu_n} \|z_n - w_n - (1-\nu_n)\eta\psi_n d_n\|$$
$$\leq \frac{1}{1-\nu} \|z_n - w_n\| + \eta\psi_n \|d_n\|$$
$$\leq \frac{1}{1-\nu} (\|z_n - x_{n+1}\| + \|w_n - x_{n+1}\|) + \eta\psi_n M_2$$
$$\leq \frac{1}{1-\nu} \left(2\|w_n - x_{n+1}\| + \sqrt{\xi_n}\right) + \eta\psi_n M_2 \to 0 \, (n \to \infty).$$

Consequently,

$$\|Tx_n - x_n\| \leq \|Tx_n - Tw_n\| + \|Tw_n - w_n\| + \|w_n - x_n\|$$
$$\leq 2\|w_n - x_n\| + \|Tw_n - w_n\| \to 0 \, (n \to \infty). \tag{9}$$

In view of (9) and Lemma 1, it follows that every weak limit point of $\{x_n\}$ is a fixed point of T. i.e., $\omega_w \{x_n\} \subset \text{Fix}(T)$. By means of Lemma 3 and the inequality (8), we get that $\{x_n\}$ converges to $P_{\text{Fix}(T)} x_0$ in norm. The proof is complete. □

Theorem 3. *Let C be a bounded closed convex subset of real Hilbert spaces H and let $T : C \to C$ be a nonexpansive mapping with $\text{Fix}(T) \neq \emptyset$. Assume that the following conditions are satisfied:*

$$\eta > 0, \delta_n \subset [\delta_1, \delta_2], \delta_1 \in (-\infty, 0], \delta_2 \in [0, \infty), \psi_n \subset [0, \infty), \lim_{n \to \infty} \psi_n = 0, \nu_n \subset (0, \nu], 0 < \nu < 1.$$

Set $x_{-1}, x_0 \in H$ arbitrarily and set $d_0 := (Tx_0 - x_0)/\eta$. Define a sequence $\{x_n\}$ by the following:

$$\begin{cases} w_n = x_n + \delta_n (x_n - x_{n-1}), \\ d_{n+1} = \frac{1}{\eta}(Tw_n - w_n) + \psi_n d_n, \\ y_n = w_n + \eta d_{n+1}, \\ z_n = \nu_n w_n + (1-\nu_n) y_n, \\ C_{n+1} = \left\{ z \in C_n : \|z_n - z\|^2 \leq \|w_n - z\|^2 - \nu_n(1-\nu_n)\|w_n - y_n\|^2 + \xi_n \right\}, \\ x_{n+1} = P_{C_{n+1}} x_0, \quad n \geq 0. \end{cases} \tag{10}$$

where the sequence $\{\xi_n\}$ is defined by $\xi_n := \eta \psi_n M_2 [\eta \psi_n M_2 + 2M_1]$, $M_1 := \text{diam } C = \sup_{x,y \in C} \|x - y\|$ and $M_2 := \max \left\{ \max_{1 \leq k \leq n_0} \|d_k\|, \frac{2}{\eta} M_1 \right\}$, where n_0 satisfies $\psi_n \leq \frac{1}{2}$ for all $n \geq n_0$. Then the iterative sequence $\{x_n\}$ defined by (10) converges to $P_{\text{Fix}(T)} x_0$ in the norm.

Proof. We divided our proof in three steps.
Step 1. Our first goal is to show that $\text{Fix}(T) \subset C_{n+1}$ for all $n \geq 0$. According to Step 1 in Theorem 2, for all $u \in \text{Fix}(T)$, we obtain

$$\|z_n - u\|^2 \leq \|w_n - u\|^2 - \nu_n(1-\nu_n)\|w_n - y_n\|^2 + \xi_n.$$

Therefore, $u \in C_{n+1}$ for each $n \geq 0$ and hence $\text{Fix}(T) \subset C_{n+1} \subset C_n$.
Step 2. As mentioned above, the next thing to do in the proof is show that $\|x_{n+1} - x_n\| \to 0$ as $n \to \infty$. Using the fact that $x_n = P_{C_n} x_0$ and $\text{Fix}(T) \subset C_n$, we have

$$\|x_n - x_0\| \leq \|u - x_0\|, \quad \text{for all } u \in \text{Fix}(T).$$

It follows that $\{x_n\}$ is bounded, in addition, we note that

$$\|x_n - x_0\| \leq \|x^* - x_0\|, \quad \text{where } x^* = P_{\text{Fix}(T)} x_0. \tag{11}$$

On the other hand, since $x_{n+1} \in C_n$, we obtain $\|x_n - x_0\| \le \|x_{n+1} - x_0\|$, which implies that $\lim_{n\to\infty} \|x_n - x_0\|$ exists. In view of (6), we have

$$\|x_n - x_{n+1}\|^2 \le \|x_{n+1} - x_0\|^2 - \|x_n - x_0\|^2,$$

which further implies that $\lim_{n\to\infty} \|x_{n+1} - x_n\| = 0$. Also, we have $\lim_{n\to\infty} \|w_n - x_n\| = 0$ and $\lim_{n\to\infty} \|w_n - x_{n+1}\| = 0$.

Step 3. Finally, we have to show that $x_n \to x^*$, where $x^* = P_{\text{Fix}(T)} x_0$. The remainder of the argument is analogous to that in Theorem 2 and is left to the reader. □

Remark 1. *We remark here that the modified inertial hybrid projection algorithm (7) (in short, MIHPA) and the modified inertial shrinking projection algorithm (10) (in short, MISPA) contain some previously known results. When $\delta_n = 0$ and $\psi_n = 0$, the MIHPA becomes the hybrid projection algorithm (in short, HPA) proposed by Nakajo and Takahashi [17] and the MISPA becomes the shrinking projection algorithm (in short, SPA) proposed by Takahashi, Takeuchi, and Kubota [18]. When $\delta_n = 0$ and $\psi_n \ne 0$, the MIHPA becomes the modified hybrid projection algorithm (in short, MHPA) proposed by Dong et al. [35], the MISPA becomes the modified shrinking projection algorithm (in short, MSPA).*

4. Numerical Experiments

In this section, we provide three numerical applications to demonstrate the computational performance of our proposed algorithms and compare them with some existing ones. All the programs are performed in MATLAB2018a on a personal computer Intel(R) Core(TM) i5-8250U CPU @ 1.60GHz 1.800 GHz, RAM 8.00 GB.

Example 1. *As an example, we consider the convex feasibility problem, for any nonempty closed convex set $C_i \subset R^N$ ($i = 0, 1, \ldots, m$), we find $x^* \in C := \bigcap_{i=0}^m C_i$, where one supposes that $C \ne \emptyset$. A mapping $T : R^N \to R^N$ is defined by $T := P_0\left(\frac{1}{m}\sum_{i=1}^m P_i\right)$, where $P_i = P_{C_i}$ stands for the metric projection onto C_i. It follows from P_i being nonexpansive that the mapping T is also nonexpansive. Furthermore, we note that $\text{Fix}(T) = \text{Fix}(P_0) \cap \bigcap_{i=1}^m \text{Fix}(P_i) = C_0 \cap \bigcap_{i=1}^m C_i = C$. In this experiment, we set C_i as a closed ball with center $c_i \in R^N$ and radius $r_i > 0$. Thus P_i can be computed with*

$$P_i(x) := \begin{cases} c_i + \frac{r_i}{\|c_i - x\|}(x - c_i), & \text{if } \|c_i - x\| > r_i; \\ x, & \text{if } \|c_i - x\| \le r_i. \end{cases}$$

Choose $r_i = 1$ ($i = 0, 1, \ldots, m$), $c_0 = [0, 0, \ldots, 0]$, $c_1 = [1, 0, \ldots, 0]$, and $c_2 = [-1, 0, \ldots, 0]$. c_i is randomly selected from $(-1/\sqrt{N}, 1/\sqrt{N})^N$ ($i = 3, \ldots, m$). We have $\text{Fix}(T) = \{0\}$ from the special choice of c_1, c_2 and r_1, r_2. In Algorithms (7) and (10), setting $m = 30$, $N = 30$, $\eta = 1$, $\psi_n = \frac{1}{100(n+1)^2}$, $v_n = 0.1$. When the iteration error $E_n = \|x_n - Tx_n\|_2 < 10^{-2}$ is satisfied, the iteration stops. We test our algorithms under different inertial parameters and initial values. Results are shown in Table 1, where "Iter." represents the number of iterations.

Table 1. Computational results for Example 1.

Algorithm	Initial Value	δ_n	0	0.1	0.2	0.3	0.4	0.5	0.6	0.7	0.8	0.9
MIHPA	rand(N,1)	Iter.	223	248	218	239	283	245	258	249	248	247
MISPA			127	137	148	159	169	163	167	187	186	190
MIHPA	ones(N,1)	Iter.	327	315	407	354	342	356	377	391	348	349
MISPA			174	189	181	199	217	208	279	250	243	256
MIHPA	10rand(N,1)	Iter.	1057	1377	1522	1494	1307	1119	1261	1098	1005	1070
MISPA			549	570	704	698	845	852	987	856	1003	975
MIHPA	−10rand(N,1)	Iter.	445	410	574	504	657	716	729	730	659	682
MISPA			316	313	350	416	423	386	427	392	516	556

Example 2. *Our another example is to consider the following variational inequality problem (in short, VI). For any nonempty closed convex set $C \subset R^N$,*

$$\text{find } x^* \in C \text{ such that } \langle f(x^*), x - x^* \rangle \geq 0, \quad \forall x \in C, \tag{12}$$

where $f : R^N \to R^N$ is a mapping. Take VI(C, f) denote the solution of VI (12). $T : R^N \to R^N$ is defined by $T := P_C(I - \gamma f)$, where $0 < \gamma < 2/L$, and L is the Lipschitz constant of the mapping f. In [39], Xu showed that T is an averaged mapping, i.e., T can be seen as the average of an identity mapping I and a nonexpansive mapping. It follows that $\text{Fix}(T) = \text{VI}(C, f)$, we can solve VI (12) by finding the fixed point of T. Taking $f : R^2 \to R^2$ as follows:

$$f(x, y) = (2x + 2y + \sin(x), -2x + 2y + \sin(y)), \quad \forall x, y \in R.$$

The feasible set C is given by $C = \{x \in R^2 | -10\mathbf{e} \leq x \leq 10\mathbf{e}\}$, where $\mathbf{e} = (1,1)^\top$. It is not hard to check that f is Lipschitz continuous with constant $L = \sqrt{26}$ and 1-strongly monotone [40]. Therefore, VI (12) has a unique solution $\mathbf{x}^* = (0,0)^\top$.

We use the Algorithm (7) (MIHPA), the Algorithm (10) (MISPA), the modified hybrid projection algorithm (MHPA), the modified shrinking projection algorithm (MSPA), the hybrid projection algorithm (HPA), and the shrinking projection algorithm (SPA) to solve Example 2. Setting $\gamma = 0.9/\sqrt{26}$, $\eta = 1$, $\psi_n = \frac{1}{100(n+1)^2}$, $v_n = 0$ (we consider that T is an average mapping). The initial values are randomly generated by the MATLAB function rand(2,1). We use $E_n = \|x_n - x^*\|_2$ to denote the iteration error of algorithms, and the maximum iteration 300 as the stopping criterion. Results are reported in Table 2, where "Iter." denotes the number of iterations.

Table 2. Computational results for Example 2.

	HPA		SPA		MHPA		MSPA		MIHPA		MISPA	
Iter.	x_n	E_n	x_n	E_n	x_n	E_n	x_n	E_n	x_n	E_n	x_n	E_n
1	(0.2944,0.8061)	0.8582	(0.2944,0.8061)	0.8582	(0.4607,0.8706)	0.9850	(0.4607,0.8706)	0.9850	(0.4607,0.8706)	0.9850	(0.4607,0.8706)	0.9850
50	(0.0049,0.0164)	0.0171	(0.0000,0.0001)	0.0001	(0.0142,0.0357)	0.0384	(0.0142,0.0264)	0.0300	(0.0094,0.0357)	0.0369	(0.0142,0.0278)	0.0312
100	(0.0006,0.0017)	0.0018	(0.0000,0.0000)	0.0000	(0.0116,0.0110)	0.0159	(0.0072,0.0133)	0.0151	(0.0096,0.0144)	0.0173	(0.0067,0.0137)	0.0153
200	(-0.0003,0.0013)	0.0014	(0.0000,0.0000)	0.0000	(0.0059,0.0053)	0.0080	(0.0034,0.0068)	0.0076	(0.0061,0.0047)	0.0077	(0.0036,0.0060)	0.0070
300	(0.0007,0.0003)	**0.0008**	(0.0000,0.0000)	**0.0000**	(0.0043,0.0030)	**0.0053**	(0.0021,0.0045)	**0.0049**	(0.0045,0.0038)	**0.0058**	(0.0018,0.0053)	**0.0056**

Example 3. *The Fermat–Weber problem is a famous model in location theory. It is can be formulated mathematically as the problem of finding $\mathbf{x} \in R^n$ that solves*

$$\min_{\mathbf{x}} \left\{ f(\mathbf{x}) := \sum_{i=1}^{m} \omega_i \|\mathbf{x} - \mathbf{a}_i\|_2 \right\}, \tag{13}$$

where $\omega_i > 0$ are given weights and $\mathbf{a}_i \in R^n$ are anchor points. It is easy to check that the objective function f in (13) is convex and coercive. Therefore, the problem has a nonempty solution set. It should be noted that f is not differentiable at the anchor points. The most famous method to solve the problem (13) is the Weiszfeld algorithm; see [41] for more discussion. Weiszfeld proposed the following fixed point algorithm: $x_{n+1} = T(x_n), n \in N$. The mapping $T : R^n \setminus \mathbf{A} \longmapsto R^n$ is defined by $T(\mathbf{x}) := \frac{1}{\sum_{i=1}^{m} \frac{\omega_i}{\|\mathbf{x} - \mathbf{a}_i\|}} \sum_{i=1}^{m} \frac{\omega_i \mathbf{a}_i}{\|\mathbf{x} - \mathbf{a}_i\|}$, where $\mathbf{A} = \{\mathbf{a}_1, \mathbf{a}_2, \ldots, \mathbf{a}_m\}$. We consider a small example with $n = 2, m = 4$ anchor points,

$$\mathbf{a}_1 = \begin{pmatrix} 0 \\ 0 \end{pmatrix}, \mathbf{a}_2 = \begin{pmatrix} 10 \\ 0 \end{pmatrix}, \mathbf{a}_3 = \begin{pmatrix} 0 \\ 10 \end{pmatrix}, \mathbf{a}_4 = \begin{pmatrix} 10 \\ 10 \end{pmatrix},$$

and $\omega_i = 1$ for all i. It follows from the special selection of anchor points $\mathbf{a}_i$ ($i = 1, 2, 3, 4$) that the optimal value of (13) is $\mathbf{x}^* = (5, 5)^\top$.

We use the same algorithms as in Example 2, and our parameter settings are as follows, setting $\eta = 1$, $\psi_n = \frac{1}{100(n+1)^2}$, $v_n = 0.1$. We use $E_n = \|x_n - x^*\|_2 < 10^{-4}$ or maximum iteration 300 as the stopping criterion. The initial values are randomly generated by the MATLAB function 10rand(2,1). Figures 1 and 2 show the convergence behavior of iterative sequence $\{x_n\}$ and iteration error E_n, respectively.

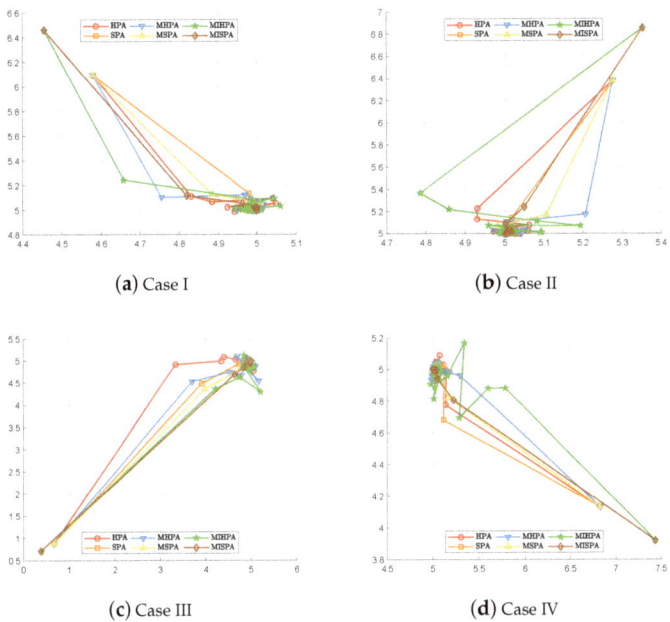

Figure 1. Convergence process at different initial values for Example 3.

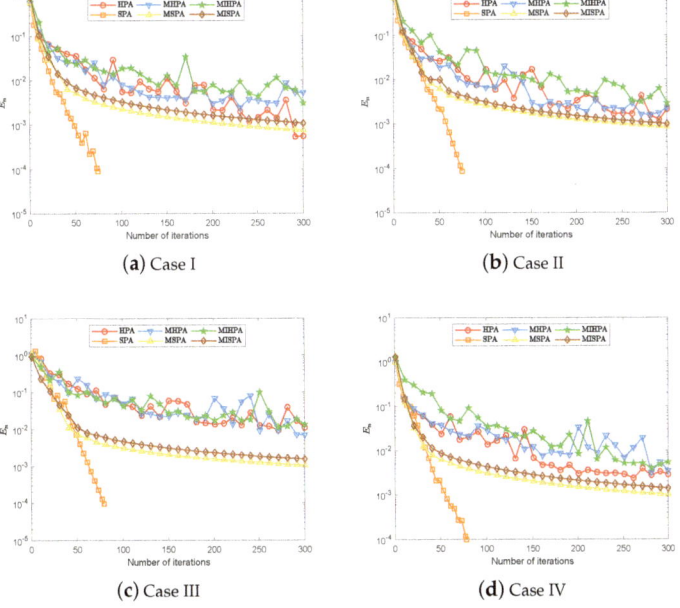

Figure 2. Convergence behavior of iteration error $\{E_n\}$ for Example 3.

Remark 2. *From Examples 1–3, we know that our proposed algorithms are effective and easy to implement. Moreover, initial values do not affect the computational performance of our algorithms. However, it should be mentioned that the MIHPA algorithm, the MISPA algorithm, the MHPA algorithm, and the MSPA algorithm will slow down the speed and accuracy of the HPA algorithm and the SPA algorithm. The acceleration may be eliminated by the projection onto the set C_n and Q_n and C_{n+1}.*

5. Conclusions

In this paper, we proposed two modified inertial hybrid and shrinking projection algorithms based on the inertial method, the Picard algorithm, the conjugate gradient method, and the projection method. We could then work with the strong convergence theorems under suitable conditions. However, numerical experiments showed that our algorithms cannot accelerate some previously known algorithms.

Author Contributions: Supervision, S.L.; Writing—original draft, B.T. and S.X. All authors have read and agreed to the published version of the manuscript.

Funding: This research received no external funding.

Acknowledgments: We greatly appreciate the reviewers for their helpful comments and suggestions.

Conflicts of Interest: The authors declare no conflict of interest.

References

1. Qin, X.; Yao, J.C. A viscosity iterative method for a split feasibility problem. *J. Nonlinear Convex Anal.* **2019**, *20*, 1497–1506.
2. Cho, S.Y. Generalized mixed equilibrium and fixed point problems in a Banach space. *J. Nonlinear Sci. Appl.* **2016**, *9*, 1083–1092. [CrossRef]
3. Nguyen, L.V.; Ansari, Q.H.; Qin, X. Linear conditioning, weak sharpness and finite convergence for equilibrium problems. *J. Glob. Optim.* **2020**. [CrossRef]
4. Dehaish, B.A.B. A regularization projection algorithm for various problems with nonlinear mappings in Hilbert spaces. *J. Inequal. Appl.* **2015**, *2015*, 1–14. [CrossRef]
5. Dehaish, B.A.B. Weak and strong convergence of algorithms for the sum of two accretive operators with applications. *J. Nonlinear Convex Anal.* **2015**, *16*, 1321–1336.
6. Qin, X.; An, N.T. Smoothing algorithms for computing the projection onto a Minkowski sum of convex sets. *Comput. Optim. Appl.* **2019**, *74*, 821–850. [CrossRef]
7. Takahahsi, W.; Yao, J.C. The split common fixed point problem for two finite families of nonlinear mappings in Hilbert spaces. *J. Nonlinear Convex Anal.* **2019**, *20*, 173–195.
8. An, N.T.; Qin, X. Solving k-center problems involving sets based on optimization techniques. *J. Glob. Optim.* **2020**, *76*, 189–209. [CrossRef]
9. Cho, S.Y.; Kang, S.M. Approximation of common solutions of variational inequalities via strict pseudocontractions. *Acta Math. Sci.* **2012**, *32*, 1607–1618. [CrossRef]
10. Takahashi, W. The shrinking projection method for a finite family of demimetric mappings with variational inequality problems in a Hilbert space. *Fixed Point Theory* **2018**, *19*, 407–419. [CrossRef]
11. Qin, X.; Cho, S.Y.; Wang, L. A regularization method for treating zero points of the sum of two monotone operators. *Fixed Point Theory Appl.* **2014**, *2014*, 75. [CrossRef]
12. Chang, S.S.; Wen, C.F.; Yao, J.C. Zero point problem of accretive operators in Banach spaces. *Bull. Malays. Math. Sci. Soc.* **2019**, *42*, 105–118. [CrossRef]
13. Tan, K.K.; Xu, H.K. Approximating fixed points of non-expansive mappings by the Ishikawa iteration process. *J. Math. Anal. Appl.* **1993**, *178*, 301–301. [CrossRef]
14. Sharma, S.; Deshpande, B. Approximation of fixed points and convergence of generalized Ishikawa iteration. *Indian J. Pure Appl. Math.* **2002**, *33*, 185–191.
15. Singh, A.; Dimri, R.C. On the convergence of Ishikawa iterates to a common fixed point for a pair of nonexpansive mappings in Banach spaces. *Math. Morav.* **2010**, *14*, 113–119. [CrossRef]

16. De la Sen, M.; Abbas, M. On best proximity results for a generalized modified Ishikawa's iterative scheme driven by perturbed 2-cyclic like-contractive self-maps in uniformly convex Banach spaces. *J. Math.* **2019**, *2019*, 1356918. [CrossRef]
17. Nakajo, K.; Takahashi, W. Strong convergence theorems for nonexpansive mappings and nonexpansive semigroups. *J. Math. Anal. Appl.* **2003**, *279*, 372–379. [CrossRef]
18. Takahashi, W.; Takeuchi, Y.; Kubota, R. Strong convergence theorems by hybrid methods for families of nonexpansive mappings in Hilbert spaces. *J. Math. Anal. Appl.* **2008**, *341*, 276–286. [CrossRef]
19. Cho, S.Y. Strong convergence analysis of a hybrid algorithm for nonlinear operators in a Banach space. *J. Appl. Anal. Comput.* **2018**, *8*, 19–31.
20. Qin, X.; Cho, S.Y.; Wang, L. Iterative algorithms with errors for zero points of m-accretive operators. *Fixed Point Theory Appl.* **2013**, *2013*, 148. [CrossRef]
21. Chang, S.S.; Wen, C.F.; Yao, J.C. Common zero point for a finite family of inclusion problems of accretive mappings in Banach spaces. *Optimization* **2018**, *67*, 1183–1196. [CrossRef]
22. Qin, X.; Cho, S.Y. Convergence analysis of a monotone projection algorithm in reflexive Banach spaces. *Acta Math. Sci.* **2017**, *37*, 488–502. [CrossRef]
23. He, S.; Dong, Q.-L. The combination projection method for solving convex feasibility problems. *Mathematics* **2018**, *6*, 249. [CrossRef]
24. Polyak, B.T. Some methods of speeding up the convergence of iteration methods. *Comput. Math. Math. Phys.* **1964**, *4*, 1–17. [CrossRef]
25. Maingé, P.E. Convergence theorems for inertial KM-type algorithms. *J. Comput. Appl. Math.* **2008**, *219*, 223–236. [CrossRef]
26. Lorenz, D.; Pock, T. An inertial forward-backward algorithm for monotone inclusions. *J. Math. Imaging Vis.* **2015**, *51*, 311–325. [CrossRef]
27. Qin, X.; Wang, L.; Yao, J.C. Inertial splitting method for maximal monotone mappings. *J. Nonlinear Convex Anal.* **2020**, in press.
28. Thong, D.V.; Hieu, D.V. Inertial extragradient algorithms for strongly pseudomonotone variational inequalities. *J. Comput. Appl. Math.* **2018**, *341*, 80–98. [CrossRef]
29. Luo, Y.L.; Tan, B. A self-adaptive inertial extragradient algorithm for solving pseudo-monotone variational inequality in Hilbert spaces. *J. Nonlinear Convex Anal.* **2020**, in press.
30. Liu, L.; Qin, X. On the strong convergence of a projection-based algorithm in Hilbert spaces. *J. Appl. Anal. Comput.* **2020**, *10*, 104–117.
31. Tan, B.; Xu, S.S.; Li, S. Inertial shrinking projection algorithms for solving hierarchical variational inequality problems. *J. Nonlinear Convex Anal.* **2020**, in press.
32. Beck, A.; Teboulle, M. A fast iterative shrinkage-thresholding algorithm for linear inverse problems. *SIAM J. Imaging Sci.* **2009**, *2*, 183–202. [CrossRef]
33. Boţ, R.I.; Csetnek, E.R.; Hendrich, C. Inertial Douglas–Rachford splitting for monotone inclusion problems. *Appl. Math. Comput.* **2015**, *256*, 472–487. [CrossRef]
34. Sakurai, K.; Iiduka, H. Acceleration of the Halpern algorithm to search for a fixed point of a nonexpansive mapping. *Fixed Point Theory Appl.* **2014**, *2014*, 202. [CrossRef]
35. Dong, Q.-L.; Yuan, H.Bi.; Cho, Y.J.; Rassias, T.M. Modified inertial Mann algorithm and inertial CQ-algorithm for nonexpansive mappings. *Optim. Lett.* **2018**, *12*, 87–102. [CrossRef]
36. Bauschke, H.H.; Combettes, P.L. *Convex Analysis and Monotone Operator Theory in Hilbert Spaces*; Springer: New York, NY, USA, 2011; Volume 48,
37. Kim, T.H.; Xu, H.K. Strong convergence of modified Mann iterations for asymptotically nonexpansive mappings and semigroups. *Nonlinear Anal.* **2006**, *64*, 1140–1152. [CrossRef]
38. Martinez-Yanes, C.; Xu, H.K. Strong convergence of the CQ method for fixed point iteration processes. *Nonlinear Anal.* **2006**, *64*, 2400–2411. [CrossRef]
39. Xu, H.K. Averaged mappings and the gradient-projection algorithm. *J. Optim. Theory Appl.* **2011**, *150*, 360–378. [CrossRef]

40. Dong, Q.-L.; Cho, Y.J.; Zhong, L.L.; Rassias, T.M. Inertial projection and contraction algorithms for variational inequalities. *J. Glob. Optim.* **2018**, *70*, 687–704. [CrossRef]
41. Beck, A.; Sabach, S. Weiszfeld's method: Old and new results. *J. Optim. Theory Appl.* **2015**, *164*, 1–40. [CrossRef]

© 2020 by the authors. Licensee MDPI, Basel, Switzerland. This article is an open access article distributed under the terms and conditions of the Creative Commons Attribution (CC BY) license (http://creativecommons.org/licenses/by/4.0/).

Article

Convergence Theorems for Modified Implicit Iterative Methods with Perturbation for Pseudocontractive Mappings

Jong Soo Jung

Department of Mathematics, Dong-a University, Busan 49315, Korea; jungjs@dau.ac.kr

Received: 14 October 2019; Accepted: 26 December 2019 ; Published: 2 January 2020

Abstract: In this paper, first, we introduce a path for a convex combination of a pseudocontractive type of mappings with a perturbed mapping and prove strong convergence of the proposed path in a real reflexive Banach space having a weakly continuous duality mapping. Second, we propose two modified implicit iterative methods with a perturbed mapping for a continuous pseudocontractive mapping in the same Banach space. Strong convergence theorems for the proposed iterative methods are established. The results in this paper substantially develop and complement the previous well-known results in this area.

Keywords: modified implicit iterative methods with perturbed mapping; pseudocontractive mapping; strongly pseudocontractive mapping; nonexpansive mapping; weakly continuous duality mapping; fixed point

1. Introduction

Let E be a real Banach space, and let E^* be the dual space of E. Let C be a nonempty closed convex subset of E. Recall that a mapping $f : C \to C$ is called *contractive* if there exists $k \in (0,1)$ such that $\|fx - fy\| \leq k\|x - y\|$, $\forall x, y \in C$ and that a mapping $S : C \to C$ is called *nonexpansive* if $\|Sx - Sy\| \leq \|x - y\|$, $\forall x, y \in C$.

Let J denote the normalized duality mapping from E into 2^{X^*} defined by

$$J(x) = \{f \in E^* : \langle x, f \rangle = \|x\|\|f\|, \|f\| = \|x\|\}, \quad x \in E,$$

where $\langle \cdot, \cdot \rangle$ denotes the generalized duality pair between E and E^*. The mapping $T : C \to C$ is called *pseudocontractive* (respectively, *strong pseudocontractive*), if there exists $j(x - y) \in J(x - y)$ such that

$$\langle Tx - Ty, j(x - y) \rangle \leq \|x - y\|^2, \quad \forall x, y \in C,$$

(respectively, $\langle Tx - Ty, j(x - y) \rangle \leq \beta \|x - y\|^2$ for some $\beta \in (0, 1)$).

The class of pseudocontractive mappings is one of the most important classes of mappings in nonlinear analysis, and it has been attracting mathematician's interest. Apart from them being a generalization of nonexpansive mappings, interest in pseudocontractive mappings stems mainly from their firm connection with the class of accretive mappings, where a mapping A with domain $D(A)$ and range $R(A)$ in E is called accretive if the inequality

$$\|x - y\| \leq \|x - y + s(Ax - Ay)\|,$$

holds for every $x, y \in D(A)$ and for all $s > 0$.

Within the past 50 years or so, many authors have been devoting their study to the existence of zeros of accretive mappings or fixed points of pseudocontractive mappings and several iterative methods for finding zeros of accretive mappings or fixed points of pseudocontractive mappings. We can refer to References [1–14] and the references in therein.

In 2007, Morales [15] introduced the following viscosity iterative method for pseudocontractive mapping:

$$x_t = tfx_t + (1-t)Tx_t, \quad t \in (0,1), \tag{1}$$

where $T : C \to E$ is a continuous pseudocontractive mapping satisfying the weakly inward condition and $f : C \to C$ is a bounded continuous strongly pseudocontractive mapping. In a reflexive Banach space with a uniformly Gâteaux differentiable norm such that every closed convex bounded subset of C has the fixed point property for nonexpansive self-mappings, he proved the strong convergence of the sequences generated by the iterative method in Equation (1) to a point q in $Fix(T)$ (the set of fixed points of T), where q is the unique solution to the following variational inequality:

$$\langle fq - q, J(p-q) \rangle \leq 0, \quad \forall p \in Fix(T). \tag{2}$$

In 2009, using the method of Reference [16], Ceng et al. [17] introduced the following modified viscosity iterative method and modified implicit viscosity iterative method with a perturbed mapping for a pseudocontractive mapping:

$$x_t = tfx_t + r_tSx_t + (1-t-r_t)Tx_t, \quad t \in (0,1), \tag{3}$$

where $0 < r_t < 1-t$, $T : C \to C$ is a continuous pseudocontractive mapping, $S : C \to C$ is a nonexpansive mapping, and $f : C \to C$ is a Lipschitz strongly pseudocontractive mapping.

$$\begin{cases} y_n = \alpha_n x_n + (1-\alpha_n)Ty_n, \\ x_{n+1} = \beta_n fy_n + \gamma_n Sy_n + (1-\beta_n-\gamma_n)y_n, \end{cases} \tag{4}$$

and

$$\begin{cases} x_n = \alpha_n y_n + (1-\alpha_n)Ty_n, \\ y_n = \beta_n fx_{n-1} + \gamma_n Sx_{n-1} + (1-\beta_n-\gamma_n)x_{n-1}, \end{cases} \tag{5}$$

where $f : C \to C$ is a contractive mapping, $x_0 \in C$ is an arbitrary initial point, and $\{\alpha_n\}$, $\{\beta_n\}$, $\{\gamma_n\} \subset (0,1]$ such that $\lim_{n\to\infty}(\gamma_n/\beta_n) = 0$ and $\beta_n + \gamma_n < 1$. In a reflexive and strictly convex Banach space with a uniformly Gâteaux differentiable norm, they proved the strong convergence of the sequences generated by the iterative methods in Equations (3)–(5) to a point q in $Fix(T)$, where q is the unique solution to the variational inequality in Equation (2). Their results developed and improved the corresponding results of Song and Chen [11], Zeng and Yao [16], Xu [18], Xu and Ori [19], and Chen et al. [20].

In this paper, as a continuation of study in this direction, in a reflexive Banach space having a weakly sequentially continuous duality mapping J_φ with gauge function φ, we consider the viscosity iterative methods in Equations (3)–(5) for a continuous pseudocontractive mapping T, a continuous bounded strongly pseudocontractive mapping f, and a nonexpansive mapping S. We establish strong convergence of the sequences generated by proposed iterative methods to a fixed point of the mapping T, which solves a variational inequality related to f. The main results develop and supplement the corresponding results of Song and Chen [11], Morales [15], Ceng et al. [17], and Xu [18] to different Banach space as well as Zeng and Yao [16], Xu and Ori [19], Chen et al. [20], and the references therein.

2. Preliminaries

Throughout the paper, we use the following notations: "$\rightharpoonup$" for weak convergence, "$\overset{*}{\rightharpoonup}$" for weak* convergence, and "$\to$" for strong convergence.

Let E be a real Banach space with the norm $\|\cdot\|$, and let E^* be its dual. The value of $x^* \in E^*$ at $x \in E$ will be denoted by $\langle x, x^* \rangle$. Let C be a nonempty closed convex subset of E, and let $T : C \to C$ be a mapping. We denote the set of fixed points of the mapping T by $Fix(T)$. That is, $Fix(T) := \{x \in C : Tx = x\}$.

Recall that a Banach space E is said to be *smooth* if for each $x \in S_E = \{x \in E : \|x\| = 1\}$, there exists a unique functional $j_x \in E^*$ such that $\langle x, j_x \rangle = \|x\|$ and $\|j_x\| = 1$ and that a Banach space E is said to be *strictly convex* [21] if the following implication holds for $x, y \in E$:

$$\|x\| \leq 1, \ \|y\| \leq 1, \ \|x - y\| > 0 \ \Rightarrow \ \left\|\frac{x+y}{2}\right\| < 1.$$

By a gauge function, we mean a continuous strictly increasing function φ defined on $\mathbb{R}^+ := [0, \infty)$ such that $\varphi(0) = 0$ and $\lim_{r \to \infty} \varphi(r) = \infty$. The mapping $J_\varphi : E \to 2^{E^*}$ defined by

$$J_\varphi(x) = \{f \in E^* : \langle x, f \rangle = \|x\|\|f\|, \|f\| = \varphi(\|x\|)\} \text{ for all } x \in E$$

is called the *duality mapping* with gauge function φ. In particular, the duality mapping with gauge function $\varphi(t) = t$ denoted by J is referred to as the *normalized duality mapping*. It is known that a Banach space E is smooth if and only if the normalized duality mapping J is single-valued. The following property of duality mapping is also well-known:

$$J_\varphi(\lambda x) = \text{sign } \lambda \left(\frac{\varphi(|\lambda| \cdot \|x\|)}{\|x\|} \right) J(x) \text{ for all } x \in E \setminus \{0\}, \ \lambda \in \mathbb{R}, \qquad (6)$$

where $\mathbb{R}$ is the set of all real numbers. The following are some elementary properties of the duality mapping J [21,22]:

(i) For $x \in E$, $J(x)$ is nonempty, bounded, closed, and convex;
(ii) $J(0) = 0$;
(iii) for $x \in E$ and a real α, $J(\alpha x) = \alpha J(x)$;
(iv) for $x, y \in E$, $f \in J(x)$ and $g \in J(y)$, $\langle x - y, f - g \rangle \geq 0$;
(v) for $x, y \in E$, $f \in J(x)$, $\|x\|^2 - \|y\|^2 \geq 2\langle x - y, f \rangle$.

We say that a Banach space E has a *weakly continuous duality mapping* if there exists a gauge function φ such that the duality mapping J_φ is single-valued and continuous from the weak topology to the weak* topology, that is, for any $\{x_n\} \in E$ with $x_n \rightharpoonup x$, $J_\varphi(x_n) \stackrel{*}{\rightharpoonup} J_\varphi(x)$. A duality mapping J_φ is weakly continuous at 0 if J_φ is single-valued and if $x_n \rightharpoonup 0$, $J_\varphi(x_n) \stackrel{*}{\rightharpoonup} 0$. For example, every l^p space ($1 < p < \infty$) has a weakly continuous duality mapping with gauge function $\varphi(t) = t^{p-1}$ [21–23]. Set

$$\Phi(t) = \int_0^t \varphi(\tau) d\tau \text{ for all } t \in \mathbb{R}^+.$$

Then it is known that $J_\varphi(x)$ is the subdifferential of the convex functional $\Phi(\|\cdot\|)$ at x. A Banach space E that has a weakly continuous duality mapping implies that E satisfies Opial's property. This means that whenever $x_n \rightharpoonup x$ and $y \neq x$, we have $\limsup_{n \to \infty} \|x_n - x\| < \limsup_{n \to \infty} \|x_n - y\|$ [21,23].

The following lemma is Lemma 2.1 of Jung [24].

Lemma 1. ([24]) *Let E be a reflexive Banach space having a weakly continuous duality mapping J_φ with gauge function φ. Let $\{x_n\}$ be a bounded sequence of E and $f : E \to E$ be a continuous mapping. Let $g : E \to \mathbb{R}$ be defined by*

$$g(z) = \limsup_{n \to \infty} \langle z - fz, J_\varphi(z - x_n) \rangle$$

for $z \in E$. Then, g is a real valued continuous function on E.

We need the following well-known lemma for the proof of our main result [21,22].

Lemma 2. *Let E be a real Banach space, and let φ be a continuous strictly increasing function on $\mathbb{R}^+$ such that $\varphi(0) = 0$ and $\lim_{r \to \infty} \varphi(r) = \infty$. Define*

$$\Phi(t) = \int_0^t \varphi(\tau) d\tau \text{ for all } t \in \mathbb{R}^+.$$

Then, the following inequalities hold:

$$\Phi(kt) \leq k\Phi(t), \ 0 < k < 1,$$

$$\Phi(\|x+y\|) \leq \Phi(\|x\|) + \langle y, j_\varphi(x+y) \rangle \text{ for all } x, y \in E,$$

where $j_\varphi(x+y) \in J_\varphi(x+y)$.

The following lemma can be found in Reference [18].

Lemma 3. ([18]) *Let $\{s_n\}$ be a sequence of nonnegative real numbers satisfying*

$$s_{n+1} \leq (1 - \lambda_n)s_n + \lambda_n \delta_n, \ n \geq 0,$$

where $\{\lambda_n\}$ and $\{\delta_n\}$ satisfy the following conditions:

(i) $\{\lambda_n\} \subset [0,1]$ and $\sum_{n=0}^\infty \lambda_n = \infty$ or, equivalently, $\prod_{n=0}^\infty (1 - \lambda_n) = 0$,
(ii) $\limsup_{n \to \infty} \delta_n \leq 0$ or $\sum_{n=0}^\infty \lambda_n |\delta_n| < \infty$,

Then, $\lim_{n \to \infty} s_n = 0$.

Let C be a nonempty closed convex subset of a real Banach space E. Recall that $S : C \to C$ is called *accretive* if $I - S$ is pseudocontractive. If $T : C \to C$ is a pseudocontractive mapping, then $I - T$ is accretive. We denote $A = J_1 = (2I - T)^{-1}$. Then, $Fix(A) = Fix(T)$ and the operator $A : R(2I - T) \to C$ is nonexpansive and single-valued, where I denotes the identity mapping.

We also need the following result which can be found in Reference [11].

Lemma 4. ([11]) *Let C be a nonempty closed convex subset of a real Banach space E, and let $T : C \to C$ be a continuous pseudocontractive mapping. We denote $A = (2I - T)^{-1}$.*

(i) *The mapping A is nonexpansive self-mapping on C, i.e., for all $x, y \in nC$, there holds*

$$\|Ax - Ay\| \leq \|x - y\|, \text{ and } Ax \in C.$$

(ii) *If $\lim_{n \to \infty} \|x_n - Tx_n\| = 0$, then $\lim_{n \to \infty} \|x_n - Ax_n\| = 0$.*

The following Lemmas, which are well-known, can be found in many books in the geometry of Banach spaces (see References [21,23]).

Lemma 5. (Demiclosedness Principle) *Let C be a nonempty closed convex subset of a Banach space E, and let $T : C \to C$ be a nonexpansive mapping. Then, $x_n \rightharpoonup x$ in C and $(I - T)x_n \to y$ imply that $(I - T)x = y$.*

Lemma 6. *If E is a Banach space such that E^* is strictly convex, then E is smooth and any duality mapping is norm-to-weak*-continuous.*

Finally, we need the following result which was given by Deimling [4].

Lemma 7. ([4]) *Let C be a nonempty closed convex subset of a Banach space E, and let $T : C \to C$ be a continuous strong pseudocontractive mapping with a pseudocontractive coefficient $\beta \in (0,1)$. Then, T has a unique fixed point in C.*

3. Convergence of Path with Perturbed Mapping

As we know, the path convergency plays an important role in proving the convergence of iterative methods to approximate fixed points. In this direction, we first prove the existence of a path for a convex combination of a pseudocontractive type of mappings with a perturbed mapping and boundedness of the path.

Proposition 1. *Let C be a nonempty closed convex subset of a real Banach space E. Let $T : C \to C$ be a continuous pseudocontractive mapping, let $S : C \to C$ be a nonexpansive mapping, and let $f : C \to C$ be a continuous strongly pseudocontractive mapping with a pseudocontractive coefficient $\beta \in (0,1)$.*

(i) *There exists a unique path $t \mapsto x_t \in C, t \in (0,1)$, satisfying*

$$x_t = tfx_t + r_t Sx_t + (1 - t - r_t)Tx_t, \tag{7}$$

provided $r_t : (0,1) \to [0, 1-t)$ is continuous and $\lim_{t \to 0}(r_t/t) = 0$.

(ii) *In particular, if T has a fixed point in C, then the path $\{x_t\}$ is bounded.*

Proof. (i) For each $t \in (0,1)$, define the mapping $T_{(S,f)} : C \to C$ as follows:

$$T_{(S,f)} = tf + r_t S + (1 - t - r_t)T,$$

where $0 < r_t < 1 - t$ and $\lim_{t \to 0}(r_t/t) = 0$. Then, it is easy to show that the mapping $T_{(S,f)}$ is a continuous strongly pseudocontractive self-mapping of C. Therefore, by Lemma 7, $T_{(S,f)}$ has a unique fixed point in C, i.e., for each given $t \in (0,1)$, there exists $x_t \in C$ such that

$$x_t = tfx_t + r_t Sx_t + (1 - t - r_t)Tx_t.$$

To show continuity, let $t, t_0 \in (0,1)$. Then, there exists $j \in J(x_t - x_{t_0})$ such that

$$\langle x_t - x_{t_0}, j \rangle = \langle tfx_t + r_t Sx_t + (1 - t - r_t)Tx_t - (t_0 fx_{t_0} + r_{t_0} Sx_{t_0} + (1 - t_0 - r_{t_0})Tx_{t_0}), j \rangle$$
$$= t\langle fx_t - fx_{t_0}, j \rangle + (t - t_0)\langle fx_{t_0}, j \rangle + r_t\langle Sx_t - Sx_{t_0}, j \rangle + (r_t - r_{t_0})\langle Sx_{t_0}, j \rangle$$
$$+ (1 - t - r_t)\langle Tx_t - Tx_{t_0}, j \rangle + ((t - t_0) + (r_t - r_{t_0}))\langle Tx_{t_0}, j \rangle,$$

and this implies that

$$\|x_t - x_{t_0}\|^2 \le t\beta\|x_t - x_{t_0}\|^2 + |t - t_0|\|fx_{t_0}\|\|x_t - x_{t_0}\|$$
$$+ r_t\|x_t - x_{t_0}\|^2 + |r_t - r_{t_0}|\|Sx_{t_0}\|\|x_t - x_{t_0}\|$$
$$+ (1 - t - r_t)\|x_t - x_{t_0}\|^2 + |t - t_0|\|Tx_{t_0}\|\|x_t - x_{t_0}\| + |r_t - r_{t_0}|\|Tx_{t_0}\|\|x_t - x_{t_0}\|.$$

and, hence,

$$\|x_t - x_{t_0}\| \le t\beta\|x_t - x_{t_0}\| + |t - t_0|\|fx_{t_0}\| + |r_t - r_{t_0}|\|Sx_{t_0}\|$$
$$+ (1 - t - r_t)\|x_t - x_{t_0}\| + |t - t_0|\|Tx_{t_0}\| + |r_t - r_{t_0}|\|Tx_{t_0}\|$$
$$= (1 - (1 - \beta)t)\|x_t - x_{t_0}\| + (\|fx_{t_0}\| + \|Tx_{t_0}\|)|t - t_0| + (\|Sx_{t_0}\| + \|Tx_{t_0}\|)|r_t - r_{t_0}|.$$

Therefore,

$$\|x_t - x_{t_0}\| \le \frac{\|fx_{t_0}\| + \|Tx_{t_0}\|}{(1 - \beta)t}|t - t_0| + \frac{\|Sx_{t_0}\| + \|Tx_{t_0}\|}{(1 - \beta)t}|r_t - r_{t_0}|,$$

which guarantees continuity.

(ii) By the same argument as in the proof of Theorem 2.1 of Reference [17], we can prove that $\{x_t\}$ defined by Equation (7) is bounded for $t \in (0, t_0)$ for some $t_0 \in (0,1)$, and so we omit its proof. □

The above path of Equation (7) is called the *modified viscosity iterative method with perturbed mapping*, where S is called the perturbed mapping.

The following result gives conditions for existence of a solution of a variational inequality:

$$\langle (I-f)q, J_\varphi(q-p) \rangle \leq 0, \quad \forall p \in Fix(T). \tag{8}$$

Theorem 1. *Let E be a Banach space such that E^* is strictly convex. Let C be a nonempty closed convex subset of a real Banach space E. Let $T : C \to C$ be a continuous pseudocontractive mapping with $Fix(T) \neq \emptyset$, let $S : C \to C$ be a nonexpansive mapping, and let $f : C \to C$ be a continuous strongly pseudocontractive mapping with a pseudocontractive coefficient $\beta \in (0,1)$. Suppose that $\{x_t\}$ defined by Equation (7) converges strongly to a point in $Fix(T)$. If we define $q := \lim_{t \to 0} x_t$, then q is a solution of the variational inequality in Equation (8).*

Proof. First, from Lemma 6, we note that E is smooth and J_φ is norm-to-weak*-continuous.
Since

$$(I-f)x_t = -\frac{1-t-r_t}{t}(I-T)x_t - \frac{r_t}{t}(I-S)x_t,$$

we have for $p \in Fix(T)$

$$\begin{aligned}\langle (I-f)x_t, J_\varphi(x_t-p) \rangle &= -\frac{1-t-r_t}{t}\langle (I-T)x_t - (I-T)p, J_\varphi(x_t-p) \rangle \\ &\quad + \frac{r_t}{t}\langle (S-I)x_t, J_\varphi(x_t-p) \rangle.\end{aligned} \tag{9}$$

Since $I - T$ is accretive and $J(x_t - p)$ is a positive-scalar multiple of $J_\varphi(x_t - p)$ (see Equation (6)), it follow from Equation (9) that

$$\begin{aligned}\langle (I-f)x_t, J_\varphi(x_t-p) \rangle &\leq \frac{r_t}{t}\langle (S-I)x_t, J_\varphi(x_t-p) \rangle \\ &\leq \frac{r_t}{t}\|(S-I)x_t\|\varphi(\|x_t-p\|).\end{aligned} \tag{10}$$

Taking the limit as $t \to 0$, by $\lim_{t \to 0} \frac{r_t}{t} = 0$, we obtain

$$\langle (I-f)q, J_\varphi(q-p) \rangle \leq 0, \quad \forall p \in Fix(T).$$

This completes the proof. □

The following lemma provides conditions under which $\{x_t\}$ defined by Equation (7) converges strongly to a point in $Fix(T)$.

Lemma 8. *Let E be a reflexive smooth Banach space having Opial's property and having some duality mapping J_φ weakly continuous at 0. Let C be a nonempty closed convex subset of E. Let $T : C \to C$ be a continuous pseudocontractive mapping with $Fix(T) \neq \emptyset$, let $S : C \to C$ be a nonexpansive mapping, and let $f : C \to C$ be a continuous bounded strongly pseudocontractive mapping with a pseudocontractive coefficient $\beta \in (0,1)$. Then, $\{x_t\}$ defined by Equation (7) converges strongly to a point in $Fix(T)$ as $t \to 0$.*

Proof. First, from Proposition 1 (ii), we know that $\{x_t : t \in (0, t_0)\}$ is bounded for $t \in (0, t_0)$ for some $t_0 \in (0,1)$.

Since f is a bounded mapping and S is a nonexpansive mapping, $\{fx_t : t \in (0, t_0)\}$ and $\{Sx_t : t \in (0, t_0)\}$ are bounded. Moreover, noting that $x_t = tfx_t + r_tSx_t + (1-t-r_t)Tx_t$, we have

$$Tx_t = \frac{1}{1-t-r_t}x_t - \frac{t}{1-t-r_t}fx_t - \frac{r_t}{1-t-r_t}Sx_t,$$

which implies that
$$\|Tx_t\| \leq \frac{1}{1-t-r_t}\|x_t\| + \frac{t}{1-t-r_t}\|fx_t\| + \frac{r_t}{1-t-r_t}\|Sx_t\|.$$

Thus, we obtain
$$\|Tx_t\| \leq 2\|x_t\| + 2t\|fx_t\| + 2r_t\|Sx_t\|, \quad \forall t \in (0, t_0)$$
and so $\{Tx_t : t \in (0, t_0)\}$ is bounded. This implies that
$$\lim_{t \to 0} \|x_t - Tx_t\| \leq \lim_{t \to 0} t\|fx_t - Tx_t\| + \lim_{t \to 0} r_t\|Sx_t - Tx_t\| = 0. \tag{11}$$

Now, let $t_m \in (0, t_0)$ for some $t_0 \in (0, 1)$ be such that $t_m \to 0$, and let $\{x_m\} := \{x_{t_m}\}$ be a subsequence of $\{x_t\}$. Then,
$$x_m = t_m f x_m + r_m S_m + (1 - t_m - r_m) T x_m.$$

Let $p \in Fix(T)$. Then, we have
$$x_m - p = t_m(fx_m - p) + r_m(Sx_m - p) + (1 - t_m - r_m)(Tx_m - Tp)$$

and
$$\begin{aligned}\|x_m - p\|\varphi(\|x_m - p\|) &= \langle x_m - p, J_\varphi(x_m - p)\rangle \\ &\leq t_m \langle fx_m - p, J_\varphi(x_m - p)\rangle + r_m \langle Sx_m - p, J_\varphi(x_m - p)\rangle \\ &\quad + (1 - t_m - r_m)\|x_m - p\|\varphi(\|x_m - p\|).\end{aligned}$$

Thus, it follows that
$$\|x_m - p\|\varphi(\|x_m - p\|) \leq \frac{t_m}{t_m + r_m}\langle fx_m - p, J_\varphi(x_m - p)\rangle + \frac{r_m}{t_m + r_m}\langle Sx_m - p, J_\varphi(x_m - p)\rangle. \tag{12}$$

Hence, we get
$$\langle p - fx_m, J_\varphi(x_m - p)\rangle \leq -\frac{t_m + r_m}{t_m}\|x_m - p\|\varphi(\|x_m - p\|) + \frac{r_m}{t_m}\langle Sx_m - p, J_\varphi(x_m - p)\rangle,$$

that is,
$$\langle p - fx_m, J_\varphi(p - x_m)\rangle \geq \frac{t_m + r_m}{t_m}\|x_m - p\|\varphi(\|x_m - p\|) + \frac{r_m}{t_m}\langle p - Sx_m, J_\varphi(x_m - p)\rangle.$$

Therefore, we have
$$\begin{aligned}\langle x_m - fx_m, J_\varphi(p - x_m)\rangle &= \langle x_m - p, J_\varphi(p - x_m)\rangle + \langle p - fx_m, J_\varphi(p - x_m)\rangle \\ &\geq -\|x_m - p\|\varphi(\|x_m - p\|) + \frac{t_m + r_m}{t_m}\|x_m - p\|\varphi(\|x_m - p\|) \\ &\quad + \frac{r_m}{t_m}\langle p - Sx_m, J_\varphi(x_m - p)\rangle \\ &= \frac{r_m}{t_m}\|x_m - p\|\varphi(\|x_m - p\|) + \frac{r_m}{t_m}\langle p - Sx_m, J_\varphi(x_m - p)\rangle.\end{aligned}$$

On the other hand, since $\{x_m\}$ is bounded and E is reflexive, $\{x_m\}$ has a weakly convergent subsequence $\{x_{m_k}\}$, say, $x_{m_k} \rightharpoonup u \in E$. From Equation (11), it follows that
$$\|x_m - Tx_m\| \leq t_m\|fx_m - Tx_m\| + r_m\|Sx_m - Tx_m\| \to 0.$$

From Lemma 4, we know that the mapping $A = (2I - T)^{-1} : C \to C$ is nonexpansive, that $Fix(A) = Fix(T)$, and that $\|x_m - Ax_m\| \to 0$. Thus, by Lemma 5, $u \in Fix(A) = Fix(T)$. Therefore, by Equation (12) and the assumption that J_φ is weakly continuous at 0, we obtain

$$\|x_{m_k} - u\|\varphi(\|x_{m_k} - u\|) \leq \frac{t_{m_k}}{t_{m_k} + r_{m_k}} \langle fx_{m_k} - u, J_\varphi(x_{m_k} - u)\rangle + \frac{r_{m_k}}{t_{m_k} + r_{m_k}} \langle Sx_{m_k} - u, J_\varphi(x_{m_k} - u)\rangle$$

$$\leq |\langle fx_{m_k} - u, J_\varphi(x_{m_k} - u)\rangle| + \frac{r_{m_k}}{t_{m_k}} |\langle Sx_{m_k} - u, J_\varphi(x_{m_k} - u)\rangle| \to 0.$$

Since φ is continuous and strictly increasing, we must have $x_{m_k} \to u$.

Now, we will show that every weakly convergent subsequence of $\{x_m\}$ has the same limit. Suppose that $x_{m_k} \rightharpoonup u$ and $x_{m_j} \rightharpoonup v$. Then, by the above proof, we have $u, v \in Fix(T)$ and $x_{m_k} \to u$ and $x_{m_j} \to v$. By Equation (12), we have the following for all $p \in Fix(T)$:

$$\|x_{m_k} - p\|\varphi(\|x_{m_k} - p\|) \leq \frac{t_{m_k}}{t_{m_k} + r_{m_k}} \langle fx_{m_k} - p, J_\varphi(x_{m_k} - p)\rangle + \frac{r_{m_k}}{t_{m_k} + r_{m_k}} \langle Sx_{m_k} - p, J_\varphi(x_{m_k} - p)\rangle$$

$$\leq \frac{t_{m_k}}{t_{m_k} + r_{m_k}} \langle fx_{m_k} - p, J_\varphi(x_{m_k} - p)\rangle + \frac{r_{m_k}}{t_{m_k}} |\langle Sx_{m_k} - p, J_\varphi(x_{m_k} - p)\rangle|$$

and

$$\|x_{m_j} - p\|\varphi(\|x_{m_j} - p\|) \leq \frac{t_{m_j}}{t_{m_j} + r_{m_j}} \langle fx_{m_j} - p, J_\varphi(x_{m_j} - p)\rangle + \frac{r_{m_j}}{t_{m_j} + r_{m_j}} \langle Sx_{m_j} - p, J_\varphi(x_{m_j} - p)\rangle$$

$$\leq \frac{t_{m_j}}{t_{m_j} + r_{m_j}} \langle fx_{m_j} - p, J_\varphi(x_{m_j} - p)\rangle + \frac{r_{m_k}}{t_{m_k}} |\langle Sx_{m_k} - p, J_\varphi(x_{m_k} - p)\rangle|.$$

Taking limits, we get

$$\Phi(\|u - v\|) = \|u - v\|\varphi(\|u - v\|) \leq \langle fu - v, J_\varphi(u - v)\rangle \tag{13}$$

and

$$\Phi(\|v - u\|) = \|v - u\|\varphi(\|v - u\|) \leq \langle fv - u, J_\varphi(v - u)\rangle. \tag{14}$$

Adding up Equations (13) and (14) yields

$$2\Phi(\|u-v\|) = 2\|u-v\|\varphi(\|u-v\|) \leq \|u-v\|\varphi(\|u-v\|) + \langle fu - fv, J_\varphi(u-v)\rangle$$
$$\leq (1+\beta)\|u-v\|\varphi(\|u-v\|) = (1+\beta)\Phi(\|u-v\|).$$

Since $\beta \in (0,1)$, this implies $\Phi(\|u-v\|) \leq 0$, that is, $u = v$. Hence, $\{x_m\}$ is strongly convergent to a point in $Fix(T)$ as $t_m \to 0$.

The same argument shows that, if $t_l \to 0$, then the subsequence $\{x_l\} := \{x_{t_l}\}$ of $\{x_t : t \in (0, t_0)\}$ for some $t_0 \in (0,1)$ is strongly convergent to the same limit. Thus, as $t \to 0$, $\{x_t\}$ converges strongly to a point in $Fix(T)$. □

Using Theorem 1 and Lemma 8, we show the existence of a unique solution of the variational inequality in Equation (8) in a reflexive Banach space having a weakly continuous duality mapping.

Theorem 2. *Let E be a reflexive Banach space having a weakly continuous duality mapping J_φ with gauge function φ, and let C be a nonempty closed convex subset of E. Let $T : C \to C$ be a continuous pseudocontractive mapping such that $Fix(T) \neq \emptyset$, let $S : C \to C$ be a nonexpansive mapping, and let $f : C \to C$ be a continuous bounded strongly pseudocontractive mapping with a pseudocontractive coefficient $\beta \in (0,1)$. Then, there exists the unique solution in $q \in Fix(T)$ of the variational inequality in Equation (8), where $q := \lim_{t \to \infty} x_t$ with x_t being defined by Equation (7).*

Proof. We notice that the definition of the weak continuity of the duality mapping J_φ implies that E is smooth. Thus, E^* is strictly convex for reflexivity of E. By Lemma 8, $\{x_t\}$ defined by Equation (7) converges strongly to a point q in $Fix(T)$ as $t \to 0$. Hence, by Theorem 1, q is the unique solution of the variational inequality in Equation (8). In fact, suppose that $q, p \in Fix(T)$ satisfy the variational inequality in Equation (8). Then, we have

$$\langle (I-f)q, J_\varphi(q-p) \rangle \leq 0 \text{ and } \langle (I-f)p, J_\varphi(p-q) \rangle \leq 0.$$

Adding these two inequalities, we have

$$(1-\beta)\Phi(\|q-p\|) = (1-\beta)\|q-p\|\varphi(\|q-p\|) \leq \langle (I-f)q - (I-f)p, J_\varphi(q-p) \rangle \leq 0,$$

and so $q = p$. □

As a direct consequence of Theorem 2, we have the following result.

Corollary 1. ([20, Theorem 3.2]) *Let E be a reflexive Banach space having a weakly continuous duality mapping J_φ with gauge function φ, and let C be a nonempty closed convex subset of E. Let $T : C \to C$ be a continuous pseudocontractive mapping such that $Fix(T) \neq \emptyset$, and let $f : C \to C$ be a continuous bounded strongly pseudocontractive mapping with a pseudocontractive coefficient $\beta \in (0,1)$. Let $\{x_t\}$ be defined by*

$$x_t = tfx_t + (1-t)Tx_t, \quad \forall t \in (0,1).$$

Then, as $t \to 0$, x_t converges strongly to a some point of T such that q is the unique solution of the variational inequality in Equation (8).

Proof. Put $S = I$ and $r_t = 0$ for all $t \in (0,1)$. Then, the result follows immediately from Theorem 2. □

Remark 1. (1) *Theorem 2 develops and supplements Theorem 2.1 of Ceng et al. [17] in the following aspects:*

  (i) *The space is replaced by the space having a weakly continuous duality mapping J_φ with gauge function φ.*
  (ii) *The Lipischiz strongly pseudocontractive mapping f in Theorem 2.1 in Reference [17] is replaced by a bounded continuous strongly pseudocontractive mapping f in Theorem 2.*

(2) *Corollary 1 complements Theorem 2.1 of Song and Chen [11] and Corollary 2.2 of Cent et al. [17] by replacing the Lipischiz strongly pseudocontractive mapping f in References [11,17] by the bounded continuous strongly pseudocontractive mapping f in Corollary 3.5 in a reflexive Banach space having a weakly continuous duality mapping J_φ with gauge function φ.*

(3) *Corollary 1 also develops Theorem 2 of Morales [15] to a reflexive Banach space having a weakly continuous duality mapping J_φ with gauge function φ.*

4. Modified Implicit Iterative Methods with Perturbed Mapping

First, we prepare the following result.

Theorem 3. *Let E be a reflexive Banach space having a weakly continuous duality mapping J_φ with gauge function φ, and let C be a nonempty closed convex subset of E. Let $T : C \to C$ be a continuous pseudocontractive mapping such that $Fix(T) \neq \emptyset$, let $S : C \to C$ be a nonexpansive mapping, and let $f : C \to C$ be a continuous bounded strongly pseudocontractive mapping with a pseudocontractive coefficient $\beta \in (0,1)$. Let $\{x_t\}$ be defined by Equation (7). If there exists a bounded sequence $\{x_n\}$ such that $\lim_{n \to \infty} \|x_n - Tx_n\| = 0$ and $q = \lim_{t \to 0} x_t$, then*

$$\limsup_{n \to \infty} \langle fq - q, J_\varphi(x_n - q) \rangle \leq 0.$$

Proof. Using the equality

$$x_t - x_n = (1 - t - r_t)(Tx_t - x_n) + t(fx_t - x_n) + r_t(Sx_t - x_n)$$

and the inequality

$$\langle Tx - Ty, J_\varphi(x - y) \rangle \le \|x - y\| \varphi(\|x - y\|), \quad \forall x, y \in C,$$

we derive

$$\begin{aligned}
\|x_t - x_n\| \varphi(\|x_t - x_n\|) &= (1 - t - r_t)\langle Tx_t - x_n, J_\varphi(x_t - x_n)\rangle + t\langle fx_t - x_n, J_\varphi(x_t - x_n)\rangle \\
&\quad + r_t \langle Sx_t - x_n, J_\varphi(x_t - x_n)\rangle \\
&= (1 - t - r_t)(\langle Tx_t - Tx_n, J_\varphi(x_t - x_n)\rangle + \langle Tx_n - x_n, J_\varphi(x_t - x_n)\rangle \\
&\quad t\langle fx_t - x_t, J_\varphi(x_t - x_n)\rangle + t\|x_t - x_n\| \varphi(\|x_t - x_n\|) \\
&\quad + r_t \langle Sx_t - x_t, J_\varphi(x_t - x_n)\rangle + r_t \|x_t - x_n\| \varphi(\|x_t - x_n\|) \\
&\le \|x_t - x_n\| \varphi(\|x_t - x_n\|) + \|Tx_n - x_n\| \varphi(\|x_t - x_n\|) \\
&\quad t\langle fx_t - x_t, J_\varphi(x_t - x_n)\rangle + r_t \|Sx_t - x_n\| \varphi(\|x_t - x_n\|)
\end{aligned}$$

and, hence,

$$\langle x_t - fx_t, J_\varphi(x_t - x_n)\rangle \le \frac{\|Tx_n - x_n\|}{t} \varphi(\|x_t - x_n\|) + \frac{r_t}{t} \|Sx_t - x_t\| \varphi(\|x_t - x_n\|).$$

Therefore, by $\limsup_{n \to \infty} \varphi(\|x_t - x_n\|) < \infty$, we have

$$\begin{aligned}
\limsup_{n \to \infty} \langle x_t - fx_t, J_\varphi(x_t - x_n)\rangle &\le \limsup_{n \to \infty} \frac{\|Tx_n - x_n\|}{t} \varphi(\|x_t - x_n\|) \\
&\quad + \limsup_{n \to \infty} \frac{r_t}{t} \|Sx_t - x_t\| \varphi(\|x_t - x_n\|) \\
&= \limsup_{n \to \infty} \frac{r_t}{t} \|Sx_t - x_t\| \varphi(\|x_t - x_n\|) \\
&= \frac{r_t}{t} \|Sx_t - x_t\| \limsup_{n \to \infty} \varphi(\|x_t - x_n\|).
\end{aligned}$$

Thus, noting that $\lim_{t \to 0} \limsup_{n \to \infty} \varphi(\|x_t - x_n\|) < \infty$, by Lemma 1, we conclude

$$\begin{aligned}
\limsup_{n \to \infty} \langle fq - q, J_\varphi(x_n - q)\rangle &= \lim_{t \to 0} \limsup_{n \to \infty} \langle fx_t - x_t, J_\varphi(x_n - x_t)\rangle \\
&\le \lim_{t \to 0} \left[\frac{r_t}{t} \|Sx_t - x_t\|\right] \lim_{t \to 0} \limsup_{n \to \infty} \varphi(\|x_t - x_n\|) \\
&= 0 \times \lim_{t \to 0} \limsup_{n \to \infty} \varphi(\|x_t - x_n\|) = 0.
\end{aligned}$$

This completes the proof. □

Theorem 4. *Let E be a reflexive Banach space having a weakly continuous duality mapping J_φ with gauge function φ, and let C be a nonempty closed convex subset of E. Let $T : C \to C$ be a continuous pseudocontractive mapping such that $Fix(T) \ne \emptyset$, let $S : C \to C$ be a nonexpansive mapping, and let $f : C \to C$ be a contractive mapping with a contractive coefficient $k \in (0,1)$. For $x_0 \in C$, let $\{x_n\}$ be defined by the following iterative scheme:*

$$\begin{cases} y_n = \alpha_n x_n + (1 - \alpha_n) T y_n \\ x_{n+1} = \beta_n f y_n + \gamma_n S y_n + (1 - \beta_n - \gamma_n) y_n, \quad \forall n \ge 0, \end{cases} \quad (15)$$

where $\{\alpha_n\}, \{\beta_n\},$ and $\{\gamma_n\}$ are three sequences in $(0,1]$ satisfying the following conditions:

(i) $\lim_{n\to\infty} \alpha_n = 0$;
(ii) $\lim_{n\to\infty} \beta_n = 0$, $\sum_{n=0}^{\infty} \beta_n = \infty$;
(iii) $\lim_{n\to\infty}(\gamma_n/\beta_n) = 0$, $\beta_n + \gamma_n \leq 1$, $\forall n \geq 0$.

Then, $\{x_n\}$ converges strongly to a fixed point x^* of T, which is the unique solution of the following variational inequality
$$\langle (I-f)x^*, J_\varphi(x^* - p)\rangle \leq 0, \quad \forall p \in Fix(T). \tag{16}$$

Proof. First, put $z_t = tfz_t + r_t Sz_t + (1-t-r_t)Tz_t$. Then, it follows from Theorem 2 that, as $t \to 0$, z_t converges strongly to some fixed point x^* of T such that x^* is the unique solution in $Fix(T)$ to the variational inequality in Equation (16).

Now, we divide the proof into several steps.

Step 1. We show that $\{x_n\}$ is bounded. To this end, let $p \in Fix(T)$. Then, we have

$$\begin{aligned}
\|y_n - p\|\varphi(\|y_n - p\|) &= \langle \alpha_n x_n + (1-\alpha_n)Ty_n - p, J_\varphi(y_n - p)\rangle \\
&\leq (1-\alpha_n)\langle Ty_n - Tp, J_\varphi(y_n - p)\rangle + \alpha_n \|x_n - p\|\varphi(\|y_n - p\|) \\
&\leq (1-\alpha_n)\|y_n - p\|\varphi(\|y_n - p\|) + \alpha_n \|x_n - p\|\varphi(\|y_n - p\|)
\end{aligned}$$

and, hence,
$$\|y_n - p\| \leq \|x_n - p\|, \quad \forall n \geq 0.$$

Thus, we obtain

$$\begin{aligned}
\|x_{n+1} - p\| &\leq \beta_n \|fy_n - p\| + \gamma_n \|Sy_n - p\| + (1 - \beta_n - \gamma_n)\|y_n - p\| \\
&\leq \beta_n(\|fy_n - fp\| + \|fp - p\|) + \gamma_n(\|Sy_n - Sp\| + \|Sp - p\|) \\
&\quad + (1 - \beta_n - \gamma_n)\|x_n - p\| \\
&\leq \beta_n k\|y_n - p\| + \beta_n \|fp - p\| + \gamma_n \|y_n - p\| + \gamma_n \|Sp - p\| \\
&\quad + (1 - \beta_n - \gamma_n)\|x_n - p\| \\
&\leq \beta_n k\|x_n - p\| + \beta_n \|fp - p\| + \gamma_n \|x_n - p\| + \gamma_n \|Sp - p\| \\
&\quad + (1 - \beta_n - \gamma_n)\|x_n - p\| \\
&= (1 - (1-k)\beta_n)\|x_n - p\| + \beta_n \|fp - p\| + \gamma_n \|Sp - p\|.
\end{aligned} \tag{17}$$

Since $\lim_{n\to\infty}(\gamma_n/\beta_n) = 0$, we may assume without loss of generality that $\gamma_n \leq \beta_n$ for all $n > 0$. Therefore, it follows from Equation (17) that

$$\begin{aligned}
\|x_{n+1} - p\| &\leq (1 - (1-k)\beta_n)\|x_n - p\| + (1-k)\beta_n \cdot \frac{1}{1-k}(\|fp - p\| + \|Sp - p\|) \\
&\leq \max\left\{\|x_n - p\|, \frac{1}{1-k}(\|fp - p\| + \|Sp - p\|)\right\}.
\end{aligned}$$

By induction, we derive

$$\|x_n - p\| \leq \max\left\{\|x_0 - p\|, \frac{1}{1-k}(\|fp - p\| + \|Sp - p\|)\right\}, \quad \forall n \geq 0.$$

This show that $\{x_n\}$ is bounded and so is $\{y_n\}$.

Step 2. We show that $\{fy_n\}$, $\{Sy_n\}$, and $\{Ty_n\}$ are bounded. Indeed, observe that

$$\|fy_n\| \leq \|fy_n - fp\| + \|fp\| \leq k\|y_n - p\| + \|fp\|$$

and

$$\|Sy_n\| \leq \|Sy_n - Sp\| + \|Sp\| \leq \|y_n - p\| + \|Sp\|.$$

Thus, $\{fy_n\}$ and $\{Sy_n\}$ are bounded. Since $\lim_{n\to\infty} \alpha_n = 0$, there exist $n_0 \geq 0$ and $a \in (0,1)$ such that $\alpha_n \leq a$ for all $n \geq n_0$. Noting that $y_n = \alpha_n x_n + (1 - \alpha_n) Ty_n$, we have

$$Ty_n = \frac{1}{1 - \alpha_n} y_n - \frac{\alpha_n}{1 - \alpha_n} x_n$$

and so

$$\|Ty_n\| \leq \frac{1}{1 - \alpha_n} \|y_n\| + \frac{\alpha_n}{1 - \alpha_n} \|x_n\| \leq \frac{1}{1 - a} \|y_n\| + \frac{a}{1 - a} \|x_n\|.$$

Consequently, the sequence $\{Ty_n\}$ is also bounded.

Step 3. We show that $\limsup_{n\to\infty} \langle fx^* - x^*, J_\varphi(y_n - x^*)\rangle \leq 0$. In fact, from condition (i) and boundedness of $\{x_n\}$ and $\{Ty_n\}$, we get

$$\|y_n - Ty_n\| = \alpha_n \|x_n - Ty_n\| \to 0 \quad (n \to \infty). \tag{18}$$

Thus, it follows from Equation (18) and Theorem 3 that $\limsup_{n\to\infty} \langle fx^* - x^*, J_\varphi(y_n - x^*)\rangle \leq 0$.

Step 4. We show that $\limsup_{n\to\infty} \langle fx^* - x^*, J_\varphi(x_{n+1} - x^*)\rangle \leq 0$. Indeed, by Equations (15) and (18), we have

$$\|x_{n+1} - y_n\| = \|\beta_n f y_n + \gamma_n Sy_n + (1 - \beta_n - \gamma_n) y_n - (\alpha_n x_n + (1 - \alpha_n) Ty_n)\|$$
$$\leq \alpha_n \|x_n - Ty_n\| + \beta_n \|f y_n - y_n\| + \gamma_n \|Sy_n - y_n\| + \|y_n - Ty_n\| \to 0 \quad (n \to \infty).$$

Since the duality mapping J_φ is single-valued and weakly continuous, we have

$$\lim_{n\to\infty} \langle fx^* - x^*, J_\varphi(x_{n+1} - x^*) - J_\varphi(y_n - x^*)\rangle = 0.$$

Therefore, we obtain from step 3 that

$$\limsup_{n\to\infty} \langle fx^* - x^*, J_\varphi(x_{n+1} - x^*)\rangle \leq \limsup_{n\to\infty} \langle fx^* - x^*, J_\varphi(y_n - x^*)\rangle$$
$$+ \limsup_{n\to\infty} \langle fx^* - x^*, J_\varphi(x_{n+1} - x^*) - J_\varphi(y_n - x^*)\rangle$$
$$= \limsup_{n\to\infty} \langle fx^* - x^*, J_\varphi(y_n - x^*)\rangle \leq 0.$$

Step 5. We show that $\lim_{n\to\infty} \|x_n - x^*\| = 0$. In fact, it follows from Equation (15) that

$$x_{n+1} - x^* = \beta_n (f y_n - fx^*) + \gamma_n (Sy_n - Sx^*) + (1 - \beta_n - \gamma_n)(y_n - x^*)$$
$$+ \beta_n (fx^* - x^*) + \gamma_n (Sx^* - x^*).$$

Therefore, using inequalities $\|y_n - x^*\| \leq \|x_n - x^*\|$, $\|fx - fy\| \leq k\|x - y\|$, and $\|Sx - Sy\| \leq \|x - y\|$ and using Lemma 2, we have

$$\Phi(\|x_{n+1} - x^*\|) \leq \Phi(\|\beta_n(fy_n - fx^*) + \gamma_n(Sy_n - Sx^*) + (1 - \beta_n - \gamma_n)(y_n - x^*)\|)$$
$$+ \beta_n \langle fx^* - x^*, J_\varphi(x_{n+1} - x^*)\rangle + \gamma_n \langle Sx^* - x^*, J_\varphi(x_{n+1} - x^*)\rangle$$
$$\leq \Phi(\beta_n k \|y_n - x^*\| + \gamma_n \|y_n - x^*\| + (1 - \beta_n - \gamma_n)\|y_n - x^*\|)$$
$$+ \beta_n \langle fx^* - x^*, J_\varphi(x_{n+1} - x^*)\rangle + \gamma_n \langle Sx^* - x^*, J_\varphi(x_{n+1} - x^*)\rangle$$
$$\leq \Phi((1 - (1 - k)\beta_n) \|x_n - x^*\|) \tag{19}$$
$$+ \beta_n \langle fx^* - x^*, J_\varphi(x_{n+1} - x^*)\rangle + \gamma_n \langle Sx^* - x^*, J_\varphi(x_{n+1} - x^*)\rangle$$
$$\leq (1 - (1 - k)\beta_n) \Phi(\|x_n - x^*\|) + \beta_n \langle fx^* - x^*, J_\varphi(x_{n+1} - x^*)\rangle$$
$$+ \gamma_n \|Sx^* - x^*\| \varphi(\|x_{n+1} - x^*\|)$$
$$\leq (1 - \lambda_n) \Phi(\|x_n - x^*\|) + \lambda_n \delta_n,$$

where $\lambda_n = (1-k)\beta_n$ and

$$\delta_n = \frac{1}{1-k}\left[\langle fx^* - x^*, J_\varphi(x_{n+1} - x^*)\rangle + \frac{\gamma_n}{\beta_n}\|Sx^* - x^*\|\varphi(\|x_{n+1} - x^*\|)\right].$$

From conditions (ii) and (iii) and from step 4, it is easily seen that $\sum_{n=0}^{\infty} \lambda_n = \infty$ and $\limsup_{n\to\infty} \delta_n \leq 0$. Thus, applying Lemma 3 to Equation (19), we conclude that $\lim_{n\to\infty} \Phi(\|x_n - x^*\|) = 0$ and, hence, $\lim_{n\to\infty} \|x_n - x^*\| = 0$. This completes the proof. □

Theorem 5. *Let E be a reflexive Banach space having a weakly continuous duality mapping J_φ with gauge function φ, and let C be a nonempty closed convex subset of E. Let $T : C \to C$ be a continuous pseudocontractive mapping such that $Fix(T) \neq \emptyset$, let $S : C \to C$ be a nonexpansive mapping, and let $f : C \to C$ be a contractive mapping with a contractive coefficient $k \in (0,1)$. For $x_0 \in C$, let $\{x_n\}$ be defined by the following iterative scheme:*

$$\begin{cases} x_n = \alpha_n y_n + (1-\alpha_n)Tx_n \\ y_n = \beta_n fx_{n-1} + \gamma_n Sx_{n-1} + (1-\beta_n - \gamma_n)x_{n-1}, \quad \forall n \geq 0, \end{cases} \quad (20)$$

where $\{\alpha_n\}$, $\{\beta_n\}$, and $\{\gamma_n\}$ are three sequences in $(0,1]$ satisfying the following conditions:

(i) $\lim_{n\to\infty} \alpha_n = 0$;
(ii) $\sum_{n=1}^{\infty} \beta_n = \infty$;
(iii) $\lim_{n\to\infty}(\gamma_n/\beta_n) = 0$, $\beta_n + \gamma_n \leq 1$, $\forall n \geq 0$.

Then, $\{x_n\}$ converges strongly to a fixed point x^* of T, which is the unique solution of the variational inequality in Equation (16).

Proof. First, as in Theorem 4, we put $z_t = tfz_t + r_t Sz_t + (1 - t - r_t)Tz_t$. Then, from Theorem 2, it follows that, as $t \to 0$, z_t converges strongly to some fixed point x^* of T such that x^* is the unique solution in $Fix(T)$ to the variational inequality in Equation (16).

Now, we divide the proof into several steps.

Step 1. We show that $\{x_n\}$ is bounded. To this end, let $p \in Fix(T)$. Then, by Equation (20), we have

$$\begin{aligned}
\|x_n - p\|\varphi(\|x_n - p\|) &= \langle \alpha_n y_n + (1-\alpha_n)Tx_n - p, J_\varphi(x_n - p)\rangle \\
&\leq (1-\alpha_n)\langle Tx_n - Tp, J_\varphi(x_n - p)\rangle + \alpha_n \|y_n - p\|\varphi(\|x_n - p\|) \\
&\leq (1-\alpha_n)\|x_n - p\|\varphi(\|x_n - p\|) + \alpha_n \|y_n - p\|\varphi(\|y_n - p\|)
\end{aligned}$$

and, hence,

$$\|x_n - p\| \leq \|y_n - p\|, \quad \forall n \geq 0.$$

Thus, we obtain

$$\begin{aligned}
\|x_n - p\| &\leq \|y_n - p\| \\
&\leq \beta_n \|fx_{n-1} - p\| + \gamma_n \|Sx_{n-1} - p\| + (1-\beta_n - \gamma_n)\|x_{n-1} - p\| \\
&\leq \beta_n(\|fx_{n-1} - fp\| + \|fp - p\|) + \gamma_n(\|Sx_{n-1} - Sp\| + \|Sp - p\|) \\
&\quad + (1-\beta_n - \gamma_n)\|x_{n-1} - p\| \\
&\leq \beta_n k \|x_{n-1} - p\| + \beta_n \|fp - p\| + \gamma_n \|x_{n-1} - p\| + \gamma_n \|Sp - p\| \\
&\quad + (1-\beta_n - \gamma_n)\|x_{n-1} - p\| \\
&= (1-(1-k)\beta_n)\|x_{n-1} - p\| + \beta_n \|fp - p\| + \gamma_n \|Sp - p\|.
\end{aligned} \quad (21)$$

Since $\lim_{n\to\infty}(\gamma_n/\beta_n) = 0$, we may assume without loss of generality that $\gamma_n \leq \beta_n$ for all $n > 0$. Therefore, it follows from Equation (21) that

$$\|x_n - p\| \leq (1 - (1-k)\beta_n)\|x_{n-1} - p\| + (1-k)\beta_n \cdot \frac{1}{1-k}(\|fp - p\| + \|Sp - p\|)$$

$$\leq \max\left\{\|x_{n-1} - p\|, \frac{1}{1-k}(\|fp - p\| + \|Sp - p\|)\right\}.$$

By induction, we derive

$$\|x_n - p\| \leq \max\left\{\|x_0 - p\|, \frac{1}{1-k}(\|fp - p\| + \|Sp - p\|)\right\}, \quad \forall n \geq 0.$$

This show that $\{x_n\}$ is bounded and so is $\{y_n\}$.

Step 2. We show that $\{fx_n\}, \{Sx_n\}$, and $\{Tx_n\}$ are bounded. Indeed, observe that

$$\|fx_n\| \leq \|fx_n - fp\| + \|fp\| \leq k\|x_n - p\| + \|fp\|$$

and

$$\|Sx_n\| \leq \|Sx_n - Sp\| + \|Sp\| \leq \|x_n - p\| + \|Sp\|.$$

Thus, $\{fx_n\}$ and $\{Sx_n\}$ are bounded. Since $\lim_{n\to\infty} \alpha_n = 0$, there exist $n_0 \geq 0$ and $a \in (0,1)$ such that $\alpha_n \leq a$ for all $n \geq n_0$. Noting that $x_n = \alpha_n y_n + (1 - \alpha_n) Tx_n$, we have

$$Tx_n = \frac{1}{1 - \alpha_n} x_n - \frac{\alpha_n}{1 - \alpha_n} y_n$$

and so

$$\|Tx_n\| \leq \frac{1}{1 - \alpha_n}\|x_n\| + \frac{\alpha_n}{1 - \alpha_n}\|y_n\| \leq \frac{1}{1-a}\|x_n\| + \frac{a}{1-a}\|y_n\|.$$

Consequently, the sequence $\{Tx_n\}$ is also bounded.

Step 3. We show that $\limsup_{n\to\infty}\langle fx^* - x^*, J_\varphi(x_n - x^*)\rangle \leq 0$. In fact, from condition (i) and boundedness of $\{x_n\}$ and $\{Tx_n\}$, we get

$$\|x_n - Tx_n\| = \alpha_n\|y_n - Tx_n\| \to 0 \quad (n \to \infty). \tag{22}$$

Thus, it follows from Equation (22) and Theorem 3 that $\limsup_{n\to\infty}\langle fx^* - x^*, J_\varphi(x_n - x^*)\rangle \leq 0$.

Step 4. We show that $\lim_{n\to\infty} \|x_n - x^*\| = 0$. In fact, using the equality

$$x_n - x^* = \alpha_n[\beta_n(fx_{n-1} - fx^*) + \gamma_n(Sx_{n-1} - Sx^*) + (1 - \beta_n - \gamma_n)(x_{n-1} - x^*)]$$
$$+ \alpha_n[\beta_n(fx^* - x^*) + \gamma_n(Sx^* - x^*)] + (1 - \alpha_n)(Tx_n - x^*)$$

by Equation (20) and the inequalities $\langle Tx - Ty, J_\varphi(x - y)\rangle \leq \|x - y\|\varphi(\|x - y\|) = \Phi(\|x - y\|)$, $\|fx - fy\| \leq k\|x - y\|$, and $\|Sx - Sy\| \leq \|x - y\|$, from Lemma 2, we derive

$$\begin{aligned}
\Phi(\|x_n - x^*\|) &= \Phi(\alpha_n\|\beta_n(fx_{n-1} - fx^*) + \gamma_n(Sx_{n-1} - Sx^*) + (1-\beta_n-\gamma_n)(x_{n-1}-x^*)\|) \\
&\quad + \alpha_n\beta_n\langle fx^* - x^*, J_\varphi(x_n - x^*)\rangle + \alpha_n\gamma_n\langle Sx^* - x^*, J_\varphi(x_n - x^*)\rangle \\
&\quad + (1-\alpha_n)\langle Tx_n - x^*, J_\varphi(x_n - x^*)\rangle \\
&\leq \alpha_n\Phi(\beta_n k\|x_{n-1} - x^*\| + \gamma_n\|x_{n-1} - x^*\| + (1-\beta_n-\gamma_n)\|x_{n-1} - x^*\|) \\
&\quad + \alpha_n\beta_n\langle fx^* - x^*, J_\varphi(x_n - x^*)\rangle + \alpha_n\gamma_n\langle Sx^* - x^*, J_\varphi(x_n - x^*)\rangle \qquad (23)\\
&\quad + (1-\alpha_n)\|x_n - x^*\|\varphi(\|x_n - x^*\|) \\
&\leq \alpha_n(1-(1-k)\beta_n)\Phi(\|x_{n-1} - x^*\|) \\
&\quad + \alpha_n\beta_n\langle fx^* - x^*, J_\varphi(x_n - x^*)\rangle + \alpha_n\gamma_n\|Sx^* - x^*\|\varphi(\|x_n - x^*\|) \\
&\quad + (1-\alpha_n)\Phi(\|x_n - x^*\|).
\end{aligned}$$

By Equation (23), we obtain

$$\begin{aligned}
\Phi(\|x_n - x^*\|) &\leq (1-(1-k)\beta_n)\Phi(\|x_{n-1} - x^*\|) + \beta_n\langle fx^* - x^*, J_\varphi(x_n - x^*)\rangle \\
&\quad + \gamma_n\|Sx^* - x^*\|\varphi(\|x_n - x^*\|) \qquad (24)\\
&\leq (1-(1-k)\beta_n)\|x_{n-1} - x^*\| + \beta_n\langle fx^* - x^*, J_\varphi(x_n - x^*)\rangle + \gamma_n\|Sx^* - x^*\|M,
\end{aligned}$$

where $M > 0$ is a constant such that $\varphi(\|x_n - x^*\|) \leq M$ for all $n \geq 1$. Put $\lambda_n = (1-k)\beta_n$ and

$$\delta_n = \frac{1}{1-k}\left[\langle fx^* - x^*, J_\varphi(x_n - x^*)\rangle + \frac{\gamma_n}{\beta_n}\|Sx^* - x^*\|M\right].$$

From conditions (ii) and (iii) and from step 3, it easily seen that $\sum_{n=0}^\infty \lambda_n = \infty$ and $\limsup_{n\to\infty}\delta_n \leq 0$. Since Equation (24) reduces to

$$\Phi(\|x_n - x^*\|) \leq (1-\lambda_n)\Phi(\|x_{n-1} - x^*\|) + \lambda_n\delta_n, \qquad (25)$$

applying Lemma 3 to Equation (25), we conclude that $\lim_{n\to\infty}\Phi(\|x_n - x^*\|) = 0$ and, hence, $\lim_{n\to\infty}\|x_n - x^*\| = 0$. This completes the proof. □

Remark 2. (1) Theorem 3 develops Theorem 2.3 of Ceng et al. [17] in the following aspects:

- (i) The space is replaced by the space having a weakly continuous duality mapping J_φ with gauge function φ.
- (ii) The Lipischiz strongly pseudocontractive mapping f in Theorem 2.3 in Reference [17] is replaced by a bounded continuous strongly pseudocontractive mapping f in Theorem 3.

(2) Theorem 4 complements Theorem 3.1 as well as Theorem 3.4 of Ceng et al. [17] in a reflexive Banach space having a weakly continuous duality mapping J_φ with gauge function φ.

(3) Theorem 5 also means that Theorem 3.2 as well as Theorem 3.5 of Ceng et al. [17] hold in a reflexive Banach space having a weakly continuous duality mapping J_φ with gauge function φ.

(4) Whenever $S = I$ and $\gamma_n = 0$ for all $n \geq 0$ in Theorem 5, it is easily seen that Theorem 3.1 Theorem 3.4 of Song and Chen [11] hold in a reflexive Banach space which has a weakly continuous duality mapping J_φ with gauge function φ.

Funding: This research was supported by the Basic Science Research Program through the National Research Foundation of Korea (NRF) funded by the Ministry of Education, Science, and Technology (2018R1D1A1B07045718).

Acknowledgments: The author thanks the anonymous reviewers for their reading and helpful comments and suggestions along with providing recent related papers, which improved the presentation of this manuscript.

Conflicts of Interest: The author declares no conflict of interest.

References

1. Chang, S.S.; Wen, C.F.; Yao, J.C. Zero point problems of accretive operators in Banach spaces. *Bull. Malays. Math. Sci. Soc.* **2019**, *3*, 105–118. [CrossRef]
2. Chidume, C.E. Global iteration schemes for strongly pseudocontractive maps. *Proc. Am. Math. Soc.* **1998**, *126*, 2641–2649. [CrossRef]
3. Chidume, C.E.; Osilike, M.O. Nonlinear accretive and pseudocontractive opeator equations in Banach spaces. *Nonlinear Anal.* **1998**, *31*, 779–789. [CrossRef]
4. Deimling, K. Zeros of accretive operators. *Manuscr. Math.* **1974**, *13*, 365–374. [CrossRef]
5. Martin, R.H. Differential equations on closed subsets of Banach spaces. *Trans. Am. Math. Soc.* **1975**, *179*, 399–414. [CrossRef]
6. Morales, C.H.; Chidume, C.E. Convergence of the steepest descent method for accretive operators. *Proc. Am. Math. Soc.* **1999**, *127*, 3677-3683. [CrossRef]
7. Morales, C.H.; Jung, J.S. Convergence of paths for pseudocontractive mappings in Banach spaces. *Proc. Am. Math. Soc.* **2000**, *128*, 3411–3419. [CrossRef]
8. Reich, S. An iterative procedure for constructing zero of accretive sets in Banach spaces. *Nonlinear Anal.* **1978**, *2*, 85–92. [CrossRef]
9. Reich, S. Strong convergence theorems for resolvents of accretive operators in Banach spaces. *J. Math. Anal. Appl.* **1980**, *75*, 287–292. [CrossRef]
10. Rezapour, S.; Zakeri, S.H. Strong convergence theorems for δ-inverse strongly accretive operators in Banach spaces. *Appl. Set-Valued Anal. Optim.* **2019**, *1*, 39–52.
11. Song, Y.S.; Chen, R.D. Convergence theorems of iterative algorithms for continuous pseudocontractive mappings. *Nonlinear Anal.* **2007**, *67*, 486–497. [CrossRef]
12. Tuyen, T.M.; Trang, N.M. Two new algorithms for finding a common zero of accretive operators in Banach spaces. *J. Nonlinear Var. Anal.* **2019**, *3*, 87–107.
13. Yao, Y.; Liou, Y.C.; Chen, R. Strong convergence of an iterative algorithm for pseudocontractive mapping in Banach spaces. *Nonlinear Anal.* **2007**, *67*, 3311–3317. [CrossRef]
14. Yuan, H. A splitting algorithm in a uniformly convex and 2-uniformly smooth Banach space. *J. Nonlinear Funct. Anal.* **2018**, *2018*, 1–12.
15. Morales, C.H. Strong convergence of path for continuous pseudo-contractive mappings. *Proc. Am. Math. Soc.* **2007**, *135*, 2831–2838. [CrossRef]
16. Zeng, L.C.; Yao, J.-C. Implicit iteration scheme with perturbed mapping for common fixed points of a finite family of nonexpansive mappings. *Nonlinear Anal.* **2006**, *64*, 2507–2515. [CrossRef]
17. Ceng, L.-C.; Petruşel, A.; Yao, J.-C. Strong convergence of modified implicit iterative algorithms with perturbed mappings for continuous pseudocontractive mappings. *Appl. Math. Comput.* **2009**, *209*, 162–176.
18. Xu, H.K. Viscosity approximation methods for nonexpansive mappings. *J. Math. Anal. Appl.* **2004**, *298*, 279–291. [CrossRef]
19. Xu, H.K.; Ori, R.G. An implicit iteration process for nonexpansive mappings. *Numer. Funct. Anal. Optim.* **2001**, *22*, 767–773. [CrossRef]
20. Chen, R.D.; Song, Y.S.; Zhou, H.Y. Convergence theorems for implicit iteration process for a finite family of continuous pseudocontractive mappings. *J. Math. Anal. Appl.* **2006**, *314*, 701–709. [CrossRef]
21. Agarwal, R.P.; O'Regan, D.; Sahu, D.R. *Fixed Point Theory for Lipschitzian-Type Mappings with Applications*; Springer: Berlin, Germany, 2009.
22. Cioranescu. I. *Geometry of Banach Spaces, Duality Mappings and Nonlinear Problems*; Kluwer Academic Publishers: Dordrecht, The Netherlands, 1990.
23. Goebel, K.; Kirk, W.A. Topics in Metric Fixed Point Theory. In *Cambridge Studies in Advanced Mathematics*; Cambridge Univirsity Press: Cambridge, UK, 1990; Volume 28.
24. Jung, J.S. Convergence of irerative algorithms for continuous pseudocontractive mappings. *Filomat* **2016**, *30*, 1767–1777. [CrossRef]

© 2020 by the author. Licensee MDPI, Basel, Switzerland. This article is an open access article distributed under the terms and conditions of the Creative Commons Attribution (CC BY) license (http://creativecommons.org/licenses/by/4.0/).

Article

Properties for ψ-Fractional Integrals Involving a General Function ψ and Applications

Jin Liang * and Yunyi Mu

School of Mathematical Sciences, Shanghai Jiao Tong University, Shanghai 200240, China; muyunyi@sjtu.edu.cn
* Correspondence: jinliang@sjtu.edu.cn

Received: 28 April 2019; Accepted: 5 June 2019; Published: 6 June 2019

Abstract: In this paper, we are concerned with the ψ-fractional integrals, which is a generalization of the well-known Riemann–Liouville fractional integrals and the Hadamard fractional integrals, and are useful in the study of various fractional integral equations, fractional differential equations, and fractional integrodifferential equations. Our main goal is to present some new properties for ψ-fractional integrals involving a general function ψ by establishing several new equalities for the ψ-fractional integrals. We also give two applications of our new equalities.

Keywords: fractional calculus; ψ-fractional integrals; fractional differential equations

1. Introduction

Fractional integrals and fractional derivatives are generalizations of classical integer-order integrals and integer-order derivatives, respectively, which have been found to be more adequate in the study of a lot of real world problems. In recent decades, various fractional-order models have been used in plasma physics, automatic control, robotics, and many other branches of science (cf., [1–24] and the references therein).

It is known that the ψ-fractional derivative operator, which was introduced in [22], extends the well-known Riemann–Liouville fractional derivative operator. Moreover, it is also easy to see that the ψ-fractional integral operator [14] extends the well-known Riemann–Liouville fractional integral operator and the Hadamard fractional integral operator (see Remark 1 below). Both the ψ-fractional derivative operator and the ψ-fractional integral operator are useful in the study of various fractional integral equations, fractional differential equations, and fractional integrodifferential equations.

The following known definitions about fractional integrals are used later.

Definition 1. [14] *Let $[t_1, t_2] \in \mathbb{R}$ and $\alpha > 0$. The Riemann–Liouville fractional integrals (left-sided and right-sided) of order α are defined by*

$$\mathcal{J}^\alpha_{t_1+} f(\mu) := \frac{1}{\Gamma(\alpha)} \int_{t_1}^{\mu} \frac{f(s)}{(\mu - s)^{1-\alpha}} ds, \ \mu > t_1$$

and

$$\mathcal{J}^\alpha_{t_2-} f(\mu) := \frac{1}{\Gamma(\alpha)} \int_{\mu}^{t_2} \frac{f(s)}{(s - \mu)^{1-\alpha}} ds, \ \mu < t_2,$$

respectively, where

$$\Gamma(t) = \int_0^\infty s^{t-1} e^{-s} ds,$$

is the Euler's gamma function.

Definition 2. [23] Let $[t_1, t_2] \in \mathbb{R}$ and $\alpha > 0$. The Hadamard fractional integrals (left-sided and right-sided) of order α are defined by

$$H_{t_1+}^\alpha f(\mu) := \frac{1}{\Gamma(\alpha)} \int_{t_1}^{\mu} \left(\ln\frac{\mu}{s}\right)^{\alpha-1} \frac{f(s)}{s} ds, \ \mu > t_1$$

and

$$H_{t_2-}^\alpha f(\mu) := \frac{1}{\Gamma(\alpha)} \int_{\mu}^{t_2} \left(\ln\frac{s}{\mu}\right)^{\alpha-1} \frac{f(s)}{s} ds, \ \mu < t_2,$$

respectively.

Definition 3. [14] Let $[t_1, t_2] \in \mathbb{R}$ and $\alpha > 0$. Suppose that $\psi(\mu) > 0$ is an increasing function on $(t_1, t_2]$, and $\psi'(\mu)$ is continuous on (t_1, t_2). The ψ-fractional integrals (left-sided and right-sided) of order α are defined by

$$I_{t_1+}^{\alpha;\psi} f(\mu) = \frac{1}{\Gamma(\alpha)} \int_{t_1}^{\mu} \psi'(s)(\psi(\mu) - \psi(s))^{\alpha-1} f(s) ds, \ \mu > t_1 \tag{1}$$

and

$$I_{t_2-}^{\alpha;\psi} f(\mu) = \frac{1}{\Gamma(\alpha)} \int_{\mu}^{t_2} \psi'(s)(\psi(s) - \psi(\mu))^{\alpha-1} f(s) ds, \ \mu < t_2, \tag{2}$$

respectively.

Remark 1.

(i) From [14], we know that, for a function f, the right-sided and left-sided Riemann–Liouville fractional integral of order α are defined by

$$\mathcal{J}_{a+}^\alpha f(x) := \frac{1}{\Gamma(\alpha)} \int_a^x \frac{f(t)}{(x-t)^{1-\alpha}} dt, \ x > a$$

and

$$\mathcal{J}_{b-}^\alpha f(x) := \frac{1}{\Gamma(\alpha)} \int_x^b \frac{f(t)}{(t-x)^{1-\alpha}} dt, \ x < b,$$

respectively. If we take $\psi(x) = x$, then it follows from (1) that

$$I_{a+}^{\alpha;x} f(x) = \frac{1}{\Gamma(\alpha)} \int_a^x (x-t)^{\alpha-1} f(t) dt = \mathcal{J}_{a+}^\alpha f(x),$$

which is the right-sided Riemann–Liouville fractional integral.

(ii) From [23], we know that, for a function f, the right-sided and left-sided Hadamard fractional integral of order α are defined by

$$H_{a+}^\alpha f(x) := \frac{1}{\Gamma(\alpha)} \int_a^x \left(\ln\frac{x}{t}\right)^{\alpha-1} \frac{f(t)}{t} dt, \ x > a$$

and

$$H_{b-}^\alpha f(x) := \frac{1}{\Gamma(\alpha)} \int_x^b \left(\ln\frac{t}{x}\right)^{\alpha-1} \frac{f(t)}{t} dt, \ x < b,$$

respectively. Hence, taking $\psi(x) = \ln x$ in (1), we have

$$\begin{aligned} I_{a+}^{\alpha;\ln x} f(x) &= \frac{1}{\Gamma(\alpha)} \int_a^x \frac{1}{t}(\ln x - \ln t)^{\alpha-1} f(t) dt \\ &= \frac{1}{\Gamma(\alpha)} \int_a^x \left(\ln\frac{x}{t}\right)^{\alpha-1} f(t) \frac{dt}{t} \\ &= H_{a+}^\alpha f(x), \end{aligned}$$

which is the right-sided Hadamard fractional integral.

Throughout this paper, we suppose that $\psi(\mu)$ is a strictly increasing function on $(0, \infty)$ and $\psi'(\mu)$ is continuous, $0 \leq t_1 < t_2$. $\zeta(\mu)$ is the inverse function of $\psi(\mu)$ and

$$\phi(\mu) := f(\mu) + f(t_1 + t_2 - \mu).$$

The rest of the paper is organized as follows. In Section 2, we give some new equalities for ψ-fractional integrals involving a general function ψ. To illustrate the applicability of our new equalities, we give two examples in Section 3 by introducing the ψ-means and presenting relationships between the arithmetic mean and the ψ-means, and by establishing a prior estimate for a class of fractional differential equations in view of the equalities established in Section 2.

2. Equalities for ψ-Fractional Integrals

Theorem 1. *Let the function $f : [t_1, t_2] \to \mathbb{R}$ be differentiable. Then, for the ψ-fractional integrals in (1) and (2), we have*

$$\frac{f(t_1) + f(t_2)}{2} - \frac{\Gamma(\alpha+1)}{2(\psi(t_2) - \psi(t_1))^\alpha}[I_{t_1^+}^{\alpha;\psi}f(t_2) + I_{t_2^-}^{\alpha;\psi}f(t_1)]$$
$$= \frac{\psi(t_2) - \psi(t_1)}{2}\int_0^1 [(1-\mu)^\alpha - \mu^\alpha]f'(\zeta((1-\mu)\psi(t_2) + \mu\psi(t_1)))$$
$$\cdot \zeta'((1-\mu)\psi(t_2) + \mu\psi(t_1))d\mu. \quad (3)$$

Proof. Write

$$I = \int_0^1 [(1-\mu)^\alpha - \mu^\alpha]f'(\zeta((1-\mu)\psi(t_2) + \mu\psi(t_1)))\zeta'((1-\mu)\psi(t_2) + \mu\psi(t_1))d\mu$$
$$= I_1 + I_2, \quad (4)$$

where

$$I_1 = \int_0^1 (1-\mu)^\alpha f'(\zeta((1-\mu)\psi(t_2) + \mu\psi(t_1)))\zeta'((1-\mu)\psi(t_2) + \mu\psi(t_1))d\mu,$$

$$I_2 = -\int_0^1 \mu^\alpha f'(\zeta((1-\mu)\psi(t_2) + \mu\psi(t_1)))\zeta'((1-\mu)\psi(t_2) + \mu\psi(t_1))d\mu.$$

Then, for I_1, we have

$$I_1 = \int_0^1 (1-\mu)^\alpha f'(\zeta((1-\mu)\psi(t_2) + \mu\psi(t_1)))\zeta'((1-\mu)\psi(t_2) + \mu\psi(t_1))d\mu$$
$$= (1-\mu)^\alpha \frac{f(\zeta(\mu\psi(t_1) + (1-\mu)\psi(t_2)))}{\psi(t_1) - \psi(t_2)}\bigg|_0^1$$
$$- \frac{\alpha}{\psi(t_2) - \psi(t_1)}\int_0^1 (1-\mu)^{\alpha-1} f(\zeta(\mu\psi(t_1) + (1-\mu)\psi(t_2)))d\mu$$
$$= \frac{f(t_2)}{\psi(t_2) - \psi(t_1)} - \frac{\alpha}{(\psi(t_2) - \psi(t_1))^{\alpha+1}}\int_{t_1}^{t_2}(\psi(\mu) - \psi(t_1))^{\alpha-1}f(\mu)\psi'(\mu)d\mu$$
$$= \frac{f(t_2)}{\psi(t_2) - \psi(t_1)} - \frac{\Gamma(\alpha+1)}{(\psi(t_2) - \psi(t_1))^{\alpha+1}}I_{t_2^-}^{\alpha;\psi}f(t_1). \quad (5)$$

For I_2, we obtain

$$\begin{aligned}
I_2 &= -\int_0^1 \mu^\alpha f'\left(\zeta((1-\mu)\psi(t_2)+\mu\psi(t_1))\right)\zeta'\left((1-\mu)\psi(t_2)+\mu\psi(t_1)\right)d\mu \\
&= \mu^\alpha \frac{f(\zeta(\mu\psi(t_1)+(1-\mu)\psi(t_2)))}{\psi(t_2)-\psi(t_1)}\bigg|_0^1 - \frac{\alpha}{\psi(t_2)-\psi(t_1)}\int_0^1 \mu^{\alpha-1} f(\zeta(\mu\psi(t_1)+(1-\mu)\psi(t_2)))d\mu \\
&= \frac{f(t_1)}{\psi(t_2)-\psi(t_1)} - \frac{\alpha}{(\psi(t_2)-\psi(t_1))^{\alpha+1}}\int_{t_1}^{t_2}(\psi(t_2)-\psi(\mu))^{\alpha-1} f(\mu)\psi'(\mu)d\mu \\
&= \frac{f(t_1)}{\psi(t_2)-\psi(t_1)} - \frac{\Gamma(\alpha+1)}{(\psi(t_2)-\psi(t_1))^{\alpha+1}} I_{t_1+}^{\alpha;\psi} f(t_2).
\end{aligned} \quad (6)$$

Thus, by (4)–(6), we get

$$I = \frac{f(t_1)+f(t_2)}{\psi(t_2)-\psi(t_1)} - \frac{\Gamma(\alpha+1)}{(\psi(t_2)-\psi(t_1))^{\alpha+1}} [I_{t_1+}^{\alpha;\psi} f(t_2) + I_{t_2-}^{\alpha;\psi} f(t_1)]. \quad (7)$$

This implies that equality (3) is true. □

Based on Theorem 1, we can obtain the following Theorems 2 and 3.

Theorem 2. *If the function $f : [t_1, t_2] \to \mathbb{R}$ is differentiable, then for the ψ-fractional integrals in (1) and (2), we have*

$$\begin{aligned}
\frac{\Gamma(\alpha+1)}{2(\psi(t_2)-\psi(t_1))^\alpha} &[I_{t_1+}^{\alpha;\psi} f(t_2) + I_{t_2-}^{\alpha;\psi} f(t_1)] - f\left(\frac{t_1+t_2}{2}\right) \\
&= \frac{t_2-t_1}{2}\int_0^1 g(\mu) f'(\mu t_1+(1-\mu)t_2)d\mu \\
&\quad - \frac{\psi(t_2)-\psi(t_1)}{2}\int_0^1 [(1-\mu)^\alpha - \mu^\alpha] f'(\zeta((1-\mu)\psi(t_2)+\mu\psi(t_1))) \\
&\quad \cdot \zeta'((1-\mu)\psi(t_2)+\mu\psi(t_1))d\mu,
\end{aligned} \quad (8)$$

where

$$g(\mu) = \begin{cases} 1, & \mu \in [0, \tfrac{1}{2}), \\ -1, & \mu \in [\tfrac{1}{2}, 1]. \end{cases}$$

Proof. Notice that

$$\begin{aligned}
&\frac{t_2-t_1}{2}\int_0^1 g(\mu) f'(\mu t_1+(1-\mu)t_2)d\mu \\
&= \frac{t_2-t_1}{2}\int_0^{\frac{1}{2}} f'(\mu t_1+(1-\mu)t_2)d\mu - \frac{t_2-t_1}{2}\int_{\frac{1}{2}}^1 f'(\mu t_1+(1-\mu)t_2)d\mu \\
&= -\frac{1}{2}f(\mu t_1+(1-\mu)t_2)\bigg|_0^{\frac{1}{2}} + \frac{1}{2}f(\mu t_1+(1-\mu)t_2)\bigg|_{\frac{1}{2}}^1 \\
&= \frac{f(t_1)+f(t_2)}{2} - f\left(\frac{t_1+t_2}{2}\right).
\end{aligned} \quad (9)$$

Combining (3) from Theorem 1 and (9), we get (8). □

Theorem 3. Let the function $f : [t_1, t_2] \to \mathbb{R}$ be differentiable. Then,

$$\frac{\Gamma(\alpha+1)}{2(\psi(t_2)-\psi(t_1))^\alpha}[I_{t_1+}^{\alpha;\psi}f(t_2) + I_{t_2-}^{\alpha;\psi}f(t_1)] - f\left(\zeta\left(\frac{\psi(t_1)+\psi(t_2)}{2}\right)\right)$$
$$= \frac{\psi(t_2)-\psi(t_1)}{2}\int_0^1 g(\mu)f'(\zeta((1-\mu)\psi(t_2) + \mu\psi(t_1)))\zeta'((1-\mu)\psi(t_2) + \mu\psi(t_1))d\mu$$
$$- \int_0^1 [(1-\mu)^\alpha - \mu^\alpha]f'(\zeta((1-\mu)\psi(t_2) + \mu\psi(t_1)))\zeta'((1-\mu)\psi(t_2) + \mu\psi(t_2))d\mu, \quad (10)$$

where

$$g(\mu) = \begin{cases} 1, & \mu \in [0, \frac{1}{2}), \\ -1, & \mu \in [\frac{1}{2}, 1]. \end{cases}$$

Proof. Observe

$$\frac{\psi(t_2)-\psi(t_1)}{2}\int_0^1 g(\mu)f'(\zeta((1-\mu)\psi(t_2) + \mu\psi(t_1)))\zeta'((1-\mu)\psi(t_2) + \mu\psi(t_1))d\mu$$
$$= \frac{\psi(t_2)-\psi(t_1)}{2}\int_0^{\frac{1}{2}} f'(\zeta((1-\mu)\psi(t_2) + \mu\psi(t_1)))\zeta'((1-\mu)\psi(t_2) + \mu\psi(t_1))d\mu$$
$$- \frac{\psi(t_2)-\psi(t_1)}{2}\int_{\frac{1}{2}}^1 f'(\zeta((1-\mu)\psi(t_2) + \mu\psi(t_1)))\zeta'((1-\mu)\psi(t_2) + \mu\psi(t_1))d\mu$$
$$= -\frac{1}{2}f(\zeta(\mu\psi(t_1) + (1-\mu)\psi(t_2)))\Big|_0^{\frac{1}{2}} + \frac{1}{2}f(\zeta(\mu\psi(t_1) + (1-\mu)\psi(t_2)))\Big|_{\frac{1}{2}}^1$$
$$= \frac{f(t_1)+f(t_2)}{2} - f\left(\zeta\left(\frac{\psi(t_1)+\psi(t_2)}{2}\right)\right). \quad (11)$$

Combining (11) of Theorem 1 and (3), we get the equality (10). □

The following result involves a point μ between t_1 and t_2.

Theorem 4. If the function $f : [t_1, t_2] \to \mathbb{R}$ is differentiable, then we have

$$\Gamma(\alpha+1)[I_{t-}^{\alpha;\psi}f(t_1) + I_{t+}^{\alpha;\psi}f(t_2)] - [f(t_1)(\psi(\mu)-\psi(t_1))^\alpha + f(t_2)(\psi(t_2)-\psi(\mu))^\alpha]$$
$$= (\psi(t_2)-\psi(\mu))^{\alpha+1}\int_0^1 (s^\alpha - 1)f'(\zeta((1-s)\psi(t_2) + s\psi(\mu)))\zeta'((1-s)\psi(t_2) + s\psi(\mu))ds$$
$$- (\psi(\mu)-\psi(t_1))^{\alpha+1}\int_0^1 (s^\alpha - 1)f'(\zeta((1-s)\psi(t_1) + s\psi(\mu)))$$
$$\cdot \zeta'((1-s)\psi(t_1) + s\psi(\mu))ds, \quad (12)$$

where $\mu \in (t_1, t_2)$.

Proof. Observe

$$\int_0^1 (s^\alpha - 1) f'(\zeta((1-s)\psi(t_2) + s\psi(\mu)))\zeta'((1-s)\psi(t_2) + s\psi(\mu))ds$$

$$= \frac{1}{\psi(\mu) - \psi(t_2)}(s^\alpha - 1)f(\zeta(s\psi(\mu) + (1-s)\psi(t_2)))\Big|_0^1$$

$$+ \frac{\alpha}{\psi(t_2) - \psi(\mu)} \int_0^1 s^{\alpha-1} f(\zeta(s\psi(\mu) + (1-s)\psi(t_2)))ds$$

$$= -\frac{f(t_2)}{\psi(t_2) - \psi(\mu)} + \frac{\alpha}{(\psi(t_2) - \psi(\mu))^{\alpha+1}} \int_t^{t_2} (\psi(t_2) - \psi(s))^{\alpha-1} f(s)\psi'(s)ds$$

$$= -\frac{f(t_2)}{\psi(t_2) - \psi(\mu)} + \frac{\Gamma(\alpha+1)}{(\psi(t_2) - \psi(\mu))^{\alpha+1}} I_{t+}^{\alpha;\psi} f(t_2), \tag{13}$$

and

$$\int_0^1 (s^\alpha - 1) f'(\zeta((1-s)\psi(t_1) + s\psi(\mu)))\zeta'((1-s)\psi(t_1) + s\psi(\mu))ds$$

$$= \frac{1}{\psi(\mu) - \psi(t_1)}(s^\alpha - 1)f(\zeta(s\psi(\mu) + (1-s)\psi(t_1)))\Big|_0^1$$

$$- \frac{\alpha}{\psi(\mu) - \psi(t_1)} \int_0^1 s^{\alpha-1} f(\zeta(s\psi(\mu) + (1-s)\psi(t_1)))ds$$

$$= \frac{f(t_1)}{\psi(\mu) - \psi(t_1)} - \frac{\alpha}{(\psi(\mu) - \psi(t_1))^{\alpha+1}} \int_{t_1}^t (\psi(s) - \psi(t_1))^{\alpha-1} f(s)\psi'(s)ds$$

$$= \frac{f(t_1)}{\psi(\mu) - \psi(t_1)} - \frac{\Gamma(\alpha+1)}{(\psi(\mu) - \psi(t_1))^{\alpha+1}} I_{t-}^{\alpha;\psi} f(t_1). \tag{14}$$

Combining (13) and (14), we get the result (12). □

Next, we will give two equalities involving function ϕ.

Theorem 5. *Let the function $f : [t_1, t_2] \to \mathbb{R}$ be differentiable. If $f \in L[t_1, t_2]$, then*

$$\frac{\phi(t_1) + \phi(t_2)}{2} - \frac{\Gamma(\alpha+1)}{2(\psi(t_2) - \psi(t_1))^\alpha} [I_{t_1+}^{\alpha;\psi}\phi(t_2) + I_{t_2-}^{\alpha;\psi}\phi(t_1)]$$

$$= \frac{t_2 - t_1}{2(\psi(t_2) - \psi(t_1))^\alpha} \int_0^1 g(s)\phi'((1-s)t_1 + t_2s)ds, \tag{15}$$

where

$$g(\mu) = (\psi((1-\mu)t_1 + t_2\mu) - \psi(t_1))^\alpha - (\psi(t_2) - \psi((1-\mu)t_1 + t_2\mu))^\alpha.$$

Proof. Write

$$I = \int_0^1 g(s)\phi'((1-s)t_1 + t_2s)ds$$

$$= \int_0^1 (\psi((1-s)t_1 + t_2s) - \psi(t_1))^\alpha \phi'((1-s)t_1 + t_2s)ds$$

$$- \int_0^1 (\psi(t_2) - \psi((1-s)t_1 + t_2s))^\alpha \phi'((1-s)t_1 + t_2s)ds$$

$$= I_1 + I_2.$$

Then, for I_1, we have

$$\begin{aligned}
I_1 &= \int_0^1 (\psi((1-s)t_1+t_2s)-\psi(t_1))^\alpha \phi'((1-s)t_1+t_2s)ds \\
&= \frac{1}{t_2-t_1}\int_{t_1}^{t_2}(\psi(s)-\psi(t_1))^\alpha d\phi(s) \\
&= \left.\frac{(\psi(s)-\psi(t_1))^\alpha \phi(s)}{t_2-t_1}\right|_{t_1}^{t_2} - \frac{\alpha}{t_2-t_1}\int_{t_1}^{t_2}\frac{\psi'(s)}{(\psi(s)-\psi(t_1))^{1-\alpha}}\phi(s)ds \\
&= \frac{(\psi(t_2)-\psi(t_1))^\alpha}{t_2-t_1}\phi(t_2) - \frac{\Gamma(\alpha+1)}{t_2-t_1}I_{t_2^-}^{\alpha;\psi}\phi(t_1).
\end{aligned} \qquad (16)$$

For I_2, we obtain

$$\begin{aligned}
I_2 &= -\int_0^1 (\psi(t_2)-\psi((1-s)t_1+t_2s))^\alpha \phi'((1-s)t_1+t_2s)ds \\
&= -\frac{1}{t_2-t_1}\int_{t_1}^{t_2}(\psi(t_2)-\psi(s))^\alpha d\phi(s) \\
&= -\left.\frac{(\psi(t_2)-\psi(s))^\alpha \phi(s)}{t_2-t_1}\right|_{t_1}^{t_2} - \frac{\alpha}{t_2-t_1}\int_{t_1}^{t_2}\frac{\psi'(s)}{(\psi(t_2)-\psi(s))^{1-\alpha}}\phi(s)ds \\
&= \frac{(\psi(t_2)-\psi(t_1))^\alpha}{t_2-t_1}\phi(t_1) - \frac{\Gamma(\alpha+1)}{t_2-t_1}I_{t_1^+}^{\alpha;\psi}\phi(t_2).
\end{aligned} \qquad (17)$$

By adding (16) and (17), we get

$$I = \frac{(\psi(t_2)-\psi(t_1))^\alpha}{t_2-t_1}[\phi(t_1)+\phi(t_2)] - \frac{\Gamma(\alpha+1)}{t_2-t_1}[I_{t_2^-}^{\alpha;\psi}\phi(t_1)+I_{t_1^+}^{\alpha;\psi}\phi(t_2)].$$

This implies that the equality (15) is true. □

Theorem 6. Let $f:[t_1,t_2]\to\mathbb{R}$ be a differentiable function and $f'\in L[t_1,t_2]$. If $h:[t_1,t_2]\to\mathbb{R}$ is integrable, then

$$\begin{aligned}
&\frac{\phi(t_1)+\phi(t_2)}{2}[I_{t_1^+}^{\alpha;\psi}h(t_2)+I_{t_2^-}^{\alpha;\psi}h(t_1)] - [I_{t_1^+}^{\alpha;\psi}(h\phi)(t_2)+I_{t_2^-}^{\alpha;\psi}(h\phi)(t_1)] \\
&= \frac{1}{2\Gamma(\alpha)}\int_{t_1}^{t_2}\left[\int_{t_1}^{t}\mathfrak{p}(s)h(s)ds - \int_{t}^{t_2}\mathfrak{p}(s)h(s)ds\right]\phi'(\mu)d\mu,
\end{aligned} \qquad (18)$$

where

$$\mathfrak{p}(\mu) = \frac{\psi'(\mu)}{(\psi(t_2)-\psi(\mu))^{1-\alpha}} + \frac{\psi'(\mu)}{(\psi(\mu)-\psi(t_1))^{1-\alpha}}.$$

Proof. Write

$$\begin{aligned}
I &= \int_{t_1}^{t_2}\left[\int_{t_1}^{t}\mathfrak{p}(s)h(s)ds - \int_{t}^{t_2}\mathfrak{p}(s)h(s)ds\right]\phi'(\mu)d\mu \\
&= \int_{t_1}^{t_2}\int_{t_1}^{t}\mathfrak{p}(s)h(s)ds\phi'(\mu)d\mu - \int_{t_1}^{t_2}\int_{t}^{t_2}\mathfrak{p}(s)h(s)ds\phi'(\mu)d\mu \\
&= I_1 + I_2.
\end{aligned}$$

Then, for I_1, we have

$$\begin{aligned}
I_1 &= \int_{t_1}^{t_2}\int_{t_1}^{t} p(s)h(s)\,ds\,\phi'(\mu)\,d\mu \\
&= \int_{t_1}^{t} p(s)h(s)\,ds\,\phi(\mu)\Big|_{t_1}^{t_2} - \int_{t_1}^{t_2} p(\mu)h(\mu)\phi(\mu)\,d\mu \\
&= \int_{t_1}^{t_2} p(s)h(s)\,ds\,\phi(t_2) - \int_{t_1}^{t_2} p(\mu)h(\mu)\phi(\mu)\,d\mu \\
&= \Gamma(\alpha)[I_{t_1+}^{\alpha;\psi}h(t_2) + I_{t_2-}^{\alpha;\psi}h(t_1)]\phi(t_2) \\
&\quad - \Gamma(\alpha)[I_{t_1+}^{\alpha;\psi}(h\phi)(t_2) + I_{t_2-}^{\alpha;\psi}(h\phi)(t_1)].
\end{aligned} \quad (19)$$

For I_2, we obtain

$$\begin{aligned}
I_2 &= -\int_{t_1}^{t_2}\int_{t}^{t_2} p(s)h(s)\,ds\,\phi'(\mu)\,d\mu \\
&= -\int_{t}^{t_2} p(s)h(s)\,ds\,\phi(\mu)\Big|_{t_1}^{t_2} - \int_{t_1}^{t_2} p(\mu)h(\mu)\phi(\mu)\,d\mu \\
&= \int_{t_1}^{t_2} p(s)h(s)\,ds\,\phi(t_1) - \int_{t_1}^{t_2} p(\mu)h(\mu)\phi(\mu)\,d\mu \\
&= \Gamma(\alpha)[I_{t_1+}^{\alpha;\psi}h(t_2) + I_{t_1-}^{\alpha;\psi}h(t_1)]\phi(t_1) \\
&\quad - \Gamma(\alpha)[I_{t_1+}^{\alpha;\psi}(h\phi)(t_2) + I_{t_2-}^{\alpha;\psi}(h\phi)(t_1)].
\end{aligned} \quad (20)$$

Combining (19) and (20), we get

$$I = \Gamma(\alpha)[I_{t_1+}^{\alpha;\psi}h(t_2) + I_{t_2-}^{\alpha;\psi}h(t_1)](\phi(t_1) + \phi(t_2)) - 2\Gamma(\alpha)[I_{t_1+}^{\alpha;\psi}(h\phi)(t_2) + I_{t_2-}^{\alpha;\psi}(h\phi)(t_1)].$$

This implies the equality (18). □

For the last result of this section, we suppose that $\psi(0) = 0$ and $\psi(1) = 1$.

Theorem 7. *Let the function $f : [\psi(t_1), \psi(t_2)] \to \mathbb{R}$ be differentiable. Then, the following equality holds:*

$$\begin{aligned}
&\frac{f(\psi(t_1)) + f(\psi(t_2))}{2} - \frac{\Gamma(\alpha+1)}{2(\psi(t_2) - \psi(t_1))^\alpha}[I_{t_1+}^{\alpha;\psi}f \circ \psi(t_2) + I_{t_2-}^{\alpha;\psi}f \circ \psi(t_1)] \\
&= \frac{\psi(t_2) - \psi(t_1)}{2}\int_0^1 [(1-\psi(\mu))^\alpha - \psi^\alpha(\mu)]\psi'(\mu) \\
&\quad \cdot f'((1-\psi(\mu))\psi(t_2) + \psi(\mu)\psi(t_1))\,d\mu,
\end{aligned} \quad (21)$$

where $f \circ \psi(\mu) = f(\psi(\mu))$.

Proof. Write

$$\begin{aligned}
I &= \int_0^1 [(1-\psi(\mu))^\alpha - \psi^\alpha(\mu)]\psi'(\mu)f'((1-\psi(\mu))\psi(t_2) + \psi(\mu)\psi(t_1))\,d\mu \\
&= \int_0^1 (1-\psi(\mu))^\alpha \psi'(\mu)f'((1-\psi(\mu))\psi(t_2) + \psi(\mu)\psi(t_1))\,d\mu \\
&\quad - \int_0^1 \psi^\alpha(\mu)\psi'(\mu)f'((1-\psi(\mu))\psi(t_2) + \psi(\mu)\psi(t_1))\,d\mu \\
&= I_1 + I_2.
\end{aligned}$$

Then, for I_1, we get

$$\begin{aligned}
I_1 &= \int_0^1 (1-\psi(\mu))^\alpha \psi'(\mu) f'((1-\psi(\mu))\psi(t_2)+\psi(\mu)\psi(t_1))d\mu \\
&= \frac{(1-\psi(\mu))^\alpha}{\psi(t_1)-\psi(t_2)} f((1-\psi(\mu))\psi(t_2)+\psi(\mu)\psi(t_1))\Big|_0^1 \\
&\quad + \int_0^1 \frac{\alpha(1-\psi(\mu))^{\alpha-1}\psi'(\mu)}{\psi(t_1)-\psi(t_2)} f((1-\psi(t)y)\psi(t_2)+\psi(\mu)\psi(t_1))d\mu \\
&= \frac{f(\psi(t_2))}{\psi(t_2)-\psi(t_1)} - \frac{\alpha}{\psi(t_2)-\psi(t_1)} \int_{t_1}^{t_2} \left(\frac{\psi(\mu)-\psi(t_1)}{\psi(t_2)-\psi(t_1)}\right)^{\alpha-1} \frac{\psi'(t)y}{\psi(t_2)-\psi(t_1)} f(\psi(\mu))d\mu \\
&= \frac{f(\psi(t_2))}{\psi(t_2)-\psi(t_1)} - \frac{\Gamma(\alpha+1)}{(\psi(t_2)-\psi(t_1))^{\alpha+1}} I_{t_2-}^{\alpha;\psi} f \circ \psi(t_1). \quad (22)
\end{aligned}$$

For I_2, we have

$$\begin{aligned}
I_2 &= -\int_0^1 \psi^\alpha(\mu)\psi'(\mu) f'((1-\psi(\mu))\psi(t_2)+\psi(\mu)\psi(t_1))d\mu \\
&= \frac{\psi^\alpha(\mu)}{\psi(t_2)-\psi(t_1)} f((1-\psi(\mu))\psi(t_2)+\psi(\mu)\psi(t_1))\Big|_0^1 \\
&\quad - \int_0^1 \frac{\alpha\psi^{\alpha-1}(\mu)\psi'(\mu)}{\psi(t_2)-\psi(t_1)} f((1-\psi(\mu))\psi(t_2)+\psi(\mu)\psi(t_1))d\mu \\
&= \frac{f(\psi(t_1))}{\psi(t_2)-\psi(t_1)} - \frac{\alpha}{\psi(t_2)-\psi(t_1)} \int_{t_1}^{t_2} \left(\frac{\psi(t_2)-\psi(\mu)}{\psi(t_2)-\psi(t_1)}\right)^{\alpha-1} \frac{\psi'(\mu)}{\psi(t_2)-\psi(t_1)} f(\psi(\mu))d\mu \\
&= \frac{f(\psi(t_1))}{\psi(t_2)-\psi(t_1)} - \frac{\Gamma(\alpha+1)}{(\psi(t_2)-\psi(t_1))^{\alpha+1}} I_{t_1+}^{\alpha;\psi} f \circ \psi(t_2). \quad (23)
\end{aligned}$$

By (22) and (23), we see that

$$I = \frac{f(\psi(t_1))+f(\psi(t_2))}{\psi(t_2)-\psi(t_1)} - \frac{\Gamma(\alpha+1)}{(\psi(t_2)-\psi(t_1))^{\alpha+1}}[I_{t_1+}^{\alpha;\psi} f \circ \psi(t_2) + I_{t_2-}^{\alpha;\psi} f \circ \psi(t_1)],$$

which means that equality (21) is true. □

3. Applications

To illustrate the applicability of the new equalities established in previous section, we give two examples in this section.

Example 1. *The arithmetic mean A is defined by*

$$A(t_1,t_2) := \frac{t_1+t_2}{2}, \quad t_1, t_2 > 0.$$

Now, we introduce the following ψ-means M_ψ and $\overline{M}_{\psi,n}$:

$$M_\psi(t_1,t_2) := \frac{\int_{t_1}^{t_2} \mu\psi'(\mu)d\mu}{\psi(t_2)-\psi(t_1)}, \quad t_1 \neq t_2, \quad (24)$$

and

$$\overline{M}_{\psi,n}(t_1,t_2) := \frac{\psi^{n+1}(t_2)-\psi^{n+1}(t_1)}{(n+1)(\psi(t_2)-\psi(t_1))}, \quad t_1 \neq t_2, \quad n \in \mathbb{N}.$$

57

As we can see from (24) that the ψ-mean $M_\psi(t_1, t_2)$ is just the following logarithmic mean [25] when $\psi(\mu) = \ln \mu$:
$$L(t_1, t_2) := \frac{t_2 - t_1}{\ln t_2 - \ln t_1}, \quad t_1 \neq t_2.$$

Moreover, we see that, when $\psi(\mu) = \mu$, the ψ-mean $M_\psi(t_1, t_2)$ is just the arithmetic mean $A(t_1, t_2)$.

The following two results, which are deduced by virtue of our new equalities in the last section, show new relationships between the arithmetic mean A and the two ψ-means above.

Theorem 8. Let $0 < t_1 < t_2$. Then,

$$|A(t_1, t_2) - M_\psi(t_1, t_2)| \leq \frac{\psi(t_2) - \psi(t_1)}{2(q+1)^{1/q}} \left(\int_0^1 [\zeta'(\mu\psi(t_1) + (1-\mu)\psi(t_2))]^{q'} d\mu \right)^{1/q'},$$

where $q > 1$ and $\frac{1}{q} + \frac{1}{q'} = 1$.

Proof. Taking $\alpha = 1$ and $f(\mu) = \mu$ in Theorem 1 and using the Hölder inequality, we obtain

$$|A(t_1, t_2) - M_\psi(t_1, t_2)|$$
$$\leq \frac{\psi(t_2) - \psi(t_1)}{2} \int_0^1 |1 - 2\mu|\zeta'(\mu\psi(t_1) + (1-\mu)\psi(t_2)) d\mu$$
$$\leq \frac{\psi(t_2) - \psi(t_1)}{2} \left(\int_0^1 |1 - 2\mu|^q d\mu \right)^{1/q} \left(\int_0^1 [\zeta'(\mu\psi(t_1) + (1-\mu)\psi(t_2))]^{q'} d\mu \right)^{1/q'}.$$

Noticing that
$$\left(\int_0^1 |1 - 2\mu|^q d\mu \right)^{1/q} = \frac{1}{(q+1)^{1/q}},$$

we get the desired result. □

Theorem 9. Let $0 < t_1 < t_2$, $\psi(0) = 0$ and $\psi(1) = 1$. Then,

$$|A(\psi^n(t_1), \psi^n(t_2)) - \overline{M}_{\psi,n}(t_1, t_2)|$$
$$\leq \frac{n(\psi(t_2) - \psi(t_1))^{1-1/q'}}{2(q+1)^{1/q}(q'(n-1)+1)^{1/q'}} \left(\psi^{q'(n-1)+1}(t_2) - \psi^{q'(n-1)+1}(t_1) \right)^{1/q'},$$

where $q > 1$ and $\frac{1}{q} + \frac{1}{q'} = 1$.

Proof. Taking $\alpha = 1$ and $f(\mu) = \mu^n$ in Theorem 7 and using the Hölder inequality, we obtain

$$|A(\psi^n(t_1), \psi^n(t_2)) - \overline{M}_{\psi,n}(t_1, t_2)|$$
$$\leq \frac{n(\psi(t_2) - \psi(t_1))}{2} \int_0^1 |1 - 2\psi(\mu)|\psi'(\mu)(\psi(\mu)\psi(t_1) + (1-\psi(\mu))\psi(t_2))^{n-1} d\mu$$
$$\leq \frac{n(\psi(t_2) - \psi(t_1))}{2} \left(\int_0^1 |1 - 2\psi(\mu)|^q \psi'(\mu) d\mu \right)^{1/q}$$
$$\cdot \left(\int_0^1 [\psi(\mu)\psi(t_1) + (1-\psi(\mu))\psi(t_2)]^{q'(n-1)} \psi'(\mu) d\mu \right)^{1/q'}.$$

Observing that

$$\left(\int_0^1 [\psi(\mu)\psi(t_1) + (1-\psi(\mu))\psi(t_2)]^{q'(n-1)}\psi'(\mu)d\mu\right)^{1/q'} = \frac{\left(\psi^{q'(n-1)+1}(t_2) - \psi^{q'(n-1)+1}(t_1)\right)^{1/q'}}{(q'(n-1)+1)^{1/q'}(\psi(t_2) - \psi(t_1))^{1/q'}},$$

we get the desired result. □

Example 2. *Consider the following fractional integrodifferential equations of Sobolev type with nonlocal conditions in $\mathbb{R}$:*

$$\begin{cases} {}^cD^{\psi;\alpha}u(t) = f\left(t, u(t), \int_a^t \rho(t,s)h(t,s,u(s))ds\right), & t \in J := [a, T], \\ u(a) = u_a - g(u), \end{cases} \tag{25}$$

where ${}^cD^{\psi;\alpha}$, $\alpha \in (0,1)$, is the ψ-Caputo fractional derivative of order α with the lower limit $a > 0$, $u_a \in \mathbb{R}$ and $g : C(J, \mathbb{R}) \to \mathbb{R}$, $f : J \times \mathbb{R} \times \mathbb{R} \to \mathbb{R}$, $\rho : \Delta \to \mathbb{R}$ and $h : \Delta \times \mathbb{R} \to \mathbb{R}$ ($\Delta = \{(t,s) \in [a,T] \times [a,T] : t \geq s\}$) are given functions.

Applying the operator $I_{a+}^{\alpha;\psi}$ to the first equation of the problem (25), we get for each $t \in (a, T]$,

$$\begin{aligned}
u(t) &= u(a) + \frac{1}{\Gamma(\alpha)} \int_a^t \psi'(s)(\psi(t) - \psi(s))^{\alpha-1} f\left(s, u(s), \int_a^s \rho(s,\tau)h(s,\tau,u(\tau))d\tau\right) ds \\
&= u_a - g(u) + \frac{1}{\Gamma(\alpha)} \int_a^t \psi'(s)(\psi(t) - \psi(s))^{\alpha-1} \\
& \quad f\left(s, u(s), \int_a^s \rho(s,\tau)h(s,\tau,u(\tau))d\tau\right) ds.
\end{aligned} \tag{26}$$

Substituting (26) into (3) of Theorem 1 with $f = u$, $t_1 = a$ and $t_2 = t$, we can obtain

$$\begin{aligned}
& u_a - g(u) + \frac{1}{2\Gamma(\alpha)} \int_a^t \psi'(s)(\psi(t) - \psi(s))^{\alpha-1} f\left(s, u(s), \int_a^s \rho(s,\tau)h(s,\tau,u(\tau))d\tau\right) ds \\
& - \frac{\alpha}{2(\psi(t) - \psi(a))^\alpha}\left[\int_a^t \psi'(s)(\psi(t) - \psi(s))^{\alpha-1} u(s)ds + \int_a^t \psi'(s)(\psi(s) - \psi(a))^{\alpha-1} u(s)ds\right] \\
& = \frac{\psi(t) - \psi(a)}{2} \int_0^1 [(1-s)^\alpha - s^\alpha]\zeta'(s\psi(a) + (1-s)\psi(t))u'(\zeta(s\psi(a) + (1-s)\psi(t)))ds.
\end{aligned}$$

Therefore, we have

$$\begin{aligned}
& \int_a^t \psi'(s)(\psi(t) - \psi(s))^{\alpha-1} u(s)ds + \int_a^t \psi'(s)(\psi(s) - \psi(a))^{\alpha-1} u(s)ds \\
& = \frac{2(\psi(t) - \psi(a))^\alpha}{\alpha}(u_a - g(u)) + \frac{(\psi(t) - \psi(a))^\alpha}{\Gamma(\alpha+1)} \int_a^t \psi'(s)(\psi(t) - \psi(s))^{\alpha-1} \\
& \quad f\left(s, u(s), \int_a^s \rho(s,\tau)h(s,\tau,u(\tau))d\tau\right) ds - \frac{(\psi(t) - \psi(a))^{\alpha+1}}{\alpha} \int_0^1 [(1-s)^\alpha - s^\alpha] \\
& \quad \zeta'(s\psi(a) + (1-s)\psi(t))u'(\zeta(s\psi(a) + (1-s)\psi(t)))ds.
\end{aligned} \tag{27}$$

Using the fact that $|a^\alpha - b^\alpha| \leq |a - b|^\alpha$ ($a, b > 0$; $0 < \alpha < 1$) and Hölder inequality to (27), we obtain the following result.

Theorem 10. *For each solution $u(t) \in C^1[a,T]$ of the problem (25), if $|u'(t)| \leq M$, then we have the following prior estimate:*

$$\left| \int_a^t \psi'(s)(\psi(t) - \psi(s))^{\alpha-1} u(s) ds + \int_a^t \psi'(s)(\psi(s) - \psi(a))^{\alpha-1} u(s) ds \right|$$
$$\leq \frac{2(\psi(t) - \psi(a))^{\alpha}}{\alpha} |u_a - g(u)|$$
$$+ \frac{(\psi(t) - \psi(a))^{\alpha}}{\Gamma(\alpha+1)} \int_a^t \psi'(s)(\psi(t) - \psi(s))^{\alpha-1} \left| f\left(s, u(s), \int_a^s \rho(s, \tau) h(s, \tau, u(\tau)) d\tau\right) \right| ds$$
$$+ \frac{M(\psi(t) - \psi(a))^{\alpha+1}}{q^{1/q} \alpha^{1+1/q}} \left(\int_0^1 [\zeta'(s\psi(a) + (1-s)\psi(t))]^{q'} ds \right)^{1/q'}, \forall t \in [a, T],$$

where $q > 1$ and $\frac{1}{q} + \frac{1}{q'} = 1$.

4. Conclusions

In this paper, we present new properties for ψ-fractional integrals involving a general function ψ by establishing several new equalities for the ψ-fractional integrals. The ψ-fractional integrals are generalizations of Riemann–Liouville fractional integrals and Hadamard fractional integrals, and our equalities are more general and new. To illustrate the applicability of our new equalities, we introduce the ψ-means and explore the relationships between the arithmetic mean and the ψ-means with the aid of our equalities. Moreover, we use our equalities to obtain an prior estimate for a class of fractional differential equations. How to study the properties of solutions to fractional equations involving ψ-Caputo fractional derivative? How to reveal other new properties about ψ-fractional integrals? How to find more applications of these properties? We will pay our attention to these problems in our future research.

Author Contributions: All the authors contributed equally and significantly in writing this paper. All authors read and approved the final manuscript.

Funding: The work was supported partly by the NSF of China (11571229).

Acknowledgments: The authors would like to thank the reviewers very much for valuable comments and suggestions.

Conflicts of Interest: The authors declare that they have no competing interests.

References

1. Anderson, J.; Moradi, S.; Rafiq, T. Nonlinear Langevin and fractional Fokker-Planck equations for anomalous diffusion by Lévy stable processes. *Entropy* **2018**, *20*, 760. [CrossRef]
2. Anguraj, A.; Karthikeyan, P.; Rivero, M.; Trujillo, J.J. On new existence results for fractional integro-differential equations with impulsive and integral conditions. *Comput. Math. Appl.* **2014**, *66*, 2587–2594. [CrossRef]
3. Chalishajar, D.N.; Karthikeyan, K. Existence and uniqueness results for boundary value problems of higher order fractional integro-differential equations involving Gronwall's inequality in Banach spaces. *Acta Math. Sci.* **2013**, *33*, 758–772. [CrossRef]
4. Chalishajar, D.N.; Karthikeyan, K. Existence of mild solutions for second order non-local impulsive neutral evolution equations with state-dependent infinite delay. *Dyn. Contin. Discrete Impuls. Syst. Ser. A Math. Anal.* **2019**, *26*, 53–68.
5. Chalishajar, D.N.; Karthikeyan, K.; Trujillo, J.J. Existence of mild solutions for fractional impulsive semilinear integro-differential equations in Banach spaces. *Commun. Appl. Nonlinear Anal.* **2012**, *19*, 45–56.
6. Diagana, T. Existence of solutions to some classes of partial fractional differential equations. *Nonlinear Anal.* **2009**, *71*, 5296–5300. [CrossRef]

7. Diagana, T.; Mophou, G.; N'Guerekata, G.M. On the existence of mild solutions to some semilinear fractional integro-differential equations. *Electron. J. Qual. Theory Differ. Equ.* **2010**, *58*, 1–17. [CrossRef]
8. El-Borai, M.M. Some probability densities and fundamental solutions of fractional evolution equations. *Chaos Solitons Fractals* **2002**, *14*, 433–440. [CrossRef]
9. Favaron, A.; Favini, A. Fractional powers and interpolation theory for multivalued linear operators and applications to degenerate differential equations. *Tsukuba J. Math.* **2011**, *35*, 259–323. [CrossRef]
10. Kamenskii, M.; Obukhovskii, V.; Petrosyan, G.; Yao, J.C. Existence and approximation of solutions to nonlocal boundary value problems for fractional differential inclusions. *Fixed Point Theory Appl.* **2019**, *2019*, 2. [CrossRef]
11. Kamenskii, M.; Obukhovskii, V.; Petrosyan, G.; Yao, J.C. Boundary value problems for semilinear differential inclusions of fractional order in a Banach space. *Appl. Anal.* **2018**, *97*, 571–591. [CrossRef]
12. Kamenskii, M.; Obukhovskii, V.; Petrosyan, G.; Yao, J.C. On approximate solutions for a class of semilinear fractional-order differential equations in Banach spaces. *Fixed Point Theory Appl.* **2017**, *2017*, 28. [CrossRef]
13. Ke, T.D.; Obukhovskii, V.; Wong, N.C.; Yao, J.C. On a class of fractional order differential inclusions with infinite delays. *Appl. Anal.* **2013**, *92*, 115–137. [CrossRef]
14. Kilbas, A.A.; Srivastava, H.M.; Trujillo, J.J. Theory and applications of fractional differential equaations. In *North-Holland Mathematics Studies*; Elsevier: Amsterdam, The Netherlands, 2006; Volume 204.
15. Li, F.; Liang, J.; Xu, H.K. Existence of mild solutions for fractional integrodifferential equations of Sobolev type with nonlocal conditions. *J. Math. Anal. Appl.* **2012**, *391*, 510–525. [CrossRef]
16. Li, F.; Liang, J.; Wang, H.W. S-asymptotically ω-periodic solution for fractional differential equations of order $q \in (0,1)$ with finite delay. *Adv. Differ. Equ.* **2017**, *2017*, 183.
17. Liang, J.; Mu, Y.; Xiao, T.J. Solutions to fractional Sobolev-type integro-differential equations in Banach spaces with operator pairs and impulsive conditions. *Banach J. Math. Anal.* **2019**, to appear.
18. Liang, J.; Mu, Y. Mild solutions to the Cauchy problem for some fractional differential equations with delay. *Axioms* **2017**, *6*, 30. [CrossRef]
19. Lv, Z.W.; Liang, J.; Xiao, T.J. Solutions to the Cauchy problem for differential equations in Banach spaces with fractional order. *Comput. Math. Appl.* **2011**, *62*, 1303–1311. [CrossRef]
20. Mophou, G.; N'Guérékata, G.M. Mild solutions for semilinear fractional differential equations. *Electron. J. Differ. Equ.* **2009**, *2009*, 1–9.
21. Mophou, G.; N'Guérékata, G.M. Existence of mild solutions for some fractional differential equations with nonlocal conditions. *Semigroup Forum* **2009**, *79*, 315–322. [CrossRef]
22. Osler, T.J. Leibniz rule for fractional derivatives and an application to infinite series. *SIAM J. Appl. Math.* **1970**, *18*, 658–674. [CrossRef]
23. Samko, S.G.; Kilbas, A.A.; Marichev, O.I. *Fractional Integrals and Derivatives, Theory and Applications*; Gordon and Breach: Amsterdam, The Netherlands, 1993.
24. Skiadas, C. *Fractional Dynamics, Anomalous Transport and Plasma Science: Lectures from CHAOS2017*; Springer: Berlin, Germany, 2018.
25. Carlson, B.C. The Logarithmic Mean. *Am. Math. Mon.* **1972**, *79*, 615–618. [CrossRef]

© 2019 by the authors. Licensee MDPI, Basel, Switzerland. This article is an open access article distributed under the terms and conditions of the Creative Commons Attribution (CC BY) license (http://creativecommons.org/licenses/by/4.0/).

Article
A Solution for Volterra Fractional Integral Equations by Hybrid Contractions

Badr Alqahtani [1], Hassen Aydi [2,3], Erdal Karapınar [3,*] and Vladimir Rakočević [4,*]

[1] Department of Mathematics, King Saud University, Riyadh 11451, Saudi Arabia
[2] Institut Supérieur d'Informatique et des Techniques de Communication, Université de Sousse, H. Sousse 4000, Tunisia
[3] China Medical University Hospital, China Medical University, Taichung 40402, Taiwan
[4] Faculty of Sciences and Mathematics, University of Niš, Višegradska 33, 18000 Niš, Serbia
* Correspondence: karapinar@mail.cmuh.org.tw or erdalkarapinar@yahoo.com (E.K.); vrakoc@sbb.rs (V.R.)

Received: 30 June 2019; Accepted: 29 July 2019; Published: 1 August 2019

Abstract: In this manuscript, we propose a solution for Volterra type fractional integral equations by using a hybrid type contraction that unifies both nonlinear and linear type inequalities in the context of metric spaces. Besides this main goal, we also aim to combine and merge several existing fixed point theorems that were formulated by linear and nonlinear contractions.

Keywords: contraction; hybrid contractions; volterra fractional integral equations; fixed point

JEL Classification: 47H10; 54H25; 46J10

1. Introduction and Preliminaries

In the last few decades, one of the most attractive research topics in nonlinear functional analysis is to solve fractional differential and fractional integral equations that can be reduced properly to standard differential equations and integral equations, respectively. In this paper, we aim to get a proper solution for Volterra type fractional integral equations by using a hybrid type contraction. For this purpose, we first initialize the new hybrid type contractions that combine linear and nonlinear inequalities.

We first recall the auxiliary functions that we shall use effectively: Let Ψ be the set of all nondecreasing functions $\Lambda : [0, \infty) \to [0, \infty)$ in a way that

(Λ_Σ) there are $k_0 \in \mathbb{N}$ and $\delta \in (0,1)$ and a convergent series $\sum_{i=1}^{\infty} v_i$ such that $v_i \geq 0$ and

$$\Lambda^{i+1}(t) \leq \delta \Lambda^k(t) + v_i, \tag{1}$$

for $i \geq i_0$ and $t \geq 0$.

Each $\Lambda \in \Phi$ is called a (c)-comparison function (see [1,2]).
The following lemma demonstrate the usability and power of such auxiliary functions:

Lemma 1 ([2]). *If $\Lambda \in \Phi$, then*

(i) *The series $\sum_{k=1}^{\infty} \Lambda^k(\sigma)$ is convergent for $\sigma \geq 0$.*
(ii) *$(\Lambda^n(\sigma))_{n \in \mathbb{N}}$ converges to 0 as $n \to \infty$ for $\sigma \geq 0$;*
(iii) *Λ is continuous at 0;*
(iv) *$\Lambda(\sigma) < \sigma$, for any $\sigma \in (0, \infty)$.*

All the way through the paper, a pair (X, d) presents a **complete metric space** if it is not mentioned otherwise. In addition, the letter T presents a self-mapping on (X, d).

In what follows, we shall state the definition of a new hybrid contraction:

Definition 1. *A mapping $T : (X, d) \to (X, d)$ is called a hybrid contraction of type A, if there is Λ in Φ so that*

$$d(T\Omega, T\omega) \leq \Lambda \left(A_T^p(\Omega, \omega) \right), \tag{2}$$

where $p \geq 0$ and $\sigma_i \geq 0, i = 1, 2, 3, 4$, such that $\sum_{i=1}^{4} \sigma_i = 1$ and

$$A_T^p(\Omega, \omega) = \begin{cases} [\sigma_1(d(\Omega, \omega))^p + \sigma_2(d(\Omega, T\Omega))^p + \sigma_3(d(\omega, T\omega))^p + \sigma_4 \left(\frac{d(\omega, T\Omega) + d(\Omega, T\omega)}{2} \right)^p]^{1/p}, \\ \qquad \text{for } p > 0, \quad \Omega, \omega \in X \\ (d(\Omega, \omega))^{\sigma_1} (d(\Omega, T\Omega))^{\sigma_2} (d(\omega, T\omega))^{\sigma_3}, \\ \qquad \text{for } p = 0, \quad \Omega, \omega \in X \setminus F_T(X), \end{cases} \tag{3}$$

where $F_T(X) = \{\varrho \in X : T\varrho = \varrho\}$.

Leu us underline some particular cases from Definition 1.

1. For $p = 1$, $\sigma_4 = 0$ and $\mu_i = \kappa \sigma_i$, for $i = 1, 2, 3$, we get a contraction of Reich-Rus-Ćirić type:

$$d(T\Omega, T\omega) \leq \mu_1 d(\Omega, \omega) + \mu_2 d(\Omega, T\Omega) + \mu_3 d(\omega, T\omega),$$

for $\Omega, \omega \in X$, where $\kappa \in [0, 1)$, see [2–4].

2. In the statement above, for $\mu_i = \frac{1}{3}$, we find particular form Reich–Rus–Ćirić type contraction,

$$d(T\Omega, T\omega) \leq \frac{1}{3}[d(\Omega, \omega) + d(\Omega, T\Omega) + d(\omega, T\omega)],$$

for $\Omega, \omega \in X$.

3. If $p = 2$, and $\sigma_1 = \sigma_2 = \sigma_3 = \frac{1}{3}$, $\sigma_4 = 0$, we find the following condition,

$$d(T\Omega, T\omega) \leq \frac{\kappa}{\sqrt{3}}[d^2(\Omega, \omega) + d^2(\Omega, T\Omega) + d^2(\omega, T\omega)]^{1/2}$$

for all $\Omega, \omega \in X$, where $\kappa \in [0, 1)$.

4. If $p = 1$ and $\sigma_2 = \sigma_3 = \frac{1}{2}$, $\sigma_1 = \sigma_4 = 0$, we have a Kannan type contraction,

$$d(T\Omega, T\omega) \leq \frac{\kappa}{2}[d(\Omega, T\Omega) + d(\omega, T\omega)],$$

for all $\Omega, \omega \in X$, see [5].

5. If $p = 2$ and $\sigma_2 = \sigma_3 = \frac{1}{2}$, $\sigma_1 = \sigma_4 = 0$, we have

$$d(T\Omega, T\omega) \leq \frac{\kappa}{\sqrt{2}}[d^2(\Omega, T\Omega) + d^2(\omega, T\omega)]^{1/2}$$

for all $\Omega, \omega \in X$.

6. If $p = 0$ and $\sigma_1 = 0$, $\sigma_2 = \delta$, $\sigma_3 = 1 - \delta$, $\sigma_4 = 0$, we get an interpolative contraction of Kannan type:

$$d(T\Omega, T\omega) \leq \kappa (d(\Omega, T\Omega))^{\delta} (d(\omega, T\omega))^{1-\delta},$$

for all $\Omega, \omega \in X \backslash F_T(X)$, where $\kappa \in [0,1)$, see [6].

7. If $p = 0$ and $\sigma_1 = \alpha$, $\sigma_2 = \beta$, $\sigma_3 = 1 - \beta - \alpha$, $\sigma_4 = 0$ with $\alpha, \beta \in (0,1)$, then

$$d(T\Omega, T\omega) \leq \kappa (d(\Omega, \omega))^\alpha (d(\Omega, T\Omega))^\beta (d(\omega, T\omega))^{1-\beta-\alpha},$$

for all $\Omega, \omega \in X \backslash F_T(X)$. It is an interpolative contraction of Reich–Rus–Ćirić type [7] (for other related interpolate contraction type mappings, see [8–11]).

In this paper, we provide some fixed point results involving the hybrid contraction (18). At the end, we give a concrete example and we resolve a Volterra fractional type integral equation.

2. Main Results

Our essential result is

Theorem 1. *Suppose that a self-mapping T on (X, d) is a hybrid contraction of type A. Then, T possesses a fixed point ρ and, for any $\varsigma_0 \in X$, the sequence $\{T^n \varsigma_0\}$ converges to ρ if either*

(C_1) *T is continuous at ρ;*
(C_2) *or, $[\sigma_2^{1/p} + \frac{\sigma_4}{2}^{1/p}] < 1$;*
(C_2) *or, $[\sigma_3^{1/p} + \frac{\sigma_4}{2}^{1/p}] < 1$.*

Proof. We shall use the standard Picard algorithm to prove the claims in the theorem. Let $\{\varsigma_n\}$ be defined by the recursive relation $\varsigma_{n+1} = T\varsigma_n$, $n \geq 0$, by taking an arbitrary point $x \in X$ and renaming it as $x = \varsigma_0$. Hereafter, we shall assume that

$$\varsigma_n \neq \varsigma_{n+1} \Leftrightarrow d(\varsigma_n, \varsigma_{n+1}) > 0 \text{ for all } n \in \mathbb{N}_0.$$

Indeed, it is easy that the converse case is trivial and terminate the proof. More precisely, if there is n_0 so that $\varsigma_{n_0} = \varsigma_{n_0+1} = T\varsigma_{n_0}$, then ς_{n_0} turns to be a fixed point of T.

Now, we shall examine the cases $p = 0$ and $p > 0$, separately. We first consider the case $p > 0$. On account of the given condition (18), we find

$$d(\varsigma_{n+1}, \varsigma_n) \leq \Lambda \left(A_T^p (\varsigma_n, \varsigma_{n-1}) \right), \quad (4)$$

where

$$A_T^p(\varsigma_n, \varsigma_{n-1}) = [\sigma_1 (d(\varsigma_n, \varsigma_{n-1}))^p + \sigma_2 (d(\varsigma_n, \varsigma_{n+1}))^p + \sigma_3 (d(\varsigma_{n-1}, \varsigma_n))^p$$
$$+ \sigma_4 \left(\frac{d(\varsigma_{n-1}, \varsigma_{n+1}) + d(\varsigma_n, \varsigma_n)}{2} \right)^p \Big]^{1/p}$$

$$= [\sigma_1 (d(\varsigma_n, \varsigma_{n-1}))^p + \sigma_2 (d(\varsigma_n, \varsigma_{n+1}))^p + \sigma_3 (d(\varsigma_{n-1}, \varsigma_n))^p$$
$$+ \sigma_4 \left(\frac{1}{2} [d(\varsigma_{n-1}, \varsigma_n) + d(\varsigma_n, \varsigma_{n+1})] \right)^p \Big]^{1/p}.$$

Suppose that $d(\varsigma_n, \varsigma_{n+1}) \geq d(\varsigma_{n-1}, \varsigma_n)$. With an elementary estimation in Label (4) from the right-hand side and keeping $\sum_{i=1}^4 \sigma_i = 1$ in mind, we find that

$$d(\varsigma_{n+1}, \varsigma_n) \leq \Lambda \left(d(\varsigma_{n+1}, \varsigma_n) \sqrt[p]{\sum_{i=1}^4 \sigma_i} \right) = \Lambda (d(\varsigma_{n+1}, \varsigma_n)) < d(\varsigma_{n+1}, \varsigma_n), \quad (5)$$

a contradiction. Attendantly, we find that $d(\varsigma_n, \varsigma_{n+1}) < d(\varsigma_{n-1}, \varsigma_n)$ and further

$$d(\varsigma_{n+1}, \varsigma_n) \leq \Lambda (d(\varsigma_{n-1}, \varsigma_n)) < d(\varsigma_{n-1}, \varsigma_n). \quad (6)$$

Inductively, from the inequalities above, we deduce

$$d(\varsigma_{n+1}, \varsigma_n) \leq \Lambda^n(d(\varsigma_1, \varsigma_0)), \text{ for all } n \in \mathbb{N}. \tag{7}$$

From Label (7) and using the triangular inequality, for all $k \geq 1$, we have

$$\begin{aligned} d(\varsigma_n, \varsigma_{n+k}) &\leq d(\varsigma_n, \varsigma_{n+1}) + \ldots + d(\varsigma_{n+k-1}, \varsigma_{n+k}) \\ &\leq \sum_{r=n}^{n+k-1} \Lambda^r(d(\varsigma_1, \varsigma_0)) \\ &\leq \sum_{r=n}^{+\infty} \Lambda^r(d(\varsigma_1, \varsigma_0)) \to 0 \text{ as } n \to \infty. \end{aligned}$$

Thus, the constructive sequence $\{\varsigma_n\}$ is Cauchy in (X, d). Taking the completeness of the metric space (X, d) into account, we conclude the existence of $\rho \in X$ such that

$$\lim_{n \to \infty} d(\varsigma_n, \rho) = 0. \tag{8}$$

Now, we shall indicate that ρ is the requested fixed point of T under the given assumptions. Suppose that $(\mathcal{C}_1)$ holds, that is, T is continuous. Then,

$$\rho = \lim_{n \to \infty} \varsigma_{n+1} = \lim_{n \to \infty} T\varsigma_n = T(\lim_{n \to \infty} \varsigma_n) = T\rho.$$

Now, we suppose that $(\mathcal{C}_2)$ holds, that is, $[\sigma_2^{1/p} + \frac{\sigma_4}{2}^{1/p}] < 1$.

$$\begin{aligned} 0 < d(T\rho, \rho) &\leq d(T\rho, \varsigma_{n+1}) + d(\varsigma_{n+1}, \rho) \tag{9} \\ &= d(T\rho, T\varsigma_{n+1}) + d(\varsigma_{n+1}, \rho) \\ &\leq \Lambda\left(\mathcal{A}_T^p(\rho, \varsigma_n)\right) + d(\varsigma_{n+1}, \rho), \\ &< \mathcal{A}_T^p(\rho, \varsigma_n) + d(\varsigma_{n+1}, \rho), \end{aligned}$$

where

$$\mathcal{A}_T^p(\rho, \varsigma_n) = \left[\sigma_1(d(\rho, \varsigma_n))^p + \sigma_2(d(\rho, T\rho))^p + \sigma_3(d(\varsigma_n, \varsigma_{n+1}))^p + \sigma_4\left(\frac{d(\varsigma_n, T\rho) + d(\rho, \varsigma_{n+1})}{2}\right)^p\right]^{1/p}.$$

As $n \to \infty$, we have

$$0 < d(T\rho, \rho) \leq \Delta d(T\rho, \rho),$$

where $\Delta := [\sigma_2^{1/p} + \frac{\sigma_4}{2}^{1/p}]$. Since $\Delta := [\sigma_2^{1/p} + \frac{\sigma_4}{2}^{1/p}] < 1$, which is a contradiction, that is, $T\rho = \rho$.

We skip the details of the case $(\mathcal{C}_3)$ since it is verbatim of the proof of the case $(\mathcal{C}_2)$. Indeed, the only the difference follows from the fact that $\mathcal{A}_T^p(\rho, \varsigma_n) \neq \mathcal{A}_T^p(\varsigma_n, \rho)$ since σ_2 not need to be equal to σ_3.

As a last step, we shall consider the case $p = 0$. Here, Label (18) and Label (3) become

$$d(T\Omega, T\omega) \leq \Lambda\left((d(\Omega, \omega))^{\sigma_1}(d(\Omega, T\Omega))^{\sigma_2}(d(\omega, T\omega))^{\sigma_3}\left[\frac{d(T\Omega, \omega) + d(\Omega, T\omega)}{2}\right]^{1-\sigma_1-\sigma_2-\sigma_3}\right) \tag{10}$$

for all $\Omega, \omega \in X \setminus F_T(X)$, where $\kappa \in [0, 1)$ and $\sigma_1, \sigma_2, \sigma_3 \in (0, 1)$. Set $\Omega = \theta_n$ and $\omega = \theta_{n-1}$ in the inequality (10), we find that

$$d\left(\theta_{n+1},\theta_n\right) = d\left(T\theta_n, T\theta_{n-1}\right) \leq \Lambda \left([d\left(\theta_n,\theta_{n-1}\right)]^{\sigma_1} [d\left(\theta_n, T\theta_n\right)]^{\sigma_2} \cdot [d\left(\theta_{n-1}, T\theta_{n-1}\right)]^{\sigma_3} \right.$$
$$\left. \cdot \left[\tfrac{1}{2}(d\left(\theta_n,\theta_n\right) + d\left(\theta_{n-1},\theta_{n+1}\right)) \right]^{1-\sigma_1-\sigma_2-\sigma_3} \right) \quad (11)$$
$$\leq \Lambda \left([d\left(\theta_n,\theta_{n-1}\right)]^{\sigma_1} \cdot [d\left(\theta_n,\theta_{n+1}\right)]^{\sigma_2} \cdot [d\left(\theta_{n-1},\theta_n\right)]^{\sigma_3} \right.$$
$$\left. \cdot \left[\tfrac{1}{2}(d\left(\theta_{n-1},\theta_n\right) + d\left(\theta_n,\theta_{n+1}\right)) \right]^{1-\sigma_1-\sigma_2-\sigma_3} \right).$$

Suppose that $d\left(\theta_{n-1},\theta_n\right) < d\left(\theta_n,\theta_{n+1}\right)$ for some $n \geq 1$. Thus,

$$\frac{1}{2}(d\left(\theta_{n-1},\theta_n\right) + d\left(\theta_n,\theta_{n+1}\right)) \leq d\left(\theta_n,\theta_{n+1}\right).$$

Consequently, inequality (11) yields that

$$[d\left(\theta_n,\theta_{n+1}\right)]^{\sigma_1+\sigma_3} \leq \Lambda \left([d\left(\theta_{n-1},\theta_n\right)]^{\sigma_1+\sigma_3} \right) < [d\left(\theta_{n-1},\theta_n\right)]^{\sigma_1+\sigma_3}. \quad (12)$$

Thus, we conclude that $d\left(\theta_{n-1},\theta_n\right) \geq d\left(\theta_n,\theta_{n+1}\right)$, which is a contradiction. Thus, we have

$$d\left(\theta_n,\theta_{n+1}\right) \leq d\left(\theta_{n-1},\theta_n\right) \quad \text{for all } n \geq 1.$$

Hence, $\{d\left(\theta_{n-1},\theta_n\right)\}$ is a non-increasing sequence with positive terms. On account of the simple observation below,

$$\frac{1}{2}(d\left(\theta_{n-1},\theta_n\right) + d\left(\theta_n,\theta_{n+1}\right)) \leq d\left(\theta_{n-1},\theta_n\right), \quad \text{for all } n \geq 1$$

together with an elementary elimination, the inequality (11) implies that

$$d\left(\theta_n,\theta_{n+1}\right) \leq \Lambda(d\left(\theta_{n-1},\theta_n\right)) < d\left(\theta_{n-1},\theta_n\right) \quad (13)$$

for all $n \in \mathbb{N}$. Since the inequality (13) is equivalent to Label (6), by following the corresponding lines, we derive that the iterated sequence $\{\theta_n\}$ is Cauchy and converges to $\theta^* \in X$ that is, $\lim_{n\to\infty} d\left(\theta_n,\theta^*\right) = 0$. Suppose that $\theta^* \neq T\theta^*$. Since $\theta_n \neq T\theta_n$ for each $n \geq 0$, by letting $x = \theta_n$ and $y = \theta^*$ in (18), we have

$$d\left(\theta_{n+1}, T\theta^*\right) = d\left(T\theta_n, T\theta^*\right) \leq \Lambda \left([d\left(\theta_n,\theta^*\right)]^{\sigma_1} \cdot [d\left(\theta_n, T\theta_n\right)]^{\sigma_2} \cdot [d\left(\theta^*, T\theta^*\right)]^{\sigma_3} \right.$$
$$\left. \cdot \left[\tfrac{1}{2}(d\left(\theta_{n+1}, T\theta^*\right) + d\left(\theta^*, T\theta_{n+1}\right)) \right]^{1-\sigma_2-\sigma_1-\sigma_3} \right). \quad (14)$$

Letting $n \to \infty$ in the inequality (14), we get $d(\theta^*, T\theta^*) = 0$, which is a contradiction. That is, $T\theta^* = \theta^*$. □

Corollary 1. *Let T be a self-mapping on (X,d). Suppose that there is $\kappa \in [0,1)$ such that*

$$d(T\Omega, T\omega) \leq \kappa \mathcal{A}_T^p(\Omega,\omega), \quad (15)$$

where $p \geq 0$. Then, there is a fixed point ρ of T if either

(C_1)   *T is continuous at such point ρ;*
(C_2)   *or, $[\sigma_2^{1/p} + \tfrac{\sigma_4}{2}^{1/p}] < 1$;*
(C_2)   *or, $[\sigma_3^{1/p} + \tfrac{\sigma_4}{2}^{1/p}] < 1$;*

Definition 2. A self-mapping T is called on (X,d) a hybrid contraction of type B, if there is $\Lambda \in \Phi$ such that

$$d(T\Omega, T\omega) \leq \Lambda\left(W_T^p(\Omega, \omega)\right), \tag{16}$$

where $p \geq 0$, $a = (\sigma_1, \sigma_2, \sigma_3)$, $\sigma_i \geq 0$, $i = 1, 2, 3$ such that $\sigma_1 + \sigma_2 + \sigma_3 = 1$ and

$$W_T^p(\Omega, \omega) = \begin{cases} [\sigma_1(d(\Omega, \omega))^p + \sigma_2(d(\Omega, T\Omega))^p + \sigma_3(d(\omega, T\omega))^p]^{1/p}, & p > 0, \Omega, \omega \in X, \\ (d(\Omega, \omega))^{\sigma_1}(d(\Omega, T\Omega))^{\sigma_2}(d(\omega, T\omega))^{\sigma_3}, & p = 0, \Omega, \omega \in X \setminus F_T(X). \end{cases} \tag{17}$$

Notice that a hybrid contraction of type A and a hybrid contraction of type B are also called a weighted contraction of type A and type B, respectively.

As corollaries of Theorem 1, we also have the following.

Corollary 2. Let T be a self-mapping on (X,d). Suppose that either T is a hybrid contraction of type B, or there is $\kappa \in [0,1)$ so that

$$d(T\Omega, T\omega) \leq \kappa W_T^p(\Omega, \omega), \tag{18}$$

where $p \geq 0$. Then, there is a fixed point ρ of T if either

(i) T is continuous at such point ρ;
(ii) or, $\sigma_2 < 1$;
(iii) or, $\sigma_3 < 1$.

Corollary 3. Let T be a self-mapping on (X,d). Suppose that:

$$d(T\Omega, T\omega) \leq \kappa d^{\sigma_1}(\Omega, \omega) \cdot d^{\sigma_2}(\Omega, T\Omega) \cdot d^{\sigma_3}(\omega, T\omega), \tag{19}$$

for all $\Omega, \omega \in X \setminus F_T(X)$, where $\kappa \in [0,1)$, $\sigma_1, \sigma_2, \sigma_3 \geq 0$ and $\sigma_1 + \sigma_2 + \sigma_3 = 1$. Then, there is a fixed point ρ of T.

Proof. Put in Corollary 2, $p = 0$ and $a = (\sigma_1, \sigma_2, \sigma_3)$. □

Remark 1. Using Corollary 3, we get Theorem 2 in [7] (for metric spaces).

Corollary 4. Let T be a self-mapping on (X,d) such that

$$d(T\Omega, T\omega) \leq \kappa \sqrt[3]{d(\Omega, \omega) \cdot d(\Omega, T\Omega) \cdot d(\omega, T\omega)}, \tag{20}$$

for all $\Omega, \omega \in X \setminus F_T(X)$, where $\kappa \in [0,1)$. Then, there is a fixed point ρ of T.

Proof. Put in Corollary 2, $p = 0$ and $a = (\frac{1}{3}, \frac{1}{3}, \frac{1}{3})$. □

Corollary 5. Let T be a self-mapping on (X,d) such that

$$d(T\Omega, T\omega) \leq \frac{\kappa}{3}[d(\Omega, \omega) + d(\Omega, T\Omega) + d(\omega, T\omega)], \tag{21}$$

for all $\Omega, \omega \in X$, where $\kappa \in [0,1)$.

Then, there is a fixed point ρ of T.

(i) T is continuous at such point $\rho \in X$;
(ii) or, $b < 3$.

Proof. Put in Corollary 2, $p = 1$ and $a = (\frac{1}{3}, \frac{1}{3}, \frac{1}{3})$. □

Corollary 6. Let T be a self-mapping on (X, d) such that

$$d(T\Omega, T\omega) \leq \frac{\kappa}{\sqrt{3}}[d^2(\Omega, \omega) + d^2(\Omega, T\Omega) + d^2(\omega, T\omega)]^{1/2}, \qquad (22)$$

for all $\Omega, \omega \in X$, where $\kappa \in [0, 1)$, then T has a fixed point in X. The sequence $\{T^n \varsigma_0\}$ converges to ρ.

(i) T is continuous at such point $\rho \in X$;
(ii) or, $b^2 < 3$.

Proof. Put in Corollary 2, $p = 2$ and $a = (\frac{1}{3}, \frac{1}{3}, \frac{1}{3})$. □

Corollary 2 is illustrated by the following.

Example 1. Choose $X = \{\tau_1, \tau_2, \tau_3, \tau_4\} \cup [0, \infty)$ (where τ_1, τ_2, τ_3 and τ_4 are negative reals). Take

1. $d(\Omega, \omega) = |\Omega - \omega|$ for $(\Omega, \omega) \in [0, \infty) \times [0, \infty)$;
2. $d(\Omega, \omega) = 0$ for $(\Omega, \omega) \in \{a, b, c, d\} \times [0, \infty)$ or $(\Omega, \omega) \in [0, \infty) \times \{\tau_1, \tau_2, \tau_3, \tau_4\}$;
3. for $(\Omega, \omega) \in \{\tau_1, \tau_2, \tau_3, \tau_4\} \times \{\tau_1, \tau_2, \tau_3, \tau_4\}$,

$d(\Omega, \omega)$	τ_1	τ_2	τ_3	τ_4
τ_1	0	1	2	4
τ_2	1	0	1	3
τ_3	2	1	0	2
τ_4	4	3	2	0

Consider $T : \begin{pmatrix} \tau_1 & \tau_2 & \tau_3 & \tau_4 \\ \tau_3 & \tau_4 & \tau_3 & \tau_4 \end{pmatrix}$ and $T\Omega = \frac{\Omega}{8}$ for $\Omega \in [0, \infty)$.

For $\Omega \in [0, \infty)$, the main theorem is satisfied straightforwardly. Thus, we examine the case $\Omega \in \{a, b, c, d\}$. Note that there is no $\kappa \in [0, 1)$ such that

$$d(T\tau_1, T\tau_2) \leq \frac{\kappa}{3}[d(\tau_1, \tau_2) + d(\tau_1, T\tau_1) + d(\tau_2, T\tau_2)],$$

namely, we have,

$$2 \leq \frac{\kappa}{3}[1 + 2 + 3].$$

Thus, Corollary 5 is not applicable.

Using (20), we have

$$d(T\tau_1, T\tau_2) \leq \kappa \sqrt[3]{d(\tau_1, \tau_2) \cdot d(\tau_1, T\tau_1) \cdot d(\tau_2, T\tau_2)},$$

i.e., $2 \leq \kappa \sqrt[3]{1 \cdot 2 \cdot 3}$, so $\kappa \geq \frac{2}{\sqrt[3]{6}} > 1$. Hence, Corollary 4 is not applicable.

Corollary 6 is applicable. In fact, for $\Omega, \omega \in X$, we have for $\kappa = \sqrt{\frac{6}{7}}$,

$$d(T\Omega, T\omega) \leq \frac{\kappa}{\sqrt{3}}[d^2(\Omega, \omega) + d^2(\Omega, T\Omega) + d^2(\omega, T\omega)]^{1/2}.$$

Here, $\{0, \tau_3, \tau_4\}$ is the set of fixed points of T.

3. Application on Volterra Fractional Integral Equations

The fractional Schrodinger equation (FSE) is known as the fundamental equation of the fractional quantum mechanics. As compared to the standard Schrodinger equation, it contains the fractional Laplacian operator instead of the usual one. This change brings profound differences in the behavior of wave function. Zhang et al. [12] investigated analytically and numerically the propagation of optical beams in the FSE with a harmonic potential. In addition, Zhang et al. [13] suggested a real physical system (the honeycomb lattice) as a possible realization of the FSE system, through utilization of the Dirac–Weyl equation, while Zhang et al. [14] investigated the dynamics of waves in the FSE with a $\mathcal{PT}$-symmetric potential. Still in fractional calculus, in this section, we study a nonlinear Volterra fractional integral equation.

Set $0 < \tau < 1$ and $J = [\sigma_0, \sigma_0 + a]$ in $\mathbb{R}$ ($a > 0$). Denote by $X = C(J, \mathbb{R})$ the set of continuous real-valued functions on J.

Now, particularly, we cosnider the following nonlinear Volterra fractional integral equation (in short, VFIE)

$$\xi(t) = \mathcal{F}(t) + \frac{1}{\Gamma(\tau)} \int_{\sigma_0}^{t} (t-s)^{\tau-1} h(s, \xi(s)) ds, \tag{23}$$

for all $t \in J$, where Γ is the gamma function, $\mathcal{F}: J \to \mathbb{R}$ and $h: J \times \mathbb{R} \to \mathbb{R}$ are continuous functions. The VFIE (23) has been investigated in the literature on fractional calculus and its applications, see [15–17].

In the following result, under some assumptions, we ensure the existence of a solution for the VFIE (23).

Theorem 2. *Suppose that*

(H1) *There are constants $M > 0$ and $N > 0$ such that*

$$|h(t, u) - h(t, v)| \leq \frac{M|u-v|}{N + |u-v|} \tag{24}$$

for all $u, v \in \mathbb{R}$;

(H2) *Such M and N verify that*

$$\frac{Ma}{\Gamma(\tau+1)} \leq N. \tag{25}$$

Then, the VFIE (23) has a solution in X.

Proof. For $\xi, \eta \in X$, consider the metric

$$d(\xi, \eta) = \sup_{t \in J} |\xi(t) - \eta(t)|.$$

Take the operator

$$T\xi(t) = \mathcal{F}(t) + \frac{1}{\Gamma(\tau)} \int_{\sigma_0}^{t} (t-s)^{\tau-1} h(s, \xi(s)) ds, \quad t \in J. \tag{26}$$

□

Clearly, T is well defined. Let $\xi, \eta \in X$, then for each $t \in J$,

$$|T\xi(t) - T\eta(t)| = \frac{1}{\Gamma(\tau)} \int_{\sigma_0}^{t} (t-s)^{\tau-1} (h(s, \xi(s)) - h(s, \eta(s))) ds$$

$$\leq \frac{1}{\Gamma(\tau)} \int_{\sigma_0}^{t} (t-s)^{\tau-1} |h(s, \xi(s)) - h(s, \eta(s))| ds$$

$$\leq \frac{Ma}{\Gamma(\tau+1)} \frac{M|\xi(s) - \eta(s)|}{N + |\xi(s) - \eta(s)||}$$

$$\leq \frac{Ma}{\Gamma(\tau+1)} \frac{M\|\xi - \eta\|}{N + \|\xi - \eta)\|}.$$

We deduce that

$$\|T\xi - T\eta\| \leq \frac{Ma}{\Gamma(\tau+1)} \frac{M\|\xi - \eta\|}{N + \|\xi - \eta\|} = \Lambda(\|\xi - \eta\|), \qquad (27)$$

where $\Lambda(t) = \frac{La}{\Gamma(\tau+1)} \frac{Mt}{N+t}$ for $t \geq 0$. By hypothesis (H2), $\Lambda \in \Phi$. Then,

$$d(T\xi, T\eta) \leq \Lambda\left(\mathcal{F}_T^p(\xi,\eta)\right), \qquad (28)$$

for $p > 0$, with $\sigma_2 = \sigma_2 = \sigma_4 = 0$ and $\sigma_1 = 1$. Applying Theorem 1, T has a fixed point in X, so the VFIE (23) has a solution in X.

4. Conclusions

The obtained results unify several existing results in a single theorem. We list some of the consequences, but it is clear that there are more consequences of our main results. Regarding the length of the paper, we skip them.

Author Contributions: B.A. analyzed and prepared the manuscript, H.A. analyzed and prepared/edited the manuscript, E.K. analyzed and prepared/edited the manuscript, V.R. analyzed and prepared the manuscript. All authors read and approved the final manuscript.

Funding: We declare that funding is not applicable for our paper.

Acknowledgments: The authors are grateful to the handling editor and reviewers for their careful reviews and useful comments. The authors would like to extend their sincere appreciation to the Deanship of Scientific Research at King Saud University for funding this group No. RG-1437-017.

Conflicts of Interest: The authors declare no conflict of interest.

References

1. Bianchini, R.M.; Grandolfi, M. Transformazioni di tipo contracttivo generalizzato in uno spazio metrico. *Atti Acad. Naz. Lincei, VII. Ser. Rend. Cl. Sci. Fis. Mat. Natur.* **1968**, *45*, 212–216.
2. Rus, I.A. *Generalized Contractions and Applications*; Cluj University Press: Cluj-Napoca, Romania, 2001.
3. Ćirić, L. A generalization of Banach's contraction principle. *Proc. Am. Math. Soc.* **1974**, *45*, 267–273.
4. Reich, S. Some remarks concerning contraction mappings. *Can. Math. Bull.* **1971**, *14*, 121–124. [CrossRef]
5. Kannan, R. Some results on fixed points. *Bull. Calcutta Math. Soc.* **1968**, *60*, 71–76.
6. Karapinar, E. Revisiting the Kannan Type Contractions via Interpolation. *Adv. Theory Nonlinear Anal. Appl.* **2018**, *2*, 85–87.
7. Karapinar, E.; Agarwal, R.; Aydi, H. Interpolative Reich-Rus-Ćirić Type Contractions on Partial Metric Spaces. *Mathematics* **2018**, *6*, 256. [CrossRef]
8. Aydi, H.; Chen, C.M.; Karapinar, E. Interpolative Ciric-Reich-Rus type contractions via the Branciari distance. *Mathematics* **2019**, *7*, 84. [CrossRef]
9. Agarwal, R.P.; Karapinar, E. Interpolative Rus-Reich-Ciric Type Contractions Via Simulation Functions. *Analele Stiintifice ale Universitatii Ovidius Constanta Seria Matematica* **2019**, in press.
10. Karapinar, E.; Alqahtani, O.; Aydi, H. On Interpolative Hardy-Rogers Type Contractions. *Symmetry* **2019**, *11*, 8. [CrossRef]
11. Aydi, H.; Karapinar, E.; de Hierro, A.F.R.L. ω-Interpolative Ciric-Reich-Rus-Type Contractions. *Mathematics* **2019**, *7*, 57. [CrossRef]
12. Zhang, Y.; Liu, X.; Belic, M.R.; Zhong, W.; Zhang, Y.; Xiao, M. Propagation Dynamics of a Light Beam in a Fractional Schrodinger Equation. *Phys. Rev. Lett.* **2015**, *115*, 180403. [CrossRef] [PubMed]
13. Zhang, D.; Zhang, Y.; Zhang, Z.; Ahmed, N.; Zhang, Y.; Li, F.; Belic, M.R.; Xiao, M. Unveiling the Link Between Fractional Schrodinger Equation and Light Propagation in Honeycomb Lattice. *Ann. Phys.* **2017**, *529*, 1700149. [CrossRef]
14. Zhang, Y.; Zhong, H.; Belic, M.R.; Zhu, Y.; Zhong, W.; Zhang, Y.; Christodoulides, D.N.; Xiao, M. $\mathcal{PT}$ symmetry in a fractional Schroodinger equation. *Laser Photonics Rev.* **2016**, *10*, 526–531. [CrossRef]

15. Baleanu, D.; Jajarmi, A.; Asad, J.H.; Blaszczyk, T. The Motion of a bead sliding on a wire in fractional sense. *Acta Phys. Pol. A* **2017**, *131*, 1561–1564. [CrossRef]
16. Baleanu, D.; Jajarmi, A.; Hajipour, M. A new formulation of the fractional optimal control problems involving Mittag-Leffler nonsingular kernel. *J. Optim. Theory Appl.* **2017**, *175*, 718–737. [CrossRef]
17. Dhage, B.C. Hybrid fixed point theory in partially ordered normed linear spaces and applications to fractional integral equations. *Differ. Equ. Appl.* **2013**, *5*, 155–184.

© 2019 by the authors. Licensee MDPI, Basel, Switzerland. This article is an open access article distributed under the terms and conditions of the Creative Commons Attribution (CC BY) license (http://creativecommons.org/licenses/by/4.0/).

Article

A General Inertial Viscosity Type Method for Nonexpansive Mappings and Its Applications in Signal Processing

Yinglin Luo [1], Meijuan Shang [2,*] and Bing Tan [1]

[1] Institute of Fundamental and Frontier Sciences, University of Electronic Science and Technology of China, Chengdu 611731, China; luoyinglink@163.com (Y.L.); bingtan72@gmail.com (B.T.)
[2] College of Science, Shijiazhuang University, Shijiazhuang 050035, China
* Correspondence: meijuanshang@163.com or 1102017@sjzc.edu.cn

Received: 9 January 2020; Accepted: 7 February 2020; Published: 20 February 2020

Abstract: In this paper, we propose viscosity algorithms with two different inertia parameters for solving fixed points of nonexpansive and strictly pseudocontractive mappings. Strong convergence theorems are obtained in Hilbert spaces and the applications to the signal processing are considered. Moreover, some numerical experiments of proposed algorithms and comparisons with existing algorithms are given to the demonstration of the efficiency of the proposed algorithms. The numerical results show that our algorithms are superior to some related algorithms.

Keywords: nonexpansive mapping; strict pseudo-contraction; variational inequality problem; inclusion problem; signal processing

MSC: 49J40; 47H05; 90C52; 47J20; 47H09

---

1. Introduction

In this paper, H denotes real Hilbert spaces with inner product $\langle \cdot, \cdot \rangle$ and norm $\|\cdot\|$. We denote the set of fixed points of an operator T by $\mathrm{Fix}(T)$, more precisely, $\mathrm{Fix}(T) := \{x \in H : Tx = x\}$.

Recall that a mapping $T : H \to H$ is said to be an η-strict pseudo-contraction if $\|Tx - Ty\|^2 - \eta\|(I - T)x - (I - T)y\|^2 \leq \|x - y\|^2, \forall x, y \in H$, where $\eta \in [0, 1)$ is a real number. A mapping $T : H \to H$ is said to be nonexpansive if $\|Tx - Ty\| \leq \|x - y\|, \forall x, y \in H$. It is evident that the class of η-strict pseudo-contractions includes the class of nonexpansive mappings, as T is nonexpansive if and only if T is 0-strict pseudo-contractive. Many classical mathematical problems can be casted into the fixed-point problem of nonexpansive mappings, such as, inclusion problem, equilibrium problem, variational inequality problem, saddle point problem, and split feasibility problem, see [1–3]. Approximating fixed points of nonexpansive mappings is an important field in many areas of pure and applied mathematics. One of the most well-known algorithms for solving such a problem is the Mann iterative algorithm [4]:

$$x_{n+1} = (1 - \theta_n)Tx_n + \theta_n x_n,$$

where θ_n is a sequence in $(0, 1)$. One knows that the iterative sequence $\{x_n\}$ converges weakly to a fixed point of T provided that $\sum_{n=0}^{\infty} \theta_n(1 - \theta_n) = +\infty$. This algorithm is slow in terms of convergence speed. Moreover, this algorithm converges is weak. To obtain more effective methods, many authors have done a lot of works in this area, see [5–8]. A mapping $f : H \to H$ is called a contraction if there exists a constant in $[0, 1)$ such that $\|f(x) - f(y)\| \leq \tau\|x - y\|, \forall x, y \in H$. One of celebrated ways to study nonexpansive operators is to use a contractive operator, which is a convex combination of the previous contractive operator and the nonexpansive operator. The viscosity type method for nonexpansive mappings is defined as follows,

$$x_{n+1} = (1 - \alpha_n)Tx_n + \alpha_n f(x_n), \tag{1}$$

where α_n is a sequence in $(0,1)$, T is the nonexpansive operator, and f is the contractive operator. In this method, a special fixed point of the nonexpansive operator is obtained by regularizing the nonexpansive operator via the contraction. This method was proposed by Attouch [9] in 1996 and further promoted by Moudafi [10] in 2000. Motivated by Moudafi, Takahashi and Takahashi [11] introduced a strong convergence theorem by the viscosity type approximation method for finding the fixed point of nonexpansive mappings in Hilbert spaces. In 2019, Qin and Yao [12] introduced a viscosity iterative method for solving a split feasibility problem. For viscosity approximation methods, one refers to [13,14]. In practical applications, one not only studies different algorithms, but also pursues the speed of these algorithms. To obtain faster convergence algorithms, many scholars have given various acceleration techniques, see, e.g., [15–19]. One of the most commonly used methods is the inertial method. In [20], Polyak introduced an inertial extrapolation based on the heavy ball method for solving the smooth convex minimization problem. Shehu et al. [21] introduced a Halpern-type algorithm with inertial terms for approximating fixed points of a nonexpansive mapping. They obtained strong convergence in real Hilbert spaces under some assumptions on the sequence of parameters. To get a more general inertial Mann algorithm for nonexpansive mappings, Dong et al. [22] introduced a general inertial Mann algorithm which includes some classical algorithms as its special cases; however, they only got the weak convergence results.

Inspired by the above works, we give two algorithms for solving fixed point problems of nonexpansive mappings via viscosity and inertial techniques in this paper. One highlight is that our algorithms, which are more consistent and efficient, are accelerated via the inertial technique and the viscosity technique. In addition, the solution also uniquely solves a monotone variational inequality. Another highlight is that we consider two different inertial parameter sequences comparing with the existing results. We establish strong convergence results in infinite dimensional Hilbert spaces without compactness. We also investigate the applications of the two proposed algorithms to variational inequality problems and inclusion problems. Furthermore, we give some numerical experiments to illustrate the convergence efficiency of our algorithms. The proposed numerical experiments show that our algorithms are superior to some related algorithms.

In this paper, Section 2 is devoted to some required prior knowledge, which will be used in this paper. In Section 3, based on viscosity type method, we propose an algorithm for solving fixed point problems of nonexpansive mappings and give an algorithm for strict pseudo-contractive mappings. In Section 4, some applications of our algorithms in real Hilbert spaces are given. Finally, some numerical experiments of our algorithms and its comparisons with other algorithms in signal processing are given in Section 5. Section 6, the last section, is the final conclusion.

2. Toolbox

In this section, we give some essential lemmas for our main convergence theorems.

Lemma 1 ([23]). *Let $\{a_n\}$ be a non-negative real sequence and $\{b_n\}$ a real sequence and $\{\alpha_n\}$ a real sequence in $(0,1)$ such that $\sum_{n=1}^{\infty} \alpha_n = \infty$. Assume that $a_{n+1} \leq \alpha_n b_n + a_n(1 - \alpha_n), \forall n \geq 1$. If, for every subsequence $\{a_{n_k}\}$ of $\{a_n\}$ satisfying $\liminf_{k \to \infty} (a_{n_k+1} - a_{n_k}) \geq 0$, $\limsup_{k \to \infty} b_{n_k} \leq 0$ holds, then $\lim_{n \to \infty} a_n = 0$.*

Lemma 2 ([24]). *Suppose that $T : H \to H$ is a nonexpansive mapping. Let $\{x_n\}$ be a vector sequence in H and let p be a vector in H. If $x_n \rightharpoonup p$ and $x_n - Tx_n \to 0$. Then $p \in \text{Fix}(T)$.*

Lemma 3 ([14]). *Let $\{\sigma_n\}$ be a non-negative real sequence such that there exists a subsequence $\{\sigma_{n_i}\}$ of $\{\sigma_n\}$ satisfying $\sigma_{n_i} < \sigma_{n_i+1}$ for all $i \in N$. Then, there exists a nondecreasing sequence $\{m_k\}$ of N such* that $\lim_{k\to\infty} m_k = \infty$ and the following properties are satisfied for all (sufficiently large) number $k \in N$: $\sigma_{m_k} \leq \sigma_{m_k+1}$ and $\sigma_k \leq \sigma_{m_k+1}$.

It is known that m_k is the largest number in the set $\{1, 2, \cdots, k\}$ such that $\sigma_{m_k} < \sigma_{m_k+1}$.

Lemma 4 ([25]). *Let $\{s_n\}$ be a sequence of non-negative real numbers such that $s_{n+1} = (1 - \beta_n)s_n + \delta_n$, $\forall \geq 0$, where $\{\beta_n\}$ is a sequence in $(0, 1)$ with $\sum_{n=0}^{\infty} \beta_n = \infty$ and $\{\delta_n\}$ satisfies $\limsup_{n\to\infty} \frac{\delta_n}{\beta_n} \leq 0$ or $\sum_{n=0}^{\infty} |\delta_n| < \infty$. Then, $\lim_{n\to\infty} s_n = 0$.*

3. Main Results

In this section, we give two strong convergence theorems for approximating the fixed points of nonexpansive mappings and strict pseudo-contractive mappings. First, we propose some assumptions which will be used in our statements.

Condition 1. *Suppose that $\{\alpha_n\}, \{\beta_n\}$ and $\{\gamma_n\}$ are three real sequences in $(0, 1)$ satisfying the following conditions.*

(1) $\sum_{n=1}^{\infty} \alpha_n = \infty$ and $\lim_{n\to\infty} \alpha_n = 0$;
(2) $\lim_{n\to\infty} \frac{\theta_n}{\alpha_n} \|x_n - x_{n-1}\| = \lim_{n\to\infty} \frac{\epsilon_n}{\alpha_n} \|x_n - x_{n-1}\| = 0$;
(3) $\alpha_n + \beta_n + \gamma_n = 1$ and $\liminf_{n\to\infty} \gamma_n \beta_n > 0$;

Remark 1. *(1) If $\theta_n = \epsilon_n = 0$, i.e., $x_n = y_n = z_n$, Algorithm 1 is the classical viscosity type algorithm without the inertial technique.*
(2) Algorithm 1 is a generalization of Shehu et al. [21]. If $f(x) = u$ and $\theta_n = \epsilon_n$, i.e., $y_n = z_n$, then it becomes the Shehu et al. Algorithm 1 with $e_n = 0$.

Algorithm 1 The viscosity type algorithm for nonexpansive mappings

Initialization: Let $x_0, x_1 \in H$ be arbitrary.
Iterative Steps: Given the current iterator x_n, calculate x_{n+1} as follows:
Step 1. Compute
$$\begin{cases} y_n = \theta_n(x_n - x_{n-1}) + x_n, \\ z_n = \epsilon_n(x_n - x_{n-1}) + x_n. \end{cases} \quad (2)$$
Step 2. Compute
$$x_{n+1} = \alpha_n f(x_n) + \beta_n y_n + \gamma_n T z_n. \quad (3)$$
Step 3. Set $n \leftarrow n + 1$ and go to Step 1.

Remark 2. *The (2) of Condition 1 is well defined, as the inertial parameters θ_n and ϵ_n in (3) can be chosen such that $0 \leq \theta_n \leq \theta_n^*$ and $0 \leq \epsilon_n \leq \epsilon_n^*$, where*

$$\theta_n^* = \begin{cases} \min\left\{\theta, \frac{\delta_n}{\|x_n - x_{n-1}\|}\right\}, & x_n \neq x_{n-1}, \\ \theta, & \text{otherwise}, \end{cases} \quad \epsilon_n^* = \begin{cases} \min\left\{\epsilon, \frac{\delta_n}{\|x_n - x_{n-1}\|}\right\}, & x_n \neq x_{n-1}, \\ \epsilon, & \text{otherwise}, \end{cases} \quad (4)$$

and $\{\delta_n\}$ is a positive sequence such that $\lim_{n\to\infty} \frac{\delta_n}{\alpha_n} = 0$. It is easy to verify that $\lim_{n\to\infty} \theta_n \|x_n - x_{n-1}\| = 0$ and $\lim_{n\to\infty} \frac{\theta_n}{\alpha_n} \|x_n - x_{n-1}\| = 0$.

Theorem 1. *Let $T : H \to H$ be a nonexpansive mapping with $\text{Fix}(T) \neq \emptyset$ and let $f : H \to H$ be a contraction with constant $k \in [0, 1)$. Suppose that $\{x_n\}$ is any sequence generated by Algorithm 1 and Condition 1 holds. Then, $\{x_n\}$ converges strongly to $p = P_{\text{Fix}(T)} \circ f(p)$.*

Proof. The proof is divided into three steps.

Step 1. One claims that $\{x_n\}$ is bounded.

Let $p \in \text{Fix}(T)$. As $y_n = \theta_n(x_n - x_{n-1}) + x_n$, one concludes

$$\|y_n - p\| \leq \theta_n \|x_n - x_{n-1}\| + \|x_n - p\|. \tag{5}$$

Similarly, one gets

$$\|z_n - p\| \leq \|x_n - p\| + \epsilon_n \|x_n - x_{n-1}\|. \tag{6}$$

From (3), one obtains

$$\begin{aligned}\|x_{n+1} - p\| &\leq \gamma_n \|p - Tz_n\| + \beta_n \|p - y_n\| + \alpha_n \|p - f(x_n)\| \\ &\leq \gamma_n \|p - z_n\| + \beta_n \|p - y_n\| + \alpha_n \|f(x_n) - f(p) + f(p) - p\| \\ &\leq (1 - \alpha_n(1-k))\|x_n - p\| \\ &\quad + \alpha_n(1-k)\big(\frac{\|f(p) - p\| + \beta_n \frac{\theta_n}{\alpha_n}\|x_n - x_{n-1}\| + \gamma_n \frac{\epsilon_n}{\alpha_n}\|x_n - x_{n-1}\|}{1-k}\big).\end{aligned} \tag{7}$$

In view of Condition 1 (2), one sees that $\sup_{n \geq 1} \frac{\theta_n}{\alpha_n}\|x_n - x_{n-1}\|$ and $\sup_{n \geq 1} \frac{\epsilon_n}{\alpha_n}\|x_n - x_{n-1}\|$ exist. Taking $M := 3 \max\big\{\|f(p) - p\|, \sup_{n \geq 1} \frac{\theta_n}{\alpha_n}\|x_n - x_{n-1}\|, \sup_{n \geq 1} \frac{\epsilon_n}{\alpha_n}\|x_n - x_{n-1}\|\big\}$, one gets from (7) that

$$\begin{aligned}\|x_{n+1} - p\| &\leq (1 - \alpha_n(1-k))\|x_n - p\| + \alpha_n(1-k)M \\ &\leq \max\{\|x_n - p\|, M\} \leq \cdots \leq \max\{\|x_1 - p\|, M\}.\end{aligned}$$

This implies that $\{x_n\}$ is bounded.

Step 2. One claims that if $\{x_n\}$ converges weakly to $z \in H$, then $z \in \text{Fix}(T)$. Letting $w_{n+1} = \alpha_n f(w_n) + \beta_n w_n + \gamma_n T w_n$, from (1), one arrives at

$$\|w_n - y_n\| \leq \theta_n \|x_n - x_{n-1}\| + \|w_n - x_n\| \tag{8}$$

and

$$\|w_n - z_n\| \leq \epsilon_n \|x_n - x_{n-1}\| + \|w_n - x_n\|. \tag{9}$$

By the definition of w_{n+1}, (8) and (9), one obtains

$$\begin{aligned}\|w_{n+1} - x_{n+1}\| &\leq \alpha_n \|f(w_n) - f(x_n)\| + \beta_n \|w_n - y_n\| + \gamma_n \|Tw_n - Tz_n\| \\ &\leq k\alpha_n \|w_n - x_n\| + \beta_n \|w_n - y_n\| + \gamma_n \|w_n - z_n\| \\ &\leq (1 - \alpha_n(1-k))\|w_n - x_n\| + (\theta_n \|x_n - x_{n-1}\| + \epsilon_n \|x_n - x_{n-1}\|).\end{aligned} \tag{10}$$

From Condition 1 and Lemma 4, one sees that (10) implies $\lim_{n \to \infty} \|w_{n+1} - x_{n+1}\| = 0$. Therefore, it follows from Step 1 that $\{w_n\}$ is bounded. By the definition of w_{n+1}, one also obtains

$$\begin{aligned}\|w_{n+1} - p\|^2 &\leq \|\alpha_n(f(w_n) - f(p)) + \beta_n(y_n - p) + \gamma_n(Ty_n - p)\|^2 + 2\alpha_n \langle f(p) - p, w_{n+1} - p\rangle \\ &\leq \alpha_n k^2 \|w_n - p\|^2 + \beta_n \|w_n - p\|^2 + \gamma_n \|Tw_n - p\|^2 - \beta_n \gamma_n \|w_n - Tw_n\|^2 \\ &\quad + 2\alpha_n \langle f(p) - p, w_{n+1} - p\rangle \\ &= (1 - \alpha_n(1 - k^2))\|w_n - p\|^2 + 2\alpha_n \langle f(p) - p, w_{n+1} - p\rangle - \beta_n \gamma_n \|w_n - Tw_n\|^2.\end{aligned} \tag{11}$$

Taking $s_n = \|w_n - p\|^2$, one sees that (11) is equivalent to

$$s_{n+1} \leq (1 - \alpha_n(1-k^2))s_n - \beta_n \gamma_n \|w_n - Tw_n\|^2 + 2\alpha_n \langle f(p) - p, w_{n+1} - p\rangle. \tag{12}$$

Now, we show $z \in \text{Fix}(T)$ by considering two possible cases on sequence $\{s_n\}$.

Case 1. Suppose that there exists a $n_0 \in N$ such that $s_{n+1} \leq s_n$ for all $n \geq n_0$. This implies that $\lim_{n \to \infty} s_n$ exists. From (12), one has

$$\beta_n \gamma_n \|w_n - Tw_n\|^2 \leq (1 - \alpha_n(1 - k^2))s_n + 2\alpha_n \langle f(p) - p, w_{n+1} - p \rangle - s_{n+1}. \tag{13}$$

As $\{w_n\}$ is bounded, from Condition 1 and (13), one deduces that

$$\lim_{n \to \infty} \beta_n \gamma_n \|w_n - Tw_n\|^2 = 0. \tag{14}$$

As $\liminf_{n \to \infty} \beta_n \gamma_n > 0$, (14) implies that

$$\lim_{n \to \infty} \|w_n - Tw_n\|^2 = 0. \tag{15}$$

As $x_n \rightharpoonup z$ and $\lim_{n \to \infty} \|w_{n+1} - x_{n+1}\| = 0$, one has $w_n \rightharpoonup z$. By using Lemma 2, one gets $z \in \text{Fix}(T)$.

Case 2. There exists a subsequence $\{s_{n_j}\}$ of such $\{s_n\}$ that $s_{n_j} < s_{n_j+1}$ for all $j \in N$. In this case, it follows from Lemma 3 that there is a nondecreasing subsequence $\{m_k\}$ of N such that $\lim_{k \to \infty} m_k \to \infty$ and the following inequalities hold for all $k \in N$:

$$s_{m_k} \leq s_{m_k+1} \text{ and } s_k \leq s_{m_k+1}. \tag{16}$$

Using a similar argument as Case 1, it is easy to get that $\lim_{k \to \infty} \|Tw_{m_k} - w_{m_k}\| = 0$. It is known that $x_n \rightharpoonup z$, which implies $x_{m_k} \rightharpoonup z$. Therefore, $z \in \text{Fix}(T)$.

Step 3. One claims that $\{x_n\}$ converges strongly to $p = P_{\text{Fix}(T)} \circ f(p)$. From (11), we deduce that

$$\|w_{n+1} - p\|^2 \leq (1 - \alpha_n(1 - k^2))\|w_n - p\|^2 + 2\alpha_n \langle f(p) - p, w_{n+1} - p \rangle. \tag{17}$$

In the following, we show that the sequence $\{\|w_n - p\|\}$ converges strongly to zero. As $\{w_n\}$ is bounded, in view of Condition 1 and Lemma 1, we only need to show that for each subsequence $\{\|w_{n_k} - p\|\}$ of $\{\|w_n - p\|\}$ such that $\liminf_{k \to \infty}(\|w_{n_k+1} - p\| - \|w_{n_k} - p\|) \geq 0$, $\limsup_{k \to \infty} \langle f(p) - p, w_{n_k+1} - p \rangle \leq 0$. For this purpose, one assumes that $\{\|w_{n_k} - p\|\}$ is a subsequence of $\{\|w_n - p\|\}$ such that $\liminf_{k \to \infty}(\|w_{n_k+1} - p\| - \|w_{n_k} - p\|) \geq 0$. This implies that

$$\liminf_{k \to \infty}(\|w_{n_k+1} - p\|^2 - \|w_{n_k} - p\|^2) = \liminf_{k \to \infty}((\|w_{n_k+1} - p\| - \|w_{n_k} - p\|) \times (\|w_{n_k+1} + p\| + \|w_{n_k} - p\|)) \geq 0. \tag{18}$$

From the definition of w_n, we obtain

$$\begin{aligned} \|w_{n_k+1} - w_{n_k}\| &\leq \|\alpha_{n_k}(f(w_{n_k}) - w_{n_k}) + \gamma_{n_k}(Tw_{n_k} - w_{n_k})\| \\ &\leq \alpha_{n_k}\|f(w_{n_k}) - w_{n_k}\| + \gamma_{n_k}\|Tw_{n_k} - w_{n_k}\| \\ &\leq \alpha_{n_k}(k\|w_{n_k} - p\| + \|f(p) - w_{n_k}\|) + \gamma_{n_k}\|Tw_{n_k} - w_{n_k}\|. \end{aligned} \tag{19}$$

Using the argument of Case 1 and Case 2 in Step 2, there exists a subsequence of $\{w_{n_k}\}$, still denoted by $\{w_{n_k}\}$, such that

$$\lim_{k \to \infty} \|Tw_{n_k} - w_{n_k}\| = 0. \tag{20}$$

By the boundedness of $\{w_n\}$, one deduces from Condition 1, (19), and (20) that

$$\lim_{n \to \infty} \|w_{n_k+1} - w_{n_k}\| = 0. \tag{21}$$

As $\{w_{n_k}\}$ is bounded, there exists a subsequence $\{w_{n_{k_j}}\}$ of $\{w_{n_k}\}$ converges weakly to some $z \in H$. This implies that

$$\limsup_{k\to\infty}\langle f(p)-p, w_{n_k}-p\rangle = \limsup_{j\to\infty}\langle f(p)-p, w_{n_{k_j}}-p\rangle = \langle f(p)-p, z-p\rangle.$$

From Step 2, one gets $z \in \text{Fix}(T)$. Since $p = P_{\text{Fix}(T)} \circ f(p)$, one arrives at

$$\limsup_{k\to\infty}\langle f(p)-p, w_{n_k}-p\rangle = \langle f(p)-p, z-p\rangle \leq 0.$$

From (21), one obtains

$$\limsup_{k\to\infty}\langle f(p)-p, w_{n_k+1}-p\rangle = \limsup_{k\to\infty}\langle f(p)-p, w_{n_k}-p\rangle + \limsup_{k\to\infty}\langle f(p)-p, w_{n_k+1}-w_{n_k}\rangle \quad (22)$$
$$= \langle f(p)-p, z-p\rangle \leq 0.$$

Therefore, one has $\|w_n - p\| \to 0$. Since $\lim_{n\to\infty}\|w_n - x_n\| = 0$, one gets $\|x_n - p\| \to 0$. □

In the following, we give a strong convergent theorem for strict pseudo-contractions.

Theorem 2. *Let $T : H \to H$ be a η-strict pseudo-contraction with $\text{Fix}(T) \neq \emptyset$ and let $f : H \to H$ be a contraction with constant $k \in [0,1)$. Suppose that $\{x_n\}$ is a vector sequence generated by Algorithm 2 and Condition 1 holds. Then, $\{x_n\}$ converges strongly to $p = P_{\text{Fix}(T)} \circ f(p)$.*

Algorithm 2 The viscosity type algorithm for strict pseudo-contractions

Initialization: Let $x_0, x_1 \in H$ be arbitrary and let $\delta \in [\eta, 1)$.
Iterative Steps: Given the current iterator x_n, calculate x_{n+1} as follows.
Step 1. Compute
$$\begin{cases} y_n = \theta_n(x_n - x_{n-1}) + x_n, \\ z_n = \epsilon_n(x_n - x_{n-1}) + x_n. \end{cases} \quad (23)$$
Step 2. Compute
$$x_{n+1} = \alpha_n f(x_n) + \beta_n y_n + \gamma_n(\delta z_n + (1-\delta)Tz_n). \quad (24)$$
Step 3. Set $n \leftarrow n+1$ and go to Step 1.

Proof. Define $Q : H \to H$ by $Qx = \delta x + (1-\delta)Tx$. It is easy to verify that $\text{Fix}(T) = \text{Fix}(Q)$. By the definition of strict pseudo-contraction, one has

$$\|Qx - Qy\|^2 = \delta\|x-y\| + (1-\delta)\|Tx - Ty\|^2 - \delta(1-\delta)\|(x-y) - (Tx-Ty)\|^2$$
$$= \delta\|x-y\| + (1-\delta)\|x-y\|^2 + \eta(1-\delta)\|(x-y)-(Tx-Ty)\|^2$$
$$\quad - \delta(1-\delta)\|(x-y)-(Tx-Ty)\|^2$$
$$\leq \|x-y\|^2 - (\delta-\eta)(1-\delta)\|(x-y)-(Tx-Ty)\|^2$$
$$\leq \|x-y\|^2.$$

Therefore, Q is nonexpansive. Then, we get the conclusions from Theorem 1 immediately. □

In the following, we give some corollaries for Theorem 1.

Recall that T is called a ρ-averaged mapping if and only if it can be written as the average of the identity mapping I and a nonexpansive mapping, that is, $T := (1-\rho)I + \rho S$, where $\rho \in (0,1)$ and $S : H \to H$ is a nonexpansive mapping. It is known that every ρ-averaged mapping is nonexpansive and $\text{Fix}(T) = \text{Fix}(S)$. A mapping $T : H \to H$ is said to be quasi-nonexpansive if, for all $p \in \text{Fix}(T)$, $\|Tx - Tp\| \leq \|x - p\|, \forall x \in H$. T is said to be strongly nonexpansive if $x_n - y_n - (Tx_n - Ty_n) \to 0$, whenever $\{x_n\}$ and $\{y_n\}$ are two sequences in H such that $\{x_n - y_n\}$ is bounded and $\|x_n - y_n\| - \|Tx_n - Ty_n\| \to 0$. T is said to be strongly quasi-nonexpansive if T is quasi-nonexpansive and

$x_n - Tx_n \to 0$ whenever $\{x_n\}$ is a bounded sequence in H such that $\|x_n - p\| - \|Tx_n - Tp\| \to 0$ for all $p \in \text{Fix}(T)$. By using Theorem 1, we obtain the following corollaries easily.

Corollary 1. *Let H be a Hilbert space and let $f : H \to H$ be a contraction with constant $k \in [0,1)$. Let $T : H \to H$ be a ρ-average mapping with $\text{Fix}(T) \neq \emptyset$. Suppose that Conditions 1 holds. Then, the sequence $\{x_n\}$ generated by Algorithm 1 converges to $p = P_{\text{Fix}(T)} \circ f(p)$ in norm.*

Corollary 2. *Let H be a Hilbert space and let $f : H \to H$ be a contraction with constant $k \in [0,1)$. Let $T : H \to H$ be a quasi-nonexpansive mapping with $\text{Fix}(T) \neq \emptyset$ and $I - T$ be demiclosed at the origin. Suppose that Conditions 1 holds. Then, the sequence $\{x_n\}$ generated by Algorithm 1 converges to $p = P_{\text{Fix}(T)} \circ f(p)$ in norm.*

Corollary 3. *Let H be a Hilbert space and let $f : H \to H$ be a contraction with constant $k \in [0,1)$. Let $T : H \to H$ be a strongly quasi-nonexpansive mapping with $\text{Fix}(T) \neq \emptyset$ and $I - T$ be demiclosed at the origin. Suppose that Conditions 1 holds. Then, the sequence $\{x_n\}$ generated by Algorithm 1 converges to $p = P_{\text{Fix}(T)} \circ f(p)$ in norm.*

4. Applications

In this section, we will give some applications of our algorithms to variational equality problems, inclusion problems and corresponding convex minimization problems.

4.1. Variational Inequality Problems

In this subsection, we consider the following variational inequality problem (for short, VIP): find $x \in C$ such that

$$\langle Ax, y - x \rangle \geq 0, \quad \forall y \in C, \tag{25}$$

where $A : H \to H$ is a single-valued operator and C is a nonempty convex closed set in H. The solutions of VIP 25 is denoted by Ω. It is known that x^* is a solution of VIP (25) if and only if $x^* = P_C(x^* - \lambda A x^*)$, where λ is an arbitrary positive constant. In recent decades, the VIP has received a lot of attention. In order to solve the VIP, various methods have been proposed, see, e.g., [26–28]. In this subsection, we will give some applications of our algorithms to the VIP (25). For this purpose, we introduce a lemma proposed by Shehu et al. [21].

Lemma 5. *Let H be a Hilbert space and let C be a nonempty convex and closed set in H. Suppose that $A : H \to H$ is a monotone L-Lipschitz operator on C and that λ is a positive number. Let $V := P_C(I - \lambda A)$ and let $S := V - \lambda(AV - A)$. Then, $I - V$ is demi-closed at the origin. Moreover, if $\lambda L < 1$, S is a strongly quasi-nonexpansive operator and $\text{Fix}(S) = \text{Fix}(V) = \Omega$.*

By using Lemma 5 and Corollary 3, we obtain the following corollary for VIP (25) immediately.

Corollary 4. *Let H be a Hilbert space and let C be a nonempty convex closed set in H. Let $f : H \to H$ be a contraction with constant $k \in [0,1)$. Let $A : H \to H$ be a monotone L-Lipschitz operator and let $\tau \in \left(0, \frac{1}{L}\right)$. Suppose that Conditions 1 holds. Then, the sequence $\{x_n\}$ generated by Algorithm 3 converges to $p = P_\Omega \circ f(p)$ in norm.*

Proof. Let $S := P_C(I - \tau A) - \tau(A(P_C(I - \tau A)) - A)$. We see from Lemma 5 that S is strongly quasi-nonexpansive and $\text{Fix}(S) = \Omega$. Then, we get the conclusions from Corollary 3 immediately. □

Algorithm 3 The viscosity type algorithm for solving variational inequality problems

Iterative Steps: Given the current iterator x_n, calculate x_{n+1} as follows.

Step 1. Compute
$$\begin{cases} y_n = x_n + \theta_n(x_n - x_{n-1}), \\ z_n = x_n + \epsilon_n(x_n - x_{n-1}). \end{cases} \quad (26)$$

Step 2. Compute
$$\begin{cases} w_n = P_C(I - \lambda A)z_n, \\ x_{n+1} = \alpha_n f(x_n) + \beta_n y_n + \gamma_n(w_n - \lambda(Aw_n - Az_n)). \end{cases} \quad (27)$$

Step 3. Set $n \leftarrow n+1$ and go to Step 1.

4.2. Inclusion Problems

Let H denote the Hilbert spaces and let $A : H \to H$ be a single-valued mapping. Then, A is said to be monotone if $\langle Ax - Ay, x - y \rangle \geq 0, \forall x, y \in H$; A is said to be α-inverse strongly monotone if $\langle Ax - Ay, x - y \rangle \geq \alpha \|A(x) - A(y)\|^2, \forall x, y \in H$. A set-valued operator $A : H \to 2^H$ is said to be monotone if $\langle x - y, u - v \rangle \geq 0, \forall x, y \in H$, where $u \in Ax$ and $v \in Ay$. Furthermore, A said to be maximal monotone if, for all $(y, v) \in Graph(A)$ and each $(x, u) \in H \times H$, $\langle x - y, u - v \rangle \geq 0$ implies that $u \in Ax$. Recall that the resolvent operator $J_r^A : H \to H$ associated operator A is defined by $J_r^A = (I + rA)^{-1}x$, where $r > 0$ and I denotes the identity operator on H. If A is a maximal monotone mapping, J_r^A is a single-valued and firmly nonexpansive mapping. Consider the following simple inclusion problem: find $x^* \in H$ such that

$$0 \in Ax^*, \quad (28)$$

where $A : H \to H$ is a maximal monotone operator. It is know that $0 \in A(x)$ if and only if $x \in \text{Fix}(J_r^A)$. By using Theorem 1, we obtain the following corollary.

Corollary 5. *Let H be a Hilbert space and let $f : H \to H$ be a contraction with constant $k \in [0,1)$. Let $A : H \to H$ be a maximal monotone operator such that $A^{-1}(0) \neq \emptyset$. Suppose that Conditions 1 holds. Then, the sequence $\{x_n\}$ generated by Algorithm 4 converges strongly to $p = P_{A^{-1}(0)} \circ f(p)$.*

Algorithm 4 The viscosity type algorithm for solving inclusion problem (28)

Initialization: Let $x_0, x_1 \in H$ be arbitrary.

Iterative Steps: Given the current iterator x_n, calculate x_{n+1} as follows.

Step 1. Compute
$$\begin{cases} y_n = x_n + \theta_n(x_n - x_{n-1}), \\ z_n = x_n + \epsilon_n(x_n - x_{n-1}). \end{cases} \quad (29)$$

Step 2. Compute
$$x_{n+1} = \alpha_n f(x_n) + \beta_n y_n + \gamma_n J_r^A(z_n). \quad (30)$$

Step 3. Set $n \leftarrow n+1$ and go to Step 1.

Proof. As $\text{Fix}(J_r^A) = A^{-1}(0)$ and J_r^A is firmly nonexpansive, one has that J_r^A is $\frac{1}{2}$-averaged. Therefore, there exists a nonexpansive mapping S such that $J_r^A = \frac{1}{2}I + \frac{1}{2}S$ and $\text{Fix}(J_r^A) = \text{Fix}(S)$. By using Corollary 1, we obtain the conclusions immediately. □

Now, we solve the following convex minimization problem.

$$\min_{x \in H} h(x), \quad (31)$$

where $h: H \to (-\infty, +\infty]$ is a proper lower semi-continuous closed convex function. The subdifferential operator $\partial h(x)$ of $h(x)$ is defined by $\partial h(x) = \{u \in H : h(y) \geq h(x) + \langle u, y - x \rangle, \forall y \in H\}$. It is known that $\partial h(x)$ is maximal monotone, and x^* is a solution of problem (31) if and only if $0 \in \partial h(x^*)$. Taking $A = \partial h(x)$, we have $J_r^A = \text{prox}_{rh}$, where $r > 0$ and prox_{rh} is defined by

$$\text{prox}_{rh}(u) = \arg\min_{x \in H} \left\{ \frac{1}{2r} \|x - u\|^2 + h(x) \right\}.$$

Corollary 6. Let H be a Hilbert space and let $f : H \to H$ be a contraction with constant $k \in [0,1)$. Let $h : H \to (-\infty, +\infty]$ be a proper closed lower semi-continuous convex function such that $\arg\min h \neq \emptyset$. Suppose that Conditions 1 holds. Then, the sequence $\{x_n\}$ generated by Algorithm 5 converges to a solution of convex minimization problem (31) in norm.

Algorithm 5 The viscosity type algorithm for solving convex minimization problems

Initialization: Let $x_0, x_1 \in H$ be arbitrary.
Iterative Steps: Given the current iterator x_n, calculate x_{n+1} as follows.
Step 1. Compute
$$\begin{cases} y_n = x_n + \theta_n(x_n - x_{n-1}), \\ z_n = x_n + \epsilon_n(x_n - x_{n-1}). \end{cases} \tag{32}$$
Step 2. Compute
$$x_{n+1} = \alpha_n f(x_n) + \beta_n y_n + \gamma_n \text{prox}_{rh}(z_n). \tag{33}$$
Step 3. Set $n \leftarrow n + 1$ and go to Step 1.

Proof. It is known that the subdifferential operator ∂h is maximal monotone since h is a proper, closed lower semi-continuous, convex function. Therefore, $\text{prox}_{rh} = J_r^{\partial h}$. Then, we get the conclusions from Corollary 5 immediately. □

In the following, we consider the following inclusion problem: find $x^* \in H$ such that

$$0 \in A(x^*) + B(x^*), \tag{34}$$

where $A : H \to H$ be an α-inverse strongly monotone mapping and let $B : H \to 2^H$ be a set-valued maximal monotone operator. It is known that $\text{Fix}(J_r^B(I - rA)) = (A + B)^{-1}(0)$. Many problems can be modelled as the inclusion problem, such as, convex programming problems, inverse problems, split feasibility problems, and minimization problems, see [29–32]. Moreover, this problem is also widely applied in machine learning, signal processing, statistical regression, and image restoration, see [33–35]. By using Theorem 1, we obtain the following corollary.

Corollary 7. Let H be a Hilbert space and let $f : H \to H$ be a contraction with constant $k \in [0,1)$. Let $A : H \to H$ be a α-inverse strongly monotone mapping with $0 < r = 2\alpha$ and let $B : H \to 2^H$ be a maximal monotone operator. Suppose that $(A + B)^{-1}(0) \neq \emptyset$ and Conditions 1 holds. Then, the sequence $\{x_n\}$ generated by Algorithm 3 converges to $p = P_{(A+B)^{-1}(0)} \circ f(p)$ in norm.

Proof. As A is inverse strongly monotone, one has that $(I - rA)$ is nonexpansive. Therefore, the operator $J_r^B(I - rA)$ is nonexpansive. Then, we get the conclusions from Theorem 1 immediately. □

5. Numerical Results

In this section, we give three numerical examples to illustrate the computational performance of our proposed algorithms. All the programs are performed in MATLAB2018a on a PC Desktop Intel(R) Core(TM) i5-8250U CPU @ 1.60 GHz 1.800 GHz, RAM 8.00 GB.

Example 1. In this example, we consider the following case that the usual gradient method is not convergent. Take the feasible set as $C := \{-5 \leq x_i \leq 5, i = 1, 2, \cdots, m\}$ and an $m \times m$ square matrix $A := (a_{ij})_{1 \leq i,j \leq m}$ whose terms are given by

$$a_{ij} = \begin{cases} 1, & \text{if } j = m+1-i \text{ and } j < i, \\ -1, & \text{if } j = m+1-i \text{ and } j > i, \\ 0, & \text{otherwise.} \end{cases}$$

One knows that zero vector $x^* = (0, \ldots, 0)$ is a solution of this problem. First, one tests the Algorithm 3 with different choices of inertial parameter θ_n and ϵ_n. Setting $f(x) = 0.5x$, $\delta_n = \frac{1}{(n+1)^2}$, $\alpha_n = \frac{n}{(n+1)^{1.1}}$, $\beta_n = \gamma_n = \frac{1-\alpha_n}{2}$, $\lambda = 0.7$, the numerical results are shown in Tables 1 and 2.

To compare the efficiency between algorithms, we consider our proposed Algorithm 3, the extragradient method (EGM) in [36], the subgradient extragradient method (SEGM) in [26], and the new inertial subgradient extragradient method (NISEGM) in [27]. The parameters are selected as follows. The initial points $x_0, x_1 \in R^m$ are generated randomly in MATLAB and we take different values of m into consideration. In EGM, SEGM, we take $\lambda = 0.7$. In Algorithm 3, we take $f(x) = 0.5x$, $\lambda = 0.7$, $\delta_n = \frac{1}{(n+1)^2}$, $\theta = 0.7$ and $\epsilon = 0.8$ in (4), $\alpha_n = \frac{n}{(n+1)^{1.1}}$, $\beta_n = \gamma_n = \frac{1-\alpha_n}{2}$. We set $\alpha_n = 0.1$, $\tau_n = \frac{n}{(n+1)^{1.1}}$, $\lambda_n = 0.8$ in NISEGM. The stopping criterion is $E_n = \|x_n - x^*\|_2 < 10^{-4}$. The results are proposed in Table 3 and Figure 1.

Table 1. Number of iterations of Algorithm 3 with $\theta = 0.5$, $m = 100$.

Initial Value		ϵ	0	0.1	0.2	0.3	0.4	0.5	0.6	0.7	0.8	0.9	1
10 × rand(m,1)	Iter.		24	23	23	23	22	22	22	21	21	21	21
100 × rand(m,1)	Iter.		27	27	26	26	26	25	25	25	25	25	25
1000 × rand(m,1)	Iter.		31	31	30	30	30	29	29	29	29	29	29

Table 2. Number of iterations of Algorithm 3 with $\epsilon = 0.7$, $m = 100$.

Initial Value		θ	0	0.1	0.2	0.3	0.4	0.5	0.6	0.7	0.8	0.9	1
10 × rand(m,1)	Iter.		24	23	23	22	22	21	21	21	21	21	22
100 × rand(m,1)	Iter.		27	27	26	26	25	25	25	25	25	25	25
1000 × rand(m,1)	Iter.		30	30	30	29	29	29	28	28	28	28	28

Remark 3. By Table 1 and Table 2, one concludes that the number of the iteration is small for the Algorithm 3 with $\theta \in [0.5, 1]$ and $\epsilon \in [0.5, 1]$.

Remark 4. (1) By numerical results of Example 1, we find that our Algorithm 3 is efficient, easy to implement and fast. Moreover, dimensions do not affect the computational performance of our algorithm.
(2) Obviously, by Example 1, we also find that our proposed Algorithm 3 outperforms the extragradient method (EGM), the subgradient extragradient method (SEGM) and the new inertial subgradient extragradient method (NISEGM) in both CPU time and number of iterations.

Table 3. Comparison between Algorithm 3, EGM, SEGM, and NISEGM in Example 1.

	Algorithm 3		Algorithm EGM		Algorithm SEGM		Algorithm NISEGM	
m	Iter.	Time (s)	Iter.	Time (s)	Iter.	Time (s)	Iter.	Time (s)
100	24	0.0102	91	0.0147	93	0.0194	84	0.0121
1000	27	0.0548	99	0.1265	101	0.1376	92	0.1136
2000	28	0.3007	101	0.7852	104	0.7018	94	0.6516
5000	29	1.6582	105	4.2879	107	4.4691	97	4.0239

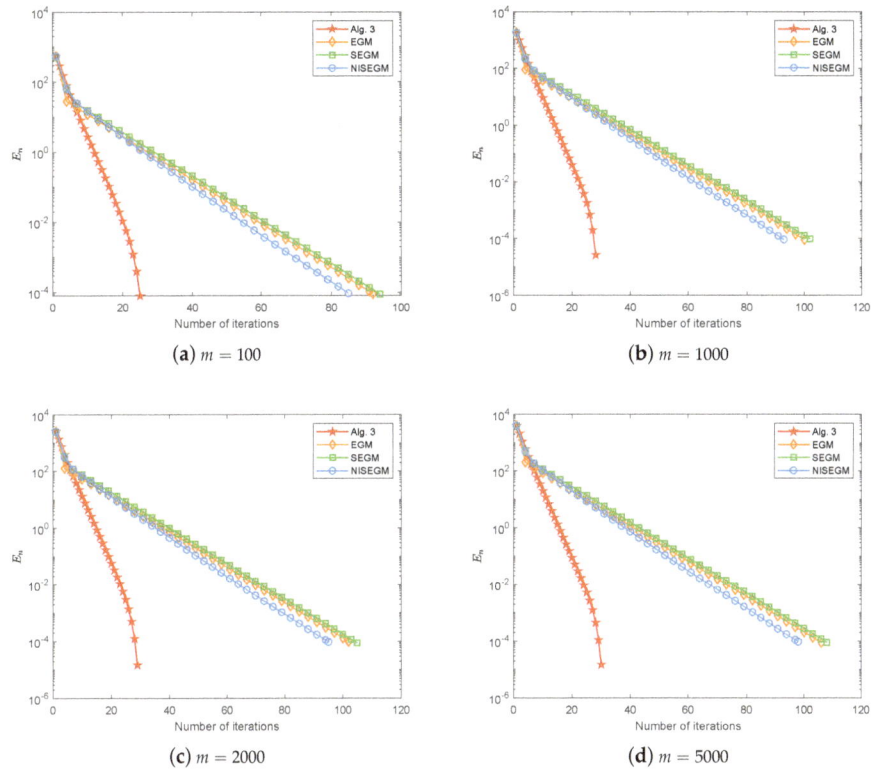

Figure 1. Convergence behavior of iteration error $\{E_n\}$ with different dimension in Example 1.

Algorithm 6 The viscosity type algorithm for solving inclusion problem (34)

Initialization: Let $x_0, x_1 \in H$ be arbitrary.
Iterative Steps: Given the current iterator x_n, calculate x_{n+1} as follows:
Step 1. Compute
$$\begin{cases} y_n = x_n + \theta_n(x_n - x_{n-1}), \\ z_n = x_n + \epsilon_n(x_n - x_{n-1}). \end{cases} \quad (35)$$
Step 2. Compute
$$x_{n+1} = \alpha_n f(x_n) + \beta_n y_n + \gamma_n J_r^B(I - rA)z_n. \quad (36)$$
Step 3. Set $n \leftarrow n+1$ and go to Step 1.

Example 2. *In this example, we consider $H = L_2([0, 2\pi])$ and the following half-space,*

$$C = \left\{ x \in L_2([0, 2\pi]) \mid \int_0^{2\pi} x(t) dt \leq 1 \right\}, \text{ and } Q = \left\{ x \in L_2([0, 2\pi]) \mid \int_0^{2\pi} |x(t) - \sin(t)|^2 dt \leq 16 \right\}.$$

Define a linear continuous operator $T : L_2([0, 2\pi]) \to L_2([0, 2\pi])$, where $(Tx)(t) := x(t)$. Then $(T^*x)(t) = x(t)$ and $\|T\| = 1$. Now, we solve the following problem,

$$\text{find } x^* \in C \quad \text{such that } Tx^* \in Q. \quad (37)$$

As $(Tx)(t) = x(t)$, (37) is actually a convex feasibility problem: find $x^* \in C \cap Q$. Moreover, it is evident that $x(t) = 0$ is a solution. Therefore, the solution set of (37) is nonempty. Take $Ax = \nabla \left(\frac{1}{2} \|Tx - P_Q Tx\|^2 \right) = T^* (I - P_Q) Tx$ and $B = \partial i_C$. Then (37) can be written in the form (34). It is clear that A is 1-Lipschitz continuous and B is maximal monotone. For our numerical computation, we can also write the projections onto set C and the projections onto set Q as follows, see [37].

$$P_C(z) = \begin{cases} \frac{1 - \int_0^{2\pi} z(t) dt}{4\pi^2} + z, & \int_0^{2\pi} z(t) dt > 1, \\ z, & \int_0^{2\pi} z(t) dt \leq 1. \end{cases}$$

and

$$P_Q(w) = \begin{cases} \sin + \frac{4}{\sqrt{\int_0^{2\pi} |w(t) - \sin(t)|^2 dt}} (w - \sin), & \int_0^{2\pi} |w(t) - \sin(t)|^2 dt > 16, \\ w, & \int_0^{2\pi} |w(t) - \sin(t)|^2 dt \leq 16. \end{cases}$$

In this numerical experiment, we consider different initial values x_0 and x_1. The error of the iterative algorithms is denoted by

$$E_n = \frac{1}{2} \|P_C(x_n) - x_n\|_2^2 + \frac{1}{2} \|P_Q(T(x_n)) - T(x_n)\|_2^2.$$

Now, we give some numerical experiment comparisons between our Algorithm 6 and the Algorithm 5.2 proposed by Shehu et al. [21]. We denote this algorithm by Shehu et al. Algorithm 5.2. In the Shehu et al. Algorithm 5.2, one sets $\lambda = 0.25$, $\epsilon_n = \frac{1}{(n+1)^2}$, $\theta = 0.5$, $\alpha_n = \frac{1}{n+1}$, $\beta_n = \gamma_n = \frac{n}{2(n+1)}$, $e_n = \frac{1}{(n+1)^2}$. In Algorithm 6, one sets $f(x) = 0.5x$, $r = 0.25$, $\delta_n = \frac{1}{(n+1)^2}$, $\theta = 0.5$, $\epsilon = 0.7$, $\alpha_n = \frac{1}{n+1}$, and $\beta_n = \gamma_n = \frac{n}{2(n+1)}$. Our stopping criterion is maximum iteration 200 or $E_n < 10^{-3}$. The results are proposed in Table 4 and Figure 2.

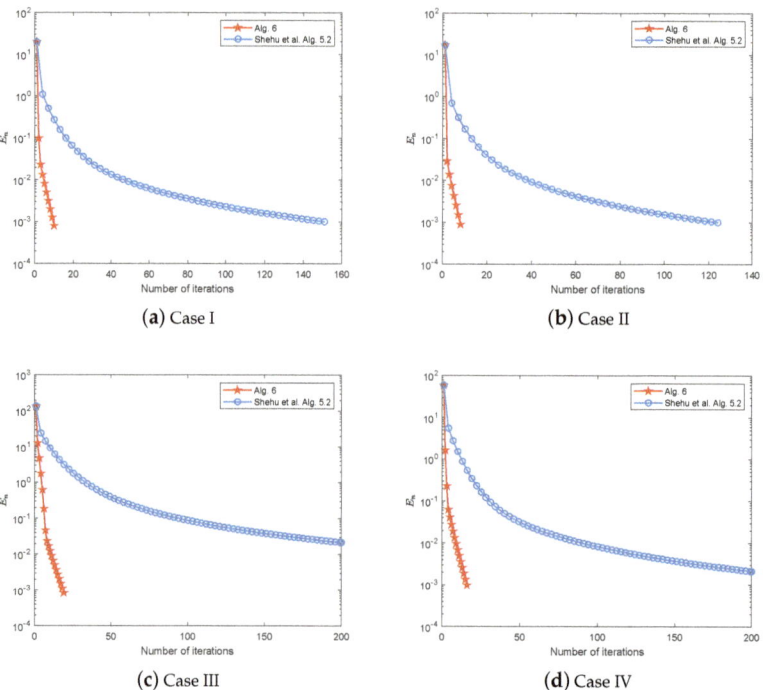

Figure 2. Convergence behavior of iteration error $\{E_n\}$ with different initial values in Example 2.

Table 4. Comparison between our Algorithm 6 and Shehu et al.'s Algorithm 5.2 in Example 2.

		Algorithm 6		Shehu et al.'s Algorithm 5.2	
Cases	Initial Values	Iter.	Time (s)	Iter.	Time (s)
I	$x_0 = \frac{t^2}{10}$ $x_1 = \frac{t^2}{10}$	9	3.4690	151	45.7379
II	$x_0 = \frac{t^2}{10}$ $x_1 = \frac{t^2}{16}$	7	2.9629	124	38.3933
III	$x_0 = \frac{t^2}{10}$ $x_1 = \frac{e^{t/2}}{2}$	18	7.0002	200	61.6568
IV	$x_0 = \frac{t^2}{10}$ $x_1 = 5\sin(2t)^2$	15	5.8423	200	62.5465

Remark 5. (1) Also, by observing numerical results of Example 2, we find that our Algorithm 6 is more efficient and faster than the Shehu et al.'s Algorithm 5.2.
(2) Our Algorithm 6 is consistent since the choice of initial value does not affect the number of iterations needed to achieve the expected results.

Example 3. In this example, we consider a linear inverse problem: $b = Ax_0 + w$, where $x_0 \in R^N$ is the (unknown) signal to recover, $w \in R^M$ is a noise vector, and $A \in R^{M \times N}$ models the acquisition device. To recover an approximation of the signal x_0, we use the Basis Pursuit denoising method. That is, one uses the ℓ_1 norm as a sparsity enforcing penalty.

$$\min_{x \in R^N} \Phi(x) = \frac{1}{2}\|b - Ax\|^2 + \lambda\|x\|_1, \tag{38}$$

where $\|x\|_1 = \sum_i |x_i|$ and λ is a parameter that is relate to noise w. It is known that (38) is referred as the least absolute selection and shrinkage operator problem, that is, the LASSO problem. The LASSO problem (38) is a special case of minimizing $F + G$, where

$$F(x) = \frac{1}{2}\|b - Ax\|^2, \quad \text{and} \quad G(x) = \lambda\|x\|_1.$$

It is easy to see that F is a smooth function with L-Lipschitz continuous gradient $\nabla F(x) = A^*(Ax - b)$, where $L = \|A^*A\|$. The ℓ_1-norm is "simple", as its proximal operator is a soft thresholding:

$$\text{prox}_{\gamma G}(x_k) = \max\left(0, 1 - \frac{\lambda\gamma}{|x_k|}\right)x_k.$$

In our experiment, we want to recover a sparse signal $x_0 \in R^N$ with k ($k \ll N$) non-zero elements. A simple linearized model of signal processing is to consider a linear operator, that is, a filtering $Ax = \varphi \star x$, where φ is a second derivative of Gaussian. We wish to solve $b = Ax_0 + w$, where w is a realization of Gaussian white noise with variance 10^{-2}. Therefore, we need to solve the (38). We compare our Algorithm 6 with another strong convergence algorithm, which was proposed by Gibali and Thong in [38]. We denote this algorithm by G-T Algorithm 1. In addition, we also compare the algorithms with the classic Forward–Backward algorithm in [33]. Our parameter settings are as follows. In all algorithms, we set regularization parameter $\lambda = \frac{1}{2}$ in (38). In the Forward–Backward algorithm, we set step size $\gamma = 1.9/L$. In G-T Algorithm 1, we set step size $\gamma = 1.9/L$, $\alpha_n = \frac{1}{n+1}$, $\beta_n = \frac{n}{2(n+1)}$ and $\mu = 0.5$. In Algorithm 6, we set step size $r = 1.9/L$, $f(x) = 0.1x$, $\theta = \epsilon = 0.9$, $\delta_n = \frac{1}{(n+1)^2}$, $\alpha_n = \frac{1}{n+1}$, $\beta_n = \frac{1}{1000(n+1)^3}$, $\gamma_n = 1 - \alpha_n - \beta_n$. We take the maximum number of iterations 5×10^4 as a common stopping criterion. In addition, we use the signal-to-noise ratio (SNR) to measure the quality of recovery, and a larger SNR means a better recovery quality. Numerical results are proposed in Table 5 and Figures 3–5. We tested the computational performance of the above algorithms in different dimension N and different sparsity k (Case I: $N = 400$, $k = 12$; Case II: $N = 400$, $k = 20$; Case III: $N = 1000$, $k = 30$; Case IV: $N = 1000$, $k = 50$). Figure 3 shows the original and noise signals in different dimension N and different sparsity k. Figure 4 shows the recovery results of different algorithms under different situation, the corresponding numerical results are shown in Table 5. Figure 5 shows the convergence behavior of $\Phi(x)$ in (38) with the number of iterations.

Table 5. Comparison the SNR between Algorithm 6, G-T Algorithm 1, and Forward–Backward in Example 3.

Cases	N	k	G-T Algorithm 1	Algorithm 6	Forward–Backward
I	400	12	16.2421	16.3742	16.3930
II	400	20	5.3994	5.4377	5.4418
III	1000	30	6.7419	6.7749	6.7792
IV	1000	50	3.2493	3.2553	3.2561

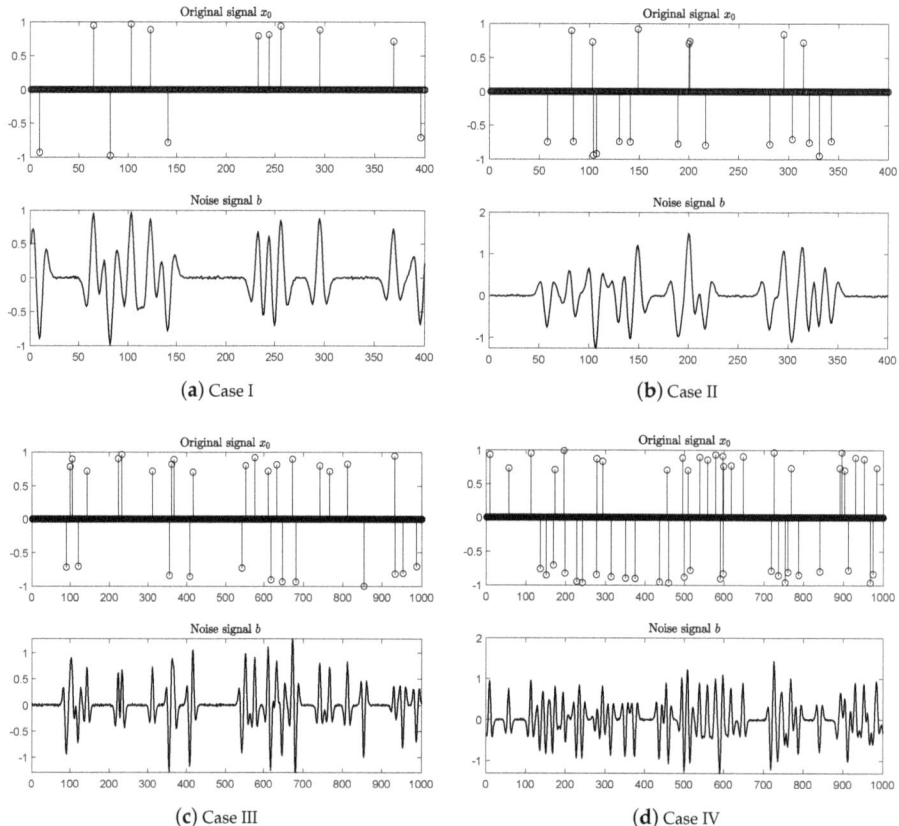

Figure 3. Original signals and noise signals at different N and k in Example 3.

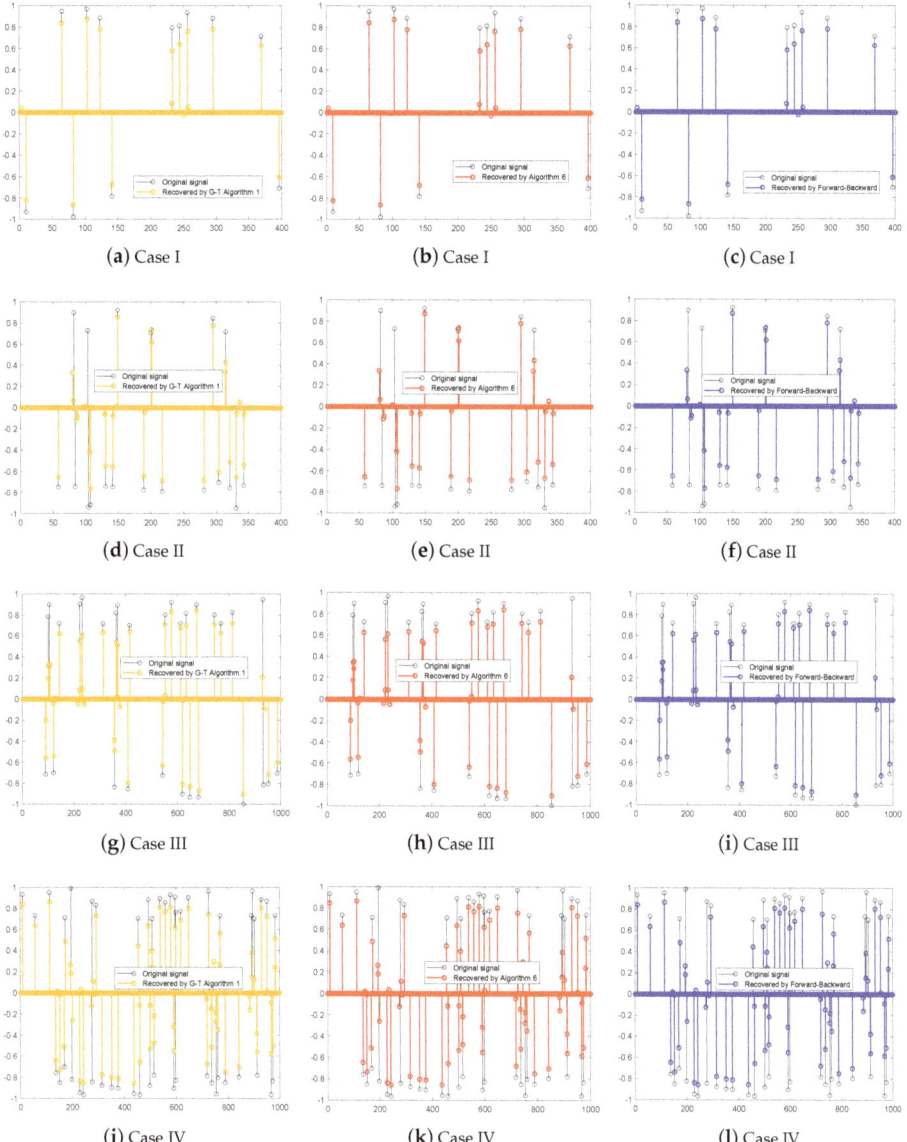

Figure 4. Recovery results under different algorithms in Example 3.

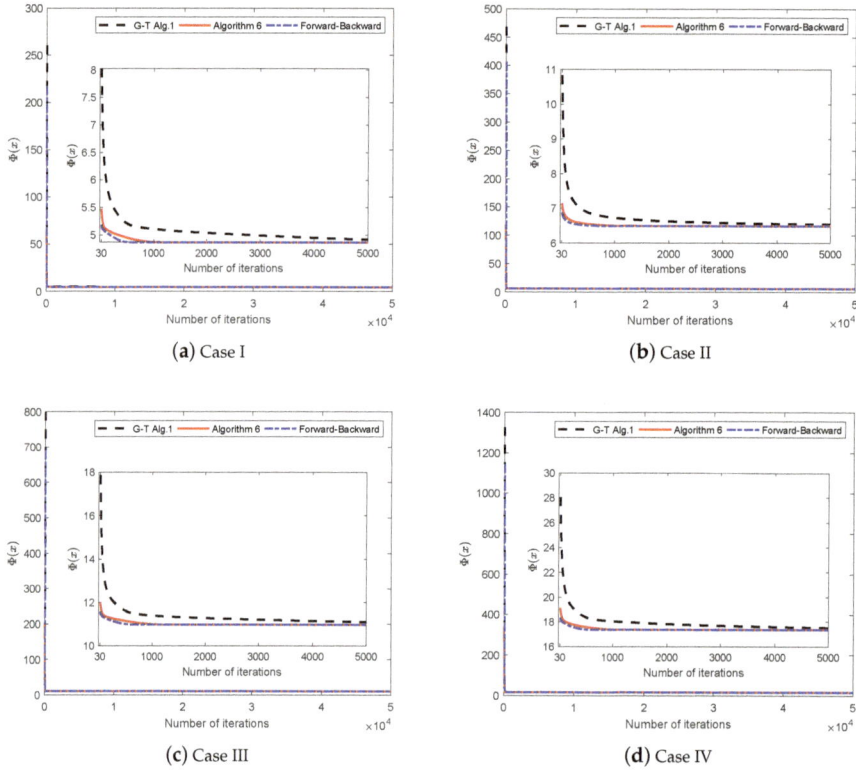

Figure 5. Convergence behavior of $\{\Phi(x)\}$ with different N and k in Example 3.

Remark 6. (1) *The LASSO problem in Example 3 also shows that our proposed algorithm is consistent and more efficient. Furthermore, dimensions and sparsity do not affect the computational performance of our proposed Algorithm 6, see Table 5 and Figures 4 and 5.*
(2) *The numerical results also show that our Algorithm 6 is superior than the algorithm proposed by Gibali and Thong [38] in terms of computational performance and accuracy.*
(3) *In addition, there is little difference between our Algorithm 6 and the classical Forward–Backward algorithm in computational performance and precision. Note that the Forward–Backward algorithm is weak convergence in the infinite dimensional Hilbert spaces; however, our proposed algorithms is strongly convergent (see Corollary 7 and Example 2).*

6. Conclusions

In this paper, we proposed a viscosity algorithm with two different inertia parameters for solving fixed-point problem of nonexpansive mappings. We also established a strong convergence theorem for strict pseudo-contractive mappings. By choosing different parameter values in inertial sequences, we analyzed the convergence behavior of our proposed algorithms. One highlight is that our algorithms are based on two different inertial parameter sequences comparing with the exiting ones. accelerated via the inertial technique and the viscosity technique. Another highlight is that, to show the effectiveness of our algorithms, we compare our algorithms with other existing algorithms in the convergence rate and applications in signal processing. Numerical experiments show that our algorithms are consistent and efficient. Finally, we remark that the framework of the space is a Hilbert space, it is of interest to further our results to the framework of Banach spaces or Hadamard manifolds.

Author Contributions: All the authors contributed equally to this work. All authors have read and agreed to the published version of the manuscript.

Funding: This paper was supported by the National Natural Science Foundation of China under Grant 11601348.

Acknowledgments: The authors are grateful to the referees for useful suggestions, which improved the contents of this paper.

Conflicts of Interest: The authors declare no conflicts of interest.

References

1. Chidume, C.E.; Romanus, O.M.; Nnyaba, U.V. An iterative algorithm for solving split equilibrium problems and split equality variational inclusions for a class of nonexpansive-type maps. *Optimization* **2018**, *67*, 1949–1962. [CrossRef]
2. Cho, S.Y.; Kang, S.M. Approximation of common solutions of variational inequalities via strict pseudocontractions. *Acta Math. Sci.* **2012**, *32*, 1607–1618. [CrossRef]
3. Qin, X.; Cho, S.Y.; Yao, J.C. Weak and strong convergence of splitting algorithms in Banach spaces. *Optimization* **2020**, *69*, 243–267. [CrossRef]
4. Mann, W.R. Mean value methods in iteration. *Proc. Amer. Math. Soc.* **1953**, *4*, 506–510. [CrossRef]
5. Chang, S.S.; Wen, C.F.; Yao, J.C. Zero point problem of accretive operators in Banach spaces. *Bull. Malays. Math. Sci. Soc.* **2019**, *42*, 105–118. [CrossRef]
6. Chang, S.S.; Wen, C.F.; Yao, J.C. Common zero point for a finite family of inclusion problems of accretive mappings in Banach spaces. *Optimization* **2018**, *67*, 1183–1196. [CrossRef]
7. Cho, S.Y.; Li, W.; Kang, S.M. Convergence analysis of an iterative algorithm for monotone operators. *J. Inequal. Appl.* **2013**, *2013*, 199. [CrossRef]
8. Qin, X.; Cho, S.Y.; Wang, L. Strong convergence of an iterative algorithm involving nonlinear mappings of nonexpansive and accretive type. *Optimization* **2018**, *67*, 1377–1388. [CrossRef]
9. Attouch, H. Viscosity approximation methods for minimization problems. *SIAM J. Optim.* **1996**, *6*, 769–806. [CrossRef]
10. Moudafi, A. Viscosity approximation methods for fixed-points problems. *J. Math. Anal. Appl.* **2000**, *241*, 46–55. [CrossRef]
11. Takahashi, S.; Takahashi, W. Viscosity approximation methods for equilibrium problems and fixed point problems in Hilbert spaces. *J. Math. Anal. Appl.* **2007**, *331*, 506–515. [CrossRef]
12. Qin, X.; Yao, J.C. A viscosity iterative method for a split feasibility problem. *J. Nonlinear Convex Anal.* **2019**, *20*, 1497–1506.
13. Qin, X.; Cho, S.Y.; Wang, L. Iterative algorithms with errors for zero points of m-accretive operators. *Fixed Point Theory Appl.* **2013**, *2013*, 148. [CrossRef]
14. Maingé, P.E. A hybrid extragradient-viscosity method for monotone operators and fixed point problems. *SIAM J. Optim.* **2008**, *47*, 1499–1515. [CrossRef]
15. Takahashi, W.; Xu, H.K.; Yao, J.C. Iterative methods for generalized split feasibility problems in Hilbert spaces. *Set-Valued Var. Anal.* **2015**, *23*, 205–221. [CrossRef]
16. Qin, X.; Cho, S.Y.; Wang, L. A regularization method for treating zero points of the sum of two monotone operators. *Fixed Point Theory Appl.* **2014**, *2014*, 75. [CrossRef]
17. Cho, S.Y.; Bin Dehaish, B.A. Weak convergence of a splitting algorithm in Hilbert spaces. *J. Appl. Anal. Comput.* **2017**, *7*, 427–438.
18. Qin, X.; Wang, L.; Yao, J.C. Inertial splitting method for maximal monotone mappings. *J. Nonlinear Convex Anal.* **2020**, in press.
19. Qin, X.; Cho, S.Y. Convergence analysis of a monotone projection algorithm in reflexive Banach spaces. *Acta Math. Sci.* **2017**, *37*, 488–502. [CrossRef]
20. Polyak, B.T. Some methods of speeding up the convergence of iteration methods. *USSR Comput. Math. Math. Phys.* **1964**, *4*, 1–17. [CrossRef]
21. Shehu, Y.; Iyiola, O.S.; Ogbuisi, F.U. Iterative method with inertial terms for nonexpansive mappings: applications to compressed sensing. *Numer Algor.* **2019**. [CrossRef]

22. Dong, Q.L.; Cho, Y.J.; Rassias, T.M. General inertial Mann algorithms and their convergence analysis for nonexpansive mappings. In *Applications of Nonlinear Analysis*; Rassias, T.M., Ed.; Springer: Berlin, Gernmany, 2018; pp. 175–191.
23. Saejung, S.; Yotkaew, P. Approximation of zeros of inverse strongly monotone operators in Banach spaces. *Nonlinear Anal.* **2012**, *75*, 742–750. [CrossRef]
24. Goebel, K.; Kirk, W.A. *Topics in Metric Fixed Point Theory*; Cambridge University Press: Cambridge, UK, 1990; Volume 28.
25. Xu, H.K. Iterative algorithms for nonlinear operators, *J. Lond. Math. Soc.* **2002**, *66*, 240–256. [CrossRef]
26. Censor, Y.; Gibali, A.; Reich, S. Strong convergence of subgradient extragradient methods for the variational inequality problem in Hilbert space. *Optim. Methods Softw.* **2011**, *26*, 827–845. [CrossRef]
27. Fan, J.; Liu, L.; Qin, X. A subgradient extragradient algorithm with inertial effects for solving strongly pseudomonotone variational inequalities. *Optimization* **2019**, 1–17. [CrossRef]
28. Malitsky, Y.V.; Semenov, V.V. A hybrid method without extrapolation step for solving variational inequality problems. *J. Glob. Optim.* **2015**, *61*, 193–202. [CrossRef]
29. Takahahsi, W.; Yao, J.C. The split common fixed point problem for two finite families of nonlinear mappings in Hilbert spaces. *J. Nonlinear Convex Anal.* **2019**, *20*, 173–195.
30. Dehaish, B.A.B. Weak and strong convergence of algorithms for the sum of two accretive operators with applications. *J. Nonlinear Convex Anal.* **2015**, *16*, 1321–1336.
31. Qin, X.; Petrusel, A.; Yao, J.C. CQ iterative algorithms for fixed points of nonexpansive mappings and split feasibility problems in Hilbert spaces. *J. Nonlinear Convex Anal.* **2018**, *19*, 157–165.
32. Cho, S.Y. Strong convergence analysis of a hybrid algorithm for nonlinear operators in a Banach space. *J. Appl. Anal. Comput.* **2018**, *8*, 19–31.
33. Combettes, P.L.; Wajs, V. Signal recovery by proximal forward-backward splitting. *Multiscale Model. Simul.* **2005**, *4*, 1168–1200. [CrossRef]
34. An, N.T.; Nam, N.M. Solving k-center problems involving sets based on optimization techniques. *J. Glob. Optim.* **2020**, *76*, 189–209. [CrossRef]
35. Qin, X.; An, N.T. Smoothing algorithms for computing the projection onto a Minkowski sum of convex sets. *Comput. Optim. Appl.* **2019** *74*, 821–850. [CrossRef]
36. Korpelevich, G.M. The extragradient method for finding saddle points and other problems. *Matecon* **1976**, *12*, 747–756.
37. Bauschke, H.H.; Combettes, P.L. *Convex Analysis and Monotone Operator Theory in Hilbert Spaces*; Springer: New York, NY, USA, 2011.
38. Gibali, A.; Thong, D.V. Tseng type methods for solving inclusion problems and its applications. *Calcolo* **2018**, *55*, 49. [CrossRef]

© 2020 by the authors. Licensee MDPI, Basel, Switzerland. This article is an open access article distributed under the terms and conditions of the Creative Commons Attribution (CC BY) license (http://creativecommons.org/licenses/by/4.0/).

Article

Inertial-Like Subgradient Extragradient Methods for Variational Inequalities and Fixed Points of Asymptotically Nonexpansive and Strictly Pseudocontractive Mappings

Lu-Chuan Ceng [1], Adrian Petruşel [2], Ching-Feng Wen [3,4,*] and Jen-Chih Yao [5]

1. Department of Mathematics, Shanghai Normal University, Shanghai 200234, China
2. Department of Mathematics, Babes-Bolyai University, Cluj-Napoca 400084, Romania
3. Center for Fundamental Science and Research Center for Nonliear Analysis and Optimization, Kaohsiung Medical University, Kaohsiung 80708, Taiwan
4. Department of Medical Research, Kaohsiung Medical University Hospital, Kaohsiung 80708, Taiwan
5. Research Center for Interneural Computing, China Medical University Hospital, Taichung 40402, Taiwan; yaojc@mail.cmu.edu.tw
* Correspondence: cfwen@kmu.edu.tw

Received: 18 August 2019; Accepted: 9 September 2019; Published: 17 September 2019

Abstract: Let VIP indicate the variational inequality problem with Lipschitzian and pseudomonotone operator and let CFPP denote the common fixed-point problem of an asymptotically nonexpansive mapping and a strictly pseudocontractive mapping in a real Hilbert space. Our object in this article is to establish strong convergence results for solving the VIP and CFPP by utilizing an inertial-like gradient-like extragradient method with line-search process. Via suitable assumptions, it is shown that the sequences generated by such a method converge strongly to a common solution of the VIP and CFPP, which also solves a hierarchical variational inequality (HVI).

Keywords: inertial-like subgradient-like extragradient method with line-search process; pseudomonotone variational inequality problem; asymptotically nonexpansive mapping; strictly pseudocontractive mapping; sequentially weak continuity

MSC: 47H05; 47H09; 47H10; 90C52

1. Introduction

Throughout this paper we assume that C is a nonempty, convex and closed subset of a real Hilbert space $(H, \|\cdot\|)$, whose inner product is denoted by $\langle \cdot, \cdot \rangle$. Moreover, let P_C denote the metric projection of H onto C.

Suppose $A : H \to H$ is a mapping. In this paper, we shall consider the following variational inequality (VI) of finding $x^* \in C$ such that

$$\langle x - x^*, Ax^* \rangle \geq 0, \quad \forall x \in C. \tag{1}$$

The set of solutions to Equation (1) is denoted by VI(C, A). In 1976, Korpelevich [1] first introduced an extragradient method, which is one of the most popular approximation ones for solving Equation (1) till now. That is, for any initial $u_0 \in C$, the sequence $\{u_n\}$ is generated by

$$\begin{cases} v_n = P_C(u_n - \tau A u_n), \\ u_{n+1} = P_C(u_n - \tau A v_n), \quad \forall n \geq 0, \end{cases} \quad (2)$$

where τ is a constant in $(0, \frac{1}{L})$ for $L > 0$ the Lipschitz constant of mapping A. In the case where VI(C, A) $\neq \emptyset$, the sequence $\{u_n\}$ constructed by Equation (2) is weakly convergent to a point in VI(C, A). Recently, light has been shed on approximation methods for solving problem Equation (1) by many researchers; see, e.g., [2–11] and references therein, to name but a few.

Let $T : C \to C$ be a mapping. We denote by Fix(T) the set of fixed points of T, i.e., Fix(T) = $\{x \in C : x = Tx\}$. T is said to be asymptotically nonexpansive if $\exists \{\theta_n\} \subset [0, +\infty)$ such that $\lim_{n \to \infty} \theta_n = 0$ and $\|T^n u - T^n v\| \leq \|u - v\| + \theta_n \|u - v\|, \forall n \geq 1, u, v \in C$. If $\theta_n \equiv 0$, then T is nonexpansive. Also, T is said to be strictly pseudocontractive if $\exists \zeta \in [0, 1)$ s.t. $\|Tu - Tv\|^2 \leq \|u - v\|^2 + \zeta \|(I - T)u - (I - T)v\|^2, \forall u, v \in C$. If $\zeta = 0$, then T reduces to a nonexpansive mapping. One knows that the class of strict pseudocontractions strictly includes the class of nonexpansive mappings. Both strict pseudocontractions and nonexpansive mappings have been studied extensively by a large number of authors via iteration approximation methods; see, e.g., [12–18] and references therein.

Let the mappings $A, B : C \to H$ be both inverse-strongly monotone and let the mapping $T : C \to C$ be asymptotically nonexpansive one with a sequence $\{\theta_n\}$. Let $f : C \to C$ be a δ-contraction with $\delta \in [0, 1)$. By using a modified extragradient method, Cai et al. [19] designed a viscosity implicit rule for finding a point in the common solution set Ω of the VIs for A and B and the FPP of T, i.e., for arbitrarily given $x_1 \in C, \{x_n\}$ is the sequence constructed by

$$\begin{cases} u_n = s_n x_n + (1 - s_n) y_n, \\ y_n = P_C(I - \lambda A) P_C(u_n - \mu B u_n), \\ x_{n+1} = P_C[(T^n y_n - \alpha_n \rho F T^n y_n) + \alpha_n f(x_n)], \end{cases}$$

where $\{\alpha_n\}, \{s_n\} \subset (0, 1]$. Under appropriate conditions imposed on $\{\alpha_n\}, \{s_n\}$, they proved that $\{x_n\}$ is convergent strongly to an element $x^* \in \Omega$ provided $\sum_{n=1}^{\infty} \|T^{n+1} y_n - T^n y_n\| < \infty$.

In the context of extragradient techniques, one has to compute metric projections two times for each computational step. Without doubt, if C is a general convex and closed set, the computation of the projection onto C might be quite consuming-time. In 2011, inspired by Korpelevich's extragradient method, Censor et al. [20] first designed the subgradient extragradient method, where a projection onto a half-space is used in place of the second projection onto C. In 2014, Kraikaew and Saejung [21] proposed the Halpern subgradient extragradient method for solving Equation (1), and proved strong convergence of the proposed method to a solution of Equation (1).

In 2018, via the inertial technique, Thong and Hieu [22] studied the inertial subgradient extragradient method, and proved weak convergence of their method to a solution of Equation (1). Very recently, they [23] constructed two inertial subgradient extragradient algorithms with linear-search process for finding a common solution of problem Equation (1) with operator A and the FPP of operator T with demiclosedness property in a real Hilbert space, where A is Lipschitzian and monotone, and T is quasi-nonexpansive. The constructed inertial subgradient extragradient algorithms (Algorithms 1 and 2) are as below:

Algorithm 1: Inertial subgradient extragradient algorithm (I) (see [[23], Algorithm 1]).

Initialization: Given $u_0, u_1 \in H$ arbitrarily. Let $\gamma > 0$, $l \in (0,1)$, $\mu \in (0,1)$.
Iterative Steps: Compute u_{n+1} in what follows:
Step 1. Put $v_n = \alpha_n(u_n - u_{n-1}) + u_n$ and calculate $y_n = P_C(v_n - \tau_n A v_n)$, where τ_n is chosen to be the largest $\tau \in \{\gamma, \gamma l, \gamma l^2, ...\}$ satisfying $\tau \|Av_n - Ay_n\| \leq \mu \|v_n - y_n\|$.
Step 2. Calculate $z_n = P_{T_n}(v_n - \tau_n A y_n)$ with $T_n := \{x \in H : \langle x - y_n, v_n - \tau_n A v_n - y_n \rangle \leq 0\}$.
Step 3. Calculate $u_{n+1} = \beta_n T z_n + (1 - \beta_n) v_n$. If $v_n = z_n = u_{n+1}$ then $v_n \in \text{Fix}(T) \cap \text{VI}(C, A)$.
Set $n := n+1$ and go to Step 1.

Algorithm 2: Inertial subgradient extragradient algorithm (II) (see [[23], Algorithm 2]).

Initialization: Given $u_0, u_1 \in H$ arbitrarily. Let $\gamma > 0$, $l \in (0,1)$, $\mu \in (0,1)$.
Iterative Steps: Calculate u_{n+1} as follows:
Step 1. Put $v_n = \alpha_n(u_n - u_{n-1}) + u_n$ and calculate $y_n = P_C(v_n - \tau_n A v_n)$, where τ_n is chosen to be the largest $\tau \in \{\gamma, \gamma l, \gamma l^2, ...\}$ satisfying $\tau \|Av_n - Ay_n\| \leq \mu \|v_n - y_n\|$.
Step 2. Calculate $z_n = P_{T_n}(v_n - \tau_n A y_n)$ with $T_n := \{x \in H : \langle x - y_n, v_n - \tau_n A v_n - y_n \rangle \leq 0\}$.
Step 3. Calculate $u_{n+1} = \beta_n T z_n + (1 - \beta_n) u_n$. If $v_n = z_n = u_n = u_{n+1}$ then $u_n \in \text{Fix}(T) \cap \text{VI}(C, A)$. Set $n := n+1$ and go to Step 1.

Under mild assumptions, they proved that the sequences generated by the proposed algorithms are weakly convergent to a point in $\text{Fix}(T) \cap \text{VI}(C, A)$. Recently, gradient-like methods have been studied extensively by many authors; see, e.g., [24–38].

Inspired by the research work of [23], we introduce two inertial-like subgradient algorithms with line-search process for solving Equation (1) with a Lipschitzian and pseudomonotone operator and the common fixed point problem (CFPP) of an asymptotically nonexpansive operator and a strictly pseudocontractive operator in H. The proposed algorithms comprehensively adopt inertial subgradient extragradient method with line-search process, viscosity approximation method, Mann iteration method and asymptotically nonexpansive mapping. Via suitable assumptions, it is shown that the sequences generated by the suggested algorithms converge strongly to a common solution of the VIP and CFPP, which also solves a hierarchical variational inequality (HVI).

2. Preliminaries

Let $x \in H$ and $\{x_n\} \subset H$. We use the notation $x_n \to x$ (resp., $x_n \rightharpoonup x$) to indicate the strong (resp., weak) convergence of $\{x_n\}$ to x. Recall that a mapping $T : C \to H$ is said to be:

(i) L-Lipschitzian (or L-Lipschitz continuous) if $\|Tx - Ty\| \leq L\|x - y\|$, $\forall x, y \in C$ for some $L > 0$;
(ii) monotone if $\langle Tu - Tv, u - v \rangle \geq 0$, $\forall u, v \in C$;
(iii) pseudomonotone if $\langle Tu, v - u \rangle \geq 0 \Rightarrow \langle Tv, v - u \rangle \geq 0$, $\forall u, v \in C$;
(iv) β-strongly monotone if $\langle Tu - Tv, u - v \rangle \geq \beta\|u - v\|^2$, $\forall u, v \in C$ for some $\beta > 0$;
(v) sequentially weakly continuous if $\forall \{u_n\} \subset C$, the relation holds: $u_n \rightharpoonup u \Rightarrow Tu_n \rightharpoonup Tu$.

For metric projections, it is well known that the following assertions hold:

(i) $\langle P_C u - P_C v, u - v \rangle \geq \|P_C u - P_C v\|^2$, $\forall u, v \in H$;
(ii) $\langle u - P_C u, v - P_C u \rangle \leq 0$, $\forall u \in H, v \in C$;
(iii) $\|u - v\|^2 \geq \|u - P_C u\|^2 + \|v - P_C u\|^2$, $\forall u \in H, v \in C$;
(iv) $\|u - v\|^2 = \|u\|^2 - \|v\|^2 - 2\langle u - v, v \rangle$, $\forall u, v \in H$;
(v) $\|\tau x + (1 - \tau) y\|^2 = \tau \|x\|^2 + (1 - \tau)\|y\|^2 - \tau(1 - \tau)\|x - y\|^2$, $\forall x, y \in H, \tau \in [0, 1]$.

Lemma 1. *[39] Assume that $A : C \to H$ is a continuous pseudomonotone mapping. Then $u^* \in C$ is a solution to the VI $\langle Au^*, v - u^* \rangle \geq 0, \forall v \in C$, iff $\langle Av, v - u^* \rangle \geq 0, \forall v \in C$.*

Lemma 2. *[40] Let the real sequence $\{t_n\} \subset [0, \infty)$ satisfy the conditions: $t_{n+1} \leq (1 - s_n)t_n + s_n b_n, \forall n \geq 1$, where $\{s_n\}$ and $\{b_n\}$ are sequences in $(-\infty, \infty)$ such that (i) $\{s_n\} \subset [0, 1]$ and $\sum_{n=1}^{\infty} s_n = \infty$, and (ii) $\limsup_{n \to \infty} b_n \leq 0$ or $\sum_{n=1}^{\infty} |s_n b_n| < \infty$. Then $\lim_{n \to \infty} t_n = 0$.*

Lemma 3. *[33] Let $T : C \to C$ be a ζ-strict pseudocontraction. If the sequence $\{u_n\} \subset C$ satisfies $u_n \rightharpoonup u \in C$ and $(I - T)u_n \to 0$, then $u \in \text{Fix}(T)$, where I is the identity operator of H.*

Lemma 4. *[33] Let $T : C \to C$ be a ζ-strictly pseudocontractive mapping. Let the real numbers $\gamma, \delta \geq 0$ satisfy $(\gamma + \delta)\zeta \leq \gamma$. Then $\|\gamma(x - y) + \delta(Tx - Ty)\| \leq (\gamma + \delta)\|x - y\|, \forall x, y \in C$.*

Lemma 5. *[41] Let the Banach space X admit a weakly continuous duality mapping, the subset $C \subset X$ be nonempty, convex and closed, and the asymptotically nonexpansive mapping $T : C \to C$ have a fixed point, i.e., $\text{Fix}(T) \neq \emptyset$. Then $I - T$ is demiclosed at zero, i.e., if the sequence $\{u_n\} \subset C$ satisfies $u_n \rightharpoonup u \in C$ and $(I - T)u_n \to 0$, then $(I - T)u = 0$, where I is the identity mapping of X.*

3. Main Results

Unless otherwise stated, we suppose the following.

- $T : H \to H$ is an asymptotically nonexpansive operator with $\{\theta_n\}$ and $S : H \to H$ is a ζ-strictly pseudocontractive mapping.
- $A : H \to H$ is sequentially weakly continuous on C, L-Lipschitzian pseudomonotone on H, and $A(C)$ is bounded.
- $f : H \to C$ is a δ-contraction with $\delta \in [0, \frac{1}{2})$.
- $\Omega = \text{Fix}(T) \cap \text{Fix}(S) \cap \text{VI}(C, A) \neq \emptyset$.
- $\{\sigma_n\} \subset [0, 1]$ and $\{\alpha_n\}, \{\beta_n\}, \{\gamma_n\}, \{\delta_n\} \subset (0, 1)$ such that
  (i) $\sup_{n \geq 1} \frac{\sigma_n}{\alpha_n} < \infty$ and $\beta_n + \gamma_n + \delta_n = 1, \forall n \geq 1$;
  (ii) $\sum_{n=1}^{\infty} \alpha_n = \infty$, $\lim_{n \to \infty} \alpha_n = \lim_{n \to \infty} \frac{\theta_n}{\alpha_n} = 0$;
  (iii) $(\gamma_n + \delta_n)\zeta \leq \gamma_n < (1 - 2\delta)\delta_n, \forall n \geq 1$ and $\liminf_{n \to \infty}((1 - 2\delta)\delta_n - \gamma_n) > 0$;
  (iv) $\limsup_{n \to \infty} \beta_n < 1$, $\liminf_{n \to \infty} \beta_n > 0$ and $\liminf_{n \to \infty} \delta_n > 0$.

We first introduce an inertial-like subgradient extragradient algorithm (Algorithm 3) with line-search process as follows:

Algorithm 3: Inertial-like subgradient extragradient algorithm (I).

Initialization: Given $x_0, x_1 \in H$ arbitrarily. Let $\gamma > 0$, $l \in (0, 1)$, $\mu \in (0, 1)$.
Iterative Steps: Compute x_{n+1} in what follows:
Step 1. Put $w_n = \sigma_n(x_n - x_{n-1}) + T^n x_n$ and calculate $y_n = P_C(I - \tau_n A)w_n$, where τ_n is chosen to be the largest $\tau \in \{\gamma, \gamma l, \gamma l^2, ...\}$ such that

$$\tau \|Aw_n - Ay_n\| \leq \mu \|w_n - y_n\|.$$

Step 2. Calculate $z_n = (1 - \alpha_n)P_{C_n}(w_n - \tau_n Ay_n) + \alpha_n f(x_n)$ with
$C_n := \{x \in H : \langle w_n - \tau_n Aw_n - y_n, x - y_n \rangle \leq 0\}$.
Step 3. Calculate

$$x_{n+1} = \gamma_n P_{C_n}(w_n - \tau_n Ay_n) + \delta_n S z_n + \beta_n T^n x_n.$$

Again set $n := n + 1$ and return to Step 1.

Lemma 6. *In Step 1 of Algorithm 3, the Armijo-like search rule*

$$\tau \|Aw_n - Ay_n\| \le \mu \|w_n - y_n\| \tag{3}$$

is well defined, and the inequality holds: $\min\{\gamma, \frac{\mu l}{L}\} \le \tau_n \le \gamma$.

Proof. Since A is L-Lipschitzian, we know that Equation (3) holds for all $\gamma l^m \le \frac{\mu}{L}$ and so τ_n is well defined. It is clear that $\tau_n \le \gamma$. Next we discuss two cases. In the case where $\tau_n = \gamma$, the inequality is valid. In the case where $\tau_n < \gamma$, from Equation (3) we derive $\|Aw_n - AP_C(w_n - \frac{\tau_n}{l}Aw_n)\| > \frac{\mu}{\tau_n}\|w_n - P_C(w_n - \frac{\tau_n}{l}Aw_n)\|$. Also, since A is L-Lipschitzian, we get $\tau_n > \frac{\mu l}{L}$. Therefore the inequality is true. □

Lemma 7. *Assume that $\{w_n\}, \{y_n\}, \{z_n\}$ are the sequences constructed by Algorithm 3. Then*

$$\begin{aligned}\|z_n - p\|^2 &\le [1 - \alpha_n(1-\delta)]\|x_n - p\|^2 + (1-\alpha_n)\Lambda_n - (1-\alpha_n)(1-\mu)\times \\ &\quad \times [\|w_n - y_n\|^2 + \|u_n - y_n\|^2] + 2\alpha_n\langle (f-I)p, z_n - p\rangle \quad \forall p \in \Omega,\end{aligned} \tag{4}$$

where $u_n := P_{C_n}(w_n - \tau_n Ay_n)$ and $\Lambda_n := \sigma_n\|x_n - x_{n-1}\|[2(1+\theta_n)\|x_n - p\| + \sigma_n\|x_n - x_{n-1}\|] + \theta_n(2+\theta_n)\|x_n - p\|^2$ for all $n \ge 1$.

Proof. We observe that

$$\begin{aligned}2\|u_n - p\|^2 &= 2\|P_{C_n}(w_n - \tau_n Ay_n) - P_{C_n}p\|^2 \le 2\langle u_n - p, w_n - \tau_n Ay_n - p\rangle \\ &= \|u_n - p\|^2 + \|w_n - p\|^2 - \|u_n - w_n\|^2 - 2\langle u_n - p, \tau_n Ay_n\rangle.\end{aligned}$$

So, it follows that $\|w_n - p\|^2 - \|u_n - w_n\|^2 - 2\langle u_n - p, \tau_n Ay_n\rangle \ge \|u_n - p\|^2$. Since A is pseudomonotone, we deduce from Equation (3) that $\langle Ay_n, y_n - p\rangle \ge 0$ and

$$\begin{aligned}\|u_n - p\|^2 &\le \|w_n - p\|^2 + 2\tau_n(\langle Ay_n, p - y_n\rangle + \langle Ay_n, y_n - u_n\rangle) - \|u_n - w_n\|^2 \\ &\le \|w_n - p\|^2 + 2\tau_n\langle Ay_n, y_n - u_n\rangle - \|u_n - w_n\|^2 \\ &= \|w_n - p\|^2 - \|y_n - w_n\|^2 + 2\langle w_n - \tau_n Ay_n - y_n, u_n - y_n\rangle - \|u_n - y_n\|^2.\end{aligned} \tag{5}$$

Since $u_n = P_{C_n}(w_n - \tau_n Ay_n)$ with $C_n := \{x \in H : 0 \ge \langle \tau_n Aw_n - w_n + y_n, y_n - x\rangle\}$, we have $\langle u_n - y_n, w_n - \tau_n Aw_n - y_n\rangle \le 0$, which together with Equation (3), implies that

$$\begin{aligned}2\langle w_n - \tau_n Ay_n - y_n, u_n - y_n\rangle &= 2\langle w_n - \tau_n Aw_n - y_n, u_n - y_n\rangle + 2\tau_n\langle Aw_n - Ay_n, u_n - y_n\rangle \\ &\le 2\mu\|w_n - y_n\|\|u_n - y_n\| \le \mu(\|w_n - y_n\|^2 + \|u_n - y_n\|^2).\end{aligned}$$

Also, from $w_n = \sigma_n(x_n - x_{n-1}) + T^n x_n$ we get

$$\begin{aligned}\|w_n - p\|^2 &= \|\sigma_n(x_n - x_{n-1}) + T^n x_n - p\|^2 \\ &\le [(1+\theta_n)\|x_n - p\| + \sigma_n\|x_n - x_{n-1}\|]^2 \\ &= (1+\theta_n)^2\|x_n - p\|^2 + \sigma_n\|x_n - x_{n-1}\|[2(1+\theta_n)\|x_n - p\| + \sigma_n\|x_n - x_{n-1}\|] \\ &= \|x_n - p\|^2 + \theta_n(2+\theta_n)\|x_n - p\|^2 + \sigma_n\|x_n - x_{n-1}\|[2(1+\theta_n)\|x_n - p\| + \sigma_n\|x_n - x_{n-1}\|] \\ &= \|x_n - p\|^2 + \Lambda_n,\end{aligned}$$

where $\Lambda_n := \theta_n(2+\theta_n)\|x_n - p\|^2 + \sigma_n\|x_n - x_{n-1}\|[2(1+\theta_n)\|x_n - p\| + \sigma_n\|x_n - x_{n-1}\|]$. Therefore, substituting the last two inequalities for Equation (5), we infer that

$$\begin{aligned}\|u_n - p\|^2 &\leq \|w_n - p\|^2 - (1-\mu)\|w_n - y_n\|^2 - (1-\mu)\|u_n - y_n\|^2 \\ &\leq \Lambda_n - (1-\mu)\|w_n - y_n\|^2 - (1-\mu)\|u_n - y_n\|^2 + \|x_n - p\|^2, \quad \forall p \in \Omega.\end{aligned} \quad (6)$$

In addition, from Algorithm 3 we have

$$z_n - p = (1-\alpha_n)(u_n - p) + \alpha_n(f-I)p + \alpha_n(f(x_n) - f(p)).$$

Since the function $h(t) = t^2, \forall t \in \mathbf{R}$ is convex, from Equation (6) we have

$$\begin{aligned}\|z_n - p\|^2 &\leq [\alpha_n\delta\|x_n - p\| + (1-\alpha_n)\|u_n - p\|]^2 + 2\alpha_n\langle(f-I)p, z_n - p\rangle \\ &\leq \alpha_n\delta\|x_n - p\|^2 + (1-\alpha_n)[\|x_n - p\|^2 + \Lambda_n - (1-\mu)\|w_n - y_n\|^2 - (1-\mu)\|u_n - y_n\|^2] \\ &\quad + 2\alpha_n\langle(f-I)p, z_n - p\rangle \\ &= [1-\alpha_n(1-\delta)]\|x_n - p\|^2 + (1-\alpha_n)\Lambda_n - (1-\alpha_n)(1-\mu)[\|w_n - y_n\|^2 + \|u_n - y_n\|^2] \\ &\quad + 2\alpha_n\langle(f-I)p, z_n - p\rangle.\end{aligned}$$

This completes the proof. □

Lemma 8. *Assume that $\{x_n\}, \{y_n\}, \{z_n\}$ are bounded vector sequences constructed by Algorithm 3. If $T^n x_n - T^{n+1}x_n \to 0$, $x_n - x_{n+1} \to 0$, $w_n - x_n \to 0$, $w_n - z_n \to 0$ and $\exists\{w_{n_k}\} \subset \{w_n\}$ such that $w_{n_k} \rightharpoonup z \in H$, then $z \in \Omega$.*

Proof. In terms of Algorithm 3, we deduce $w_n - x_n = T^n x_n' - x_n + \sigma_n(x_n - x_{n-1}), \forall n \geq 1$, and hence $\|T^n x_n - x_n\| \leq \|w_n - x_n\| + \sigma_n\|x_n - x_{n-1}\| \leq \|w_n - x_n\| + \|x_n - x_{n-1}\|$. Using the conditions $x_n - x_{n+1} \to 0$ and $w_n - x_n \to 0$, we get

$$\lim_{n\to\infty}\|T^n x_n - x_n\| = 0. \quad (7)$$

Combining the assumptions $w_n - x_n \to 0$ and $w_n - z_n \to 0$ yields

$$\|z_n - x_n\| \leq \|w_n - z_n\| + \|w_n - x_n\| \to 0, \quad (n \to \infty).$$

Then, from Equation (4) it follows that

$$\begin{aligned}(1-\alpha_n)(1-\mu)&[\|w_n - y_n\|^2 + \|u_n - y_n\|^2] \\ &\leq [1-\alpha_n(1-\delta)]\|x_n - p\|^2 + (1-\alpha_n)\Lambda_n - \|z_n - p\|^2 + 2\alpha_n\langle(f-I)p, z_n - p\rangle \\ &\leq \|x_n - p\|^2 - \|z_n - p\|^2 + \Lambda_n + 2\alpha_n\|(f-I)p\|\|z_n - p\| \\ &\leq \|x_n - z_n\|(\|x_n - p\| + \|z_n - p\|) + \Lambda_n + 2\alpha_n\|(f-I)p\|\|z_n - p\|,\end{aligned}$$

where $\Lambda_n := \theta_n(2+\theta_n)\|x_n - p\|^2 + \sigma_n\|x_n - x_{n-1}\|[2(1+\theta_n)\|x_n - p\| + \sigma_n\|x_n - x_{n-1}\|]$. Since $\alpha_n \to 0$, $\Lambda_n \to 0$ and $x_n - z_n \to 0$, from the boundedness of $\{x_n\}, \{z_n\}$ we get

$$\lim_{n\to\infty}\|w_n - y_n\| = 0 \quad \text{and} \quad \lim_{n\to\infty}\|u_n - y_n\| = 0.$$

Thus as $n \to \infty$,

$$\|w_n - u_n\| \leq \|w_n - y_n\| + \|y_n - u_n\| \to 0 \quad \text{and} \quad \|x_n - u_n\| \leq \|x_n - w_n\| + \|w_n - u_n\| \to 0.$$

Furthermore, using Algorithm 3 we have $x_{n+1} - z_n = \gamma_n(u_n - z_n) + \delta_n(Sz_n - z_n) + \beta_n(T^n x_n - z_n)$, which hence implies

$$\begin{aligned}\delta_n \|Sz_n - z_n\| &= \|x_{n+1} - z_n - \beta_n(T^n x_n - z_n) - \gamma_n(u_n - z_n)\| \\ &= \|x_{n+1} - x_n + \delta_n(x_n - z_n) - \gamma_n(u_n - x_n) - \beta_n(T^n x_n - x_n)\| \\ &\leq \|x_{n+1} - x_n\| + \|x_n - z_n\| + \|u_n - x_n\| + \|T^n x_n - x_n\|.\end{aligned}$$

Note that $x_n - x_{n+1} \to 0$, $z_n - x_n \to 0$, $x_n - u_n \to 0$, $x_n - T^n x_n \to 0$ and $\liminf_{n\to\infty} \delta_n > 0$. So we obtain

$$\lim_{n\to\infty} \|z_n - Sz_n\| = 0. \tag{8}$$

Noticing $y_n = P_C(I - \tau_n A)w_n$, we have $\langle x - y_n, w_n - \tau_n A w_n - y_n \rangle \leq 0$, $\forall x \in C$, and hence

$$\langle w_n - y_n, x - y_n \rangle + \tau_n \langle A w_n, y_n - w_n \rangle \leq \tau_n \langle A w_n, x - w_n \rangle, \quad \forall x \in C. \tag{9}$$

Since A is Lipschitzian, we infer from the boundedness of $\{w_n\}$ that $\{Aw_n\}$ is bounded. From $w_n - y_n \to 0$, we get the boundedness of $\{y_{n_k}\}$. Taking into account $\tau_n \geq \min\{\gamma, \frac{\mu l}{L}\}$, from Equation (9) we have $\liminf_{k\to\infty} \langle Aw_{n_k}, x - w_{n_k} \rangle \geq 0$, $\forall x \in C$. Moreover, note that $\langle Ay_n, x - y_n \rangle = \langle Ay_n - Aw_n, x - w_n \rangle + \langle Aw_n, x - w_n \rangle + \langle Ay_n, w_n - y_n \rangle$. Since A is L-Lipschitzian, from $w_n - y_n \to 0$ we get $Aw_n - Ay_n \to 0$. According to Equation (9) we have $\liminf_{k\to\infty} \langle Ay_{n_k}, x - y_{n_k} \rangle \geq 0$, $\forall x \in C$.

We claim $x_n - Tx_n \to 0$ below. Indeed, note that

$$\begin{aligned}\|Tx_n - x_n\| &\leq \|Tx_n - T^{n+1} x_n\| + \|T^{n+1} x_n - T^n x_n\| + \|T^n x_n - x_n\| \\ &\leq (2 + \theta_1)\|x_n - T^n x_n\| + \|T^{n+1} x_n - T^n x_n\|.\end{aligned}$$

Hence from Equation (7) and the assumption $T^n x_n - T^{n+1} x_n \to 0$ we get

$$\lim_{n\to\infty} \|x_n - Tx_n\| = 0. \tag{10}$$

We now choose a sequence $\{\varepsilon_k\} \subset (0,1)$ such that $\varepsilon_k \downarrow 0$ as $k \to \infty$. For each $k \geq 1$, we denote by m_k the smallest natural number satisfying

$$\langle Ay_{n_j}, x - y_{n_j} \rangle + \varepsilon_k \geq 0, \quad \forall j \geq m_k.$$

From the decreasing property of $\{\varepsilon_k\}$, it is easy to see that $\{m_k\}$ is increasing. Considering that $\{y_{m_k}\} \subset C$ implies $Ay_{m_k} \neq 0$, $\forall k \geq 1$, we put

$$\mu_{m_k} = \frac{Ay_{m_k}}{\|Ay_{m_k}\|^2}.$$

So we have $\langle Ay_{m_k}, \mu_{m_k} \rangle = 1$, $\forall k \geq 1$. Thus, from Equation (9), we have $\langle x + \varepsilon_k \mu_{m_k} - y_{m_k}, Ay_{m_k} \rangle \geq 0$, $\forall k \geq 1$. Also, since A is pseudomonotone, we get

$$\langle A(x + \varepsilon_k \mu_{m_k}), x + \varepsilon_k \mu_{m_k} - y_{m_k} \rangle \geq 0, \quad \forall k \geq 1.$$

Consequently,

$$\langle x - y_{m_k}, Ax \rangle \geq \langle x + \varepsilon_k \mu_{m_k} - y_{m_k}, Ax - A(x + \varepsilon_k \mu_{m_k}) \rangle - \varepsilon_k \langle \mu_{m_k}, Ax \rangle, \quad \forall k \geq 1. \tag{11}$$

We show $\lim_{k\to\infty} \varepsilon_k \mu_{m_k} = 0$. In fact, since $w_{n_k} \rightharpoonup z$ and $w_n - y_n \to 0$, we get $y_{n_k} \rightharpoonup z$. So, $\{y_n\} \subset C$ guarantees $z \in C$. Also, since A is sequentially weakly continuous on C, we deduce that $Ay_{n_k} \rightharpoonup Az$. So, we get $Az \neq 0$. It follows that $0 < \|Az\| \leq \liminf_{k\to\infty} \|Ay_{n_k}\|$. Since $\{y_{m_k}\} \subset \{y_{n_k}\}$ and $\varepsilon_k \downarrow 0$ as $k \to \infty$, we obtain that

$$0 \leq \limsup_{k\to\infty} \|\varepsilon_k \mu_{m_k}\| = \limsup_{k\to\infty} \frac{\varepsilon_k}{\|Ay_{m_k}\|} \leq \frac{\limsup_{k\to\infty} \varepsilon_k}{\liminf_{k\to\infty} \|Ay_{n_k}\|} = 0.$$

Thus $\varepsilon_k \mu_{m_k} \to 0$.

The last step is to show $z \in \Omega$. Indeed, we have $x_{n_k} \rightharpoonup z$. From Equation (10) we also have $x_{n_k} - Tx_{n_k} \to 0$. Note that Lemma 5 yields the demiclosedness of $I - T$ at zero. Thus $z \in \text{Fix}(T)$. Moreover, since $w_n - z_n \to 0$ and $w_{n_k} \rightharpoonup z$, we have $z_{n_k} \rightharpoonup z$. From Equation (8) we get $z_{n_k} - Sz_{n_k} \to 0$. By Lemma 5 we know that $I - S$ is demiclosed at zero, and hence we have $(I - S)z = 0$, i.e., $z \in \text{Fix}(S)$. In addition, taking $k \to \infty$, we infer that the right hand side of Equation (11) converges to zero by the Lipschitzian property of A, the boundedness of $\{y_{m_k}\}, \{\mu_{m_k}\}$, and the limit $\lim_{k\to\infty} \varepsilon_k \mu_{m_k} = 0$. Therefore, $\langle Ax, x - z \rangle = \liminf_{k\to\infty} \langle Ax, x - y_{m_k} \rangle \geq 0, \forall x \in C$. From Lemma 3 we get $z \in \text{VI}(C, A)$, and hence $z \in \Omega$. This completes the proof. □

Theorem 1. *Let $\{x_n\}$ be the sequence constructed by Algorithm 3. Suppose that $T^n x_n - T^{n+1} x_n \to 0$. Then*

$$x_n \to x^* \in \Omega \iff \begin{cases} x_n - x_{n+1} \to 0, \\ x_n - T^n x_n \to 0, \\ \sup_{n\geq 1} \|(T^n - f)x_n\| < \infty, \end{cases}$$

where $x^* \in \Omega$ is only a solution of the HVI: $\langle (f - I)x^*, p - x^* \rangle \leq 0, \forall p \in \Omega$.

Proof. Without loss of generality, we may assume that $\{\beta_n\} \subset [a, b] \subset (0, 1)$. We can claim that $P_\Omega \circ f$ is a contractive map. Banach's Contraction Principle ensures that it has a unique fixed point, i.e., $P_\Omega f(x^*) = x^*$. So, there exists a unique solution $x^* \in \Omega$ to the HVI

$$\langle (I - f)x^*, p - x^* \rangle \geq 0, \quad \forall p \in \Omega. \tag{12}$$

It is clear that the necessity of the theorem is valid. In fact, if $x_n \to x^* \in \Omega$, then as $n \to \infty$, we obtain that $\|x_n - x_{n+1}\| \to 0$, $\|x_n - T^n x_n\| \leq \|x_n - x^*\| + \|x^* - T^n x_n\| \leq (2 + \theta_n)\|x_n - x^*\| \to 0$, and

$$\sup_{n\geq 1} \|T^n x_n - f(x_n)\| \leq \sup_{n\geq 1} (\|T^n x_n - x^*\| + \|x^* - f(x^*)\| + \|f(x^*) - f(x_n)\|)$$
$$\leq \sup_{n\geq 1} [(1 + \theta_n)\|x_n - x^*\| + \|x^* - f(x^*)\| + \delta\|x^* - x_n\|]$$
$$\leq \sup_{n\geq 1} [(2 + \theta_n)\|x_n - x^*\| + \|x^* - f(x^*)\|] < \infty.$$

We now assume that $\lim_{n\to\infty}(\|x_n - x_{n+1}\| + \|x_n - T^n x_n\|) = 0$ and $\sup_{n\geq 1} \|(T^n - f)x_n\| < \infty$, and prove the sufficiency by the following steps.

Step 1. We claim the boundedness of $\{x_n\}$. In fact, take a fixed $p \in \Omega$ arbitrarily. From Equation (6) we get

$$\|w_n - p\|^2 - (1 - \mu)\|w_n - y_n\|^2 - (1 - \mu)\|u_n - y_n\|^2 \geq \|u_n - p\|^2, \tag{13}$$

which hence yields

$$\|w_n - p\| \geq \|u_n - p\|, \quad \forall n \geq 1. \tag{14}$$

By the definition of w_n, we have

$$\begin{aligned}\|w_n - p\| &\leq (1+\theta_n)\|x_n - p\| + \sigma_n\|x_n - x_{n-1}\| \\ &= (1+\theta_n)\|x_n - p\| + \alpha_n \cdot \frac{\sigma_n}{\alpha_n}\|x_n - x_{n-1}\|.\end{aligned} \quad (15)$$

From $\sup_{n\geq 1} \frac{\sigma_n}{\alpha_n} < \infty$ and $\sup_{n\geq 1} \|x_n - x_{n-1}\| < \infty$, we deduce that $\sup_{n\geq 1} \frac{\sigma_n}{\alpha_n}\|x_n - x_{n-1}\| < \infty$, which immediately implies that $\exists M_1 > 0$ s.t.

$$M_1 \geq \frac{\sigma_n}{\alpha_n}\|x_n - x_{n-1}\|, \quad \forall n \geq 1. \quad (16)$$

From Equations (14)–(16), we obtain

$$\|u_n - p\| \leq \|w_n - p\| \leq (1+\theta_n)\|x_n - p\| + \alpha_n M_1, \quad \forall n \geq 1. \quad (17)$$

Note that $A(C)$ is bounded, $y_n = P_C(I - \tau_n)Aw_n$, $f(H) \subset C \subset C_n$ and $u_n = P_{C_n}(w_n - \tau_n Ay_n)$. Hence, we know that $\{Ay_n\}$ is a bounded sequence. So, from $\sup_{n\geq 1}\|(T^n - f)x_n\| < \infty$, it follows that

$$\begin{aligned}\|u_n - f(x_n)\| &= \|P_{C_n}(w_n - \tau_n Ay_n) - P_{C_n}f(x_n)\| \leq \|w_n - \tau_n Ay_n - f(x_n)\| \\ &\leq \|w_n - T^n x_n\| + \|T^n x_n - f(x_n)\| + \tau_n\|Ay_n\| \\ &\leq \|x_n - x_{n-1}\| + \|(T^n - f)x_n\| + \gamma\|Ay_n\| \leq M_0,\end{aligned}$$

where $\sup_{n\geq 1}(\|x_n - x_{n-1}\| + \|(T^n - f)x_n\| + \gamma\|Ay_n\|) \leq M_0$ for some $M_0 > 0$. Taking into account $\lim_{n\to\infty} \frac{\theta_n(2+\theta_n)}{\alpha_n(1-\beta_n)} = 0$, we know that $\exists n_0 \geq 1$ such that

$$\theta_n(2+\theta_n) \leq \frac{\alpha_n(1-\beta_n)(1-\delta)}{2} \ (\leq \frac{\alpha_n(1-\delta)}{2}), \quad \forall n \geq n_0.$$

So, from Algorithm 3 and Equation (17) it follows that for all $n \geq n_0$,

$$\begin{aligned}\|z_n - p\| &\leq \alpha_n\delta\|x_n - p\| + (1-\alpha_n)\|u_n - p\| + \alpha_n\|(f-I)p\| \\ &\leq [1 - \alpha_n(1-\delta) + \theta_n]\|x_n - p\| + \alpha_n(M_1 + \|(f-I)p\|) \\ &\leq [1 - \frac{\alpha_n(1-\delta)}{2}]\|x_n - p\| + \alpha_n(M_1 + \|(f-I)p\|),\end{aligned}$$

which together with Lemma 4 and $(\gamma_n + \delta_n)\zeta \leq \gamma_n$, implies that for all $n \geq n_0$,

$$\begin{aligned}\|x_{n+1} - p\| &= \|\beta_n(T^n x_n - p) + \gamma_n(z_n - p) + \delta_n(Sz_n - p) + \gamma_n(u_n - z_n)\| \\ &\leq \beta_n(1+\theta_n)\|x_n - p\| + (1-\beta_n)\|z_n - p\| + \gamma_n\alpha_n\|u_n - f(x_n)\| \\ &\leq \beta_n(1+\theta_n)\|x_n - p\| + (1-\beta_n)[(1 - \frac{\alpha_n(1-\delta)}{2})\|x_n - p\| + \alpha_n(M_0 + M_1 + \|(f-I)p\|)] \\ &\leq [1 - \frac{\alpha_n(1-\beta_n)(1-\delta)}{2} + \beta_n\frac{\alpha_n(1-\beta_n)(1-\delta)}{2}]\|x_n - p\| + \alpha_n(1-\beta_n)(M_0 + M_1 + \|(f-I)p\|) \\ &= [1 - \frac{\alpha_n(1-\beta_n)^2(1-\delta)}{2}]\|x_n - p\| + \frac{\alpha_n(1-\beta_n)^2(1-\delta)}{2} \cdot \frac{2(M_0+M_1+\|(f-I)p\|)}{(1-\delta)(1-\beta_n)}.\end{aligned}$$

By induction, we obtain $\|x_n - p\| \leq \max\{\|x_{n_0} - p\|, \frac{2(M_0+M_1+\|(f-I)p\|)}{(1-\delta)(1-b)}\}, \forall n \geq n_0$. Therefore, we derive the boundedness of $\{x_n\}$ and hence the one of sequences $\{u_n\}, \{w_n\}, \{y_n\}, \{z_n\}, \{f(x_n)\}, \{Sz_n\}, \{T^n x_n\}$.

Step 2. We claim that $\exists M_4 > 0$ s.t.

$$(1-\alpha_n)(1-\beta_n)(1-\mu)[\|w_n - y_n\|^2 + \|u_n - y_n\|^2] \leq \|x_n - p\|^2 - \|x_{n+1} - p\|^2 + \alpha_n M_4, \quad \forall n \geq n_0.$$

In fact, using Lemmas 4 and 7 and the convexity of $\|\cdot\|^2$, we get

$$\begin{aligned}
\|x_{n+1} - p\|^2 &= \|\beta_n(T^n x_n - p) + \gamma_n(z_n - p) + \delta_n(Sz_n - p) + \gamma_n(u_n - z_n)\|^2 \\
&\leq \beta_n \|T^n x_n - p\|^2 + (1 - \beta_n)\|\tfrac{1}{1-\beta_n}[\gamma_n(z_n - p) + \delta_n(Sz_n - p)]\|^2 \\
&\quad + 2(1 - \beta_n)\alpha_n \|u_n - f(x_n)\| \|x_{n+1} - p\| \\
&\leq \beta_n \|T^n x_n - p\|^2 + (1 - \beta_n)\{[1 - \alpha_n(1 - \delta)]\|x_n - p\|^2 + (1 - \alpha_n)\Lambda_n \\
&\quad - (1 - \alpha_n)(1 - \mu)[\|w_n - y_n\|^2 + \|u_n - y_n\|^2] + 2\alpha_n \langle (f - I)p, z_n - p \rangle\} \\
&\quad + 2(1 - \beta_n)\alpha_n \|u_n - f(x_n)\| \|x_{n+1} - p\| \\
&\leq \beta_n \|T^n x_n - p\|^2 + (1 - \beta_n)\{[1 - \alpha_n(1 - \delta)]\|x_n - p\|^2 + (1 - \alpha_n)\Lambda_n \\
&\quad - (1 - \alpha_n)(1 - \mu)[\|w_n - y_n\|^2 + \|u_n - y_n\|^2] + \alpha_n M_2\},
\end{aligned} \tag{18}$$

where

$$\Lambda_n := \theta_n(2 + \theta_n)\|x_n - p\|^2 + \sigma_n \|x_n - x_{n-1}\|[2(1 + \theta_n)\|x_n - p\| + \sigma_n \|x_n - x_{n-1}\|],$$

and

$$\sup_{n \geq 1} 2(\|(f - I)p\| \|z_n - p\| + \|u_n - f(x_n)\| \|x_{n+1} - p\|) \leq M_2$$

for some $M_2 > 0$. Also, from Equation (16) we have

$$\begin{aligned}
\Lambda_n &= \theta_n(2 + \theta_n)\|x_n - p\|^2 + \sigma_n \|x_n - x_{n-1}\|[2(1 + \theta_n)\|x_n - p\| + \sigma_n \|x_n - x_{n-1}\|] \\
&\leq \theta_n(2 + \theta_n)\|x_n - p\|^2 + \alpha_n M_1 [2(1 + \theta_n)\|x_n - p\| + \alpha_n M_1] \\
&= \alpha_n \{\tfrac{\theta_n}{\alpha_n}(2 + \theta_n)\|x_n - p\|^2 + M_1 [2(1 + \theta_n)\|x_n - p\| + \alpha_n M_1]\} \leq \alpha_n M_3,
\end{aligned} \tag{19}$$

where

$$\sup_{n \geq 1} \{\tfrac{\theta_n}{\alpha_n}(2 + \theta_n)\|x_n - p\|^2 + M_1 [2(1 + \theta_n)\|x_n - p\| + \alpha_n M_1]\} \leq M_3$$

for some $M_3 > 0$. Note that

$$\theta_n(2 + \theta_n) \leq \frac{\alpha_n(1 - \beta_n)(1 - \delta)}{2}, \quad \forall n \geq n_0.$$

Substituting Equation (19) for Equation (18), we obtain that for all $n \geq n_0$,

$$\begin{aligned}
\|x_{n+1} - p\|^2 &\leq \beta_n(1 + \theta_n)^2 \|x_n - p\|^2 + (1 - \beta_n)\{[1 - \alpha_n(1 - \delta)]\|x_n - p\|^2 + (1 - \alpha_n)\alpha_n M_3 \\
&\quad - (1 - \alpha_n)(1 - \mu)[\|w_n - y_n\|^2 + \|u_n - y_n\|^2] + \alpha_n M_2\} \\
&\leq [1 - \tfrac{\alpha_n(1 - \beta_n)(1 - \delta)}{2}]\|x_n - p\|^2 + \alpha_n M_3 \\
&\quad - (1 - \alpha_n)(1 - \beta_n)(1 - \mu)[\|w_n - y_n\|^2 + \|u_n - y_n\|^2] + \alpha_n M_2 \\
&\leq \|x_n - p\|^2 - (1 - \alpha_n)(1 - \beta_n)(1 - \mu)[\|w_n - y_n\|^2 + \|u_n - y_n\|^2] + \alpha_n M_4,
\end{aligned}$$

where $M_4 := M_2 + M_3$. This immediately implies that for all $n \geq n_0$,

$$(1 - \alpha_n)(1 - \beta_n)(1 - \mu)[\|w_n - y_n\|^2 + \|u_n - y_n\|^2] \leq \|x_n - p\|^2 - \|x_{n+1} - p\|^2 + \alpha_n M_4. \tag{20}$$

Step 3. We claim that $\exists M > 0$ s.t.

$$\begin{aligned}
\|x_{n+1} - p\|^2 &\leq [1 - \tfrac{(1-2\delta)\delta_n - \gamma_n}{1 - \alpha_n \gamma_n}\alpha_n]\|x_n - p\|^2 + \tfrac{[(1-2\delta)\delta_n - \gamma_n]\alpha_n}{1 - \alpha_n \gamma_n} \cdot \{\tfrac{2\gamma_n}{(1-2\delta)\delta_n - \gamma_n}\|f(x_n) - p\| \|z_n - x_{n+1}\| \\
&\quad + \tfrac{2\delta_n}{(1-2\delta)\delta_n - \gamma_n}\|f(x_n) - p\| \|z_n - x_n\| + \tfrac{2\delta_n}{(1-2\delta)\delta_n - \gamma_n}\langle f(p) - p, x_n - p\rangle \\
&\quad + \tfrac{\gamma_n + \delta_n}{(1-2\delta)\delta_n - \gamma_n}(\tfrac{\theta_n}{\alpha_n} \cdot \tfrac{2M^2}{1-b} + \tfrac{\sigma_n}{\alpha_n}\|x_n - x_{n-1}\|3M)\}.
\end{aligned}$$

In fact, we get

$$\|w_n - p\|^2 \leq [(1+\theta_n)\|x_n - p\| + \sigma_n\|x_n - x_{n-1}\|]^2$$
$$= \|x_n - p\|^2 + \theta_n(2+\theta_n)\|x_n - p\|^2 + \sigma_n\|x_n - x_{n-1}\|[2(1+\theta_n)\|x_n - p\| + \sigma_n\|x_n - x_{n-1}\|] \quad (21)$$
$$\leq \|x_n - p\|^2 + \theta_n 2M^2 + \sigma_n\|x_n - x_{n-1}\|3M,$$

where $M \geq \sup_{n\geq 1}\{(1+\theta_n)\|x_n - p\|, \sigma_n\|x_n - x_{n-1}\|\}$ for some $M > 0$. From Algorithm 3 and the convexity of $\|\cdot\|^2$, we have

$$\|x_{n+1} - p\|^2 = \|\beta_n(T^n x_n - p) + \gamma_n(z_n - p) + \delta_n(Sz_n - p) + \gamma_n(u_n - z_n)\|^2$$
$$\leq \|\beta_n(T^n x_n - p) + \gamma_n(z_n - p) + \delta_n(Sz_n - p)\|^2 + 2\gamma_n\alpha_n\langle u_n - f(x_n), x_{n+1} - p\rangle$$
$$\leq \beta_n\|T^n x_n - p\|^2 + (1-\beta_n)\|\tfrac{1}{1-\beta_n}[\gamma_n(z_n - p) + \delta_n(Sz_n - p)]\|^2$$
$$+ 2\gamma_n\alpha_n\langle u_n - p, x_{n+1} - p\rangle + 2\gamma_n\alpha_n\langle p - f(x_n), x_{n+1} - p\rangle,$$

which together with Lemma 4, leads to

$$\|x_{n+1} - p\|^2 \leq \beta_n(1+\theta_n)^2\|x_n - p\|^2 + (1-\beta_n)\|z_n - p\|^2 + 2\gamma_n\alpha_n\|u_n - p\|\|x_{n+1} - p\|$$
$$+ 2\gamma_n\alpha_n\langle p - f(x_n), x_{n+1} - p\rangle$$
$$\leq \beta_n(1+\theta_n)^2\|x_n - p\|^2 + (1-\beta_n)[(1-\alpha_n)\|u_n - p\|^2 + 2\alpha_n\langle f(x_n) - p, z_n - p\rangle]$$
$$+ \gamma_n\alpha_n(\|u_n - p\|^2 + \|x_{n+1} - p\|^2) + 2\gamma_n\alpha_n\langle p - f(x_n), x_{n+1} - p\rangle.$$

From Equations (17) and (21) we know that

$$\|u_n - p\|^2 \leq \|x_n - p\|^2 + \theta_n 2M^2 + \sigma_n\|x_n - x_{n-1}\|3M.$$

Hence, we have

$$\|x_{n+1} - p\|^2 \leq [1 - \alpha_n(1-\beta_n)]\|x_n - p\|^2 + \beta_n\theta_n 2M^2 + (1-\beta_n)(1-\alpha_n)(\theta_n 2M^2$$
$$+ \sigma_n\|x_n - x_{n-1}\|3M) + 2\alpha_n\delta_n\langle f(x_n) - p, z_n - p\rangle + \gamma_n\alpha_n(\|x_n - p\|^2$$
$$+ \|x_{n+1} - p\|^2) + (1-\beta_n)\alpha_n(\theta_n 2M^2 + \sigma_n\|x_n - x_{n-1}\|3M)$$
$$+ 2\gamma_n\alpha_n\langle f(x_n) - p, z_n - x_{n+1}\rangle$$
$$\leq [1 - \alpha_n(1-\beta_n)]\|x_n - p\|^2 + 2\gamma_n\alpha_n\|f(x_n) - p\|\|z_n - x_{n+1}\|$$
$$+ 2\alpha_n\delta_n\delta\|x_n - p\|^2 + 2\alpha_n\delta_n\langle f(p) - p, x_n - p\rangle + 2\alpha_n\delta_n\|f(x_n) - p\|\|z_n - x_n\|$$
$$+ \gamma_n\alpha_n(\|x_n - p\|^2 + \|x_{n+1} - p\|^2) + (1-\beta_n)(\tfrac{\theta_n 2M^2}{1-\beta_n} + \sigma_n\|x_n - x_{n-1}\|3M),$$

which immediately yields

$$\|x_{n+1} - p\|^2$$
$$\leq [1 - \tfrac{(1-2\delta)\delta_n - \gamma_n}{1-\alpha_n\gamma_n}\alpha_n]\|x_n - p\|^2 + \tfrac{[(1-2\delta)\delta_n - \gamma_n]\alpha_n}{1-\alpha_n\gamma_n} \cdot \{\tfrac{2\gamma_n}{(1-2\delta)\delta_n - \gamma_n}\|f(x_n) - p\|\|z_n - x_{n+1}\|$$
$$+ \tfrac{2\delta_n}{(1-2\delta)\delta_n - \gamma_n}\|f(x_n) - p\|\|z_n - x_n\| + \tfrac{2\delta_n}{(1-2\delta)\delta_n - \gamma_n}\langle f(p) - p, x_n - p\rangle \quad (22)$$
$$+ \tfrac{\gamma_n + \delta_n}{(1-2\delta)\delta_n - \gamma_n}(\tfrac{\theta_n}{\alpha_n} \cdot \tfrac{2M^2}{1-b} + \tfrac{\sigma_n}{\alpha_n}\|x_n - x_{n-1}\|3M)\}.$$

Step 4. We claim the strong convergence of $\{x_n\}$ to a unique solution $x^* \in \Omega$ to the HVI Equation (12). In fact, setting $p = x^*$, from Equation (22) we know that

$$\|x_{n+1} - x^*\|^2 \leq [1 - \tfrac{(1-2\delta)\delta_n - \gamma_n}{1 - \alpha_n \gamma_n} \alpha_n] \|x_n - x^*\|^2 + \tfrac{[(1-2\delta)\delta_n - \gamma_n]\alpha_n}{1 - \alpha_n \gamma_n} \cdot \{ \tfrac{2\gamma_n}{(1-2\delta)\delta_n - \gamma_n} \|f(x_n) - x^*\| \|z_n - x_{n+1}\|$$
$$+ \tfrac{2\delta_n}{(1-2\delta)\delta_n - \gamma_n} \|f(x_n) - x^*\| \|z_n - x_n\| + \tfrac{2\delta_n}{(1-2\delta)\delta_n - \gamma_n} \langle f(x^*) - x^*, x_n - x^* \rangle$$
$$+ \tfrac{\gamma_n + \delta_n}{(1-2\delta)\delta_n - \gamma_n} (\tfrac{\theta_n}{\alpha_n} \cdot \tfrac{2M^2}{1-b} + \tfrac{\sigma_n}{\alpha_n} \|x_n - x_{n-1}\| 3M) \}.$$

According to Lemma 4, it is sufficient to prove that $\limsup_{n \to \infty} \langle (f - I)x^*, x_n - x^* \rangle \leq 0$. Since $x_n - x_{n+1} \to 0$, $\alpha_n \to 0$ and $\{\beta_n\} \subset [a, b] \subset (0, 1)$, from Equation (20) we get

$$\limsup_{n \to \infty} (1 - \alpha_n)(1 - b)(1 - \mu)[\|w_n - y_n\|^2 + \|u_n - y_n\|^2]$$
$$\leq \limsup_{n \to \infty} [\|x_n - p\|^2 - \|x_{n+1} - p\|^2 + \alpha_n M_4]$$
$$\leq \limsup_{n \to \infty} (\|x_n - p\| + \|x_{n+1} - p\|) \|x_n - x_{n+1}\| = 0,$$

which hence leads to

$$\lim_{n \to \infty} \|w_n - y_n\| = \lim_{n \to \infty} \|u_n - y_n\| = 0. \tag{23}$$

Obviously, the assumptions $\|x_n - x_{n+1}\| \to 0$ and $\|x_n - T^n x_n\| \to 0$ guarantee that $\|w_n - x_n\| \leq \|T^n x_n - x_n\| + \|x_n - x_{n-1}\| \to 0$ $(n \to \infty)$. Thus,

$$\|x_n - y_n\| \leq \|x_n - w_n\| + \|w_n - y_n\| \to 0, \quad (n \to \infty).$$

Since $z_n = (1 - \alpha_n)u_n + \alpha_n f(x_n)$ with $u_n := P_{C_n}(w_n - \tau_n Ay_n)$, from Equation (23) and the boundedness of $\{x_n\}, \{u_n\}$, we get

$$\|z_n - y_n\| \leq \alpha_n(\|f(x_n)\| + \|u_n\|) + \|u_n - y_n\| \to 0, \quad (n \to \infty), \tag{24}$$

and hence

$$\|z_n - x_n\| \leq \|z_n - y_n\| + \|y_n - x_n\| \to 0, \quad (n \to \infty).$$

Obviously, combining Equations (23) and (24) guarantees that

$$\|w_n - z_n\| \leq \|w_n - y_n\| + \|y_n - z_n\| \to 0, \quad (n \to \infty).$$

Since $\{x_n\}$ is bounded, we know that $\exists \{x_{n_k}\} \subset \{x_n\}$ s.t.

$$\limsup_{n \to \infty} \langle (f - I)x^*, x_n - x^* \rangle = \lim_{k \to \infty} \langle (f - I)x^*, x_{n_k} - x^* \rangle. \tag{25}$$

Next, we may suppose that $x_{n_k} \rightharpoonup \tilde{x}$. Hence from Equation (25) we get

$$\limsup_{n \to \infty} \langle (f - I)x^*, x_n - x^* \rangle = \lim_{k \to \infty} \langle (f - I)x^*, x_{n_k} - x^* \rangle = \langle (f - I)x^*, \tilde{x} - x^* \rangle. \tag{26}$$

From $w_n - x_n \to 0$ and $x_{n_k} \rightharpoonup \tilde{x}$ it follows that $w_{n_k} \rightharpoonup \tilde{x}$.

Since $T^n x_n - T^{n+1} x_n \to 0$, $x_n - x_{n+1} \to 0$, $w_n - x_n \to 0$, $w_n - z_n \to 0$ and $w_{n_k} \rightharpoonup \tilde{x}$, from Lemma 8 we conclude that $\tilde{x} \in \Omega$. Therefore, from Equations (12) and (26) we infer that

$$\limsup_{n \to \infty} \langle (f - I)x^*, x_n - x^* \rangle = \langle (f - I)x^*, \tilde{x} - x^* \rangle \leq 0.$$

Note that

$$\sum_{n=0}^{\infty} \frac{(1 - 2\delta)\delta_n - \gamma_n}{1 - \alpha_n \gamma_n} \alpha_n = \infty.$$

It is clear that

$$\limsup_{n \to \infty} \{ \frac{2\gamma_n}{(1-2\delta)\delta_n - \gamma_n} \| f(x_n) - x^* \| \| z_n - x_{n+1} \| + \frac{2\delta_n}{(1-2\delta)\delta_n - \gamma_n} \| f(x_n) - x^* \| \| z_n - x_n \|$$
$$+ \frac{2\delta_n}{(1-2\delta)\delta_n - \gamma_n} \langle f(x^*) - x^*, x_n - x^* \rangle + \frac{\gamma_n + \delta_n}{(1-2\delta)\delta_n - \gamma_n} (\frac{\theta_n}{\alpha_n} \cdot \frac{2M^2}{1-b} + \frac{\sigma_n}{\alpha_n} \| x_n - x_{n-1} \| 3M) \} \leq 0.$$

Consequently, all conditions of Lemma 4 are satisfied, and hence we immediately deduce that $x_n \to x^*$. This completes the proof. □

Next, we introduce another inertial-like subgradient extragradient algorithm (Algorithm 4) with line-search process as the following.

It is remarkable that Lemmas 6–8 are still valid for Algorithm 4.

Algorithm 4: Inertial-like subgradient extragradient algorithm (II).

Initialization: Given $x_0, x_1 \in H$ arbitrarily. Let $\gamma > 0$, $l \in (0,1)$, $\mu \in (0,1)$.
Iterative Steps: Compute x_{n+1} in what follows:
Step 1. Put $w_n = \sigma_n(x_n - x_{n-1}) + T^n x_n$ and calculate $y_n = P_C(w_n - \tau_n A w_n)$, where τ_n is chosen to be the largest $\tau \in \{\gamma, \gamma l, \gamma l^2, ...\}$ such that

$$\tau \| A w_n - A y_n \| \leq \mu \| w_n - y_n \|.$$

Step 2. Calculate $z_n = (1 - \alpha_n) P_{C_n}(w_n - \tau_n A y_n) + \alpha_n f(x_n)$ with
$C_n := \{ x \in H : \langle w_n - \tau_n A w_n - y_n, x - y_n \rangle \leq 0 \}$.
Step 3. Calculate

$$x_{n+1} = \gamma_n P_{C_n}(w_n - \tau_n A y_n) + \delta_n S z_n + \beta_n T^n w_n.$$

Again set $n := n + 1$ and return to Step 1.

Theorem 2. Let $\{x_n\}$ be the sequence constructed by Algorithm 4. Suppose that $T^n x_n - T^{n+1} x_n \to 0$. Then

$$x_n \to x^* \in \Omega \Leftrightarrow \begin{cases} x_n - x_{n+1} \to 0, \\ x_n - T^n x_n \to 0, \\ \sup_{n \geq 1} \| (T^n - f) x_n \| < \infty, \end{cases}$$

where $x^* \in \Omega$ is only a solution of the HVI: $\langle (I - f)x^*, p - x^* \rangle \geq 0$, $\forall p \in \Omega$.

Proof. Using the same reasoning as in the proof of Theorem 1, we know that there is only a solution $x^* \in \Omega$ of Equation (12), and that the necessity of the theorem is true.

We claim the sufficiency of the theorem below. For the purpose, we suppose that $\lim_{n\to\infty}(\|x_n - x_{n+1}\| + \|x_n - T^n x_n\|) = 0$ and $\sup_{n\geq 1}\|(T^n - f)x_n\| < \infty$. Then we prove the sufficiency by the following steps.

Step 1. We claim the boundedness of $\{x_n\}$. In fact, using the same reasoning as in Step 1 of the proof of Theorem 1, we obtain that inequalities Equations (13)–(17) hold. Noticing $\lim_{n\to\infty} \frac{\theta_n(2+\theta_n)}{\alpha_n(1-\beta_n)} = 0$, we infer that $\exists n_0 \geq 1$ s.t.

$$\theta_n(2+\theta_n) \leq \frac{\alpha_n(1-\beta_n)(1-\delta)}{2} \left(\leq \frac{\alpha_n(1-\delta)}{2}\right), \quad \forall n \geq n_0.$$

So, from Algorithm 4 and Equation (17) it follows that for all $n \geq n_0$,

$$\|z_n - p\| \leq \alpha_n\delta\|x_n - p\| + (1-\alpha_n)[(1+\theta_n)\|x_n - p\| + \alpha_n M_1] + \alpha_n\|(f-I)p\|$$
$$\leq [1 - \frac{\alpha_n(1-\delta)}{2}]\|x_n - p\| + \alpha_n(M_1 + \|(f-I)p\|),$$

which together with Lemma 4 and $(\gamma_n + \delta_n)\zeta \leq \gamma_n$, implies that for all $n \geq n_0$,

$$\|x_{n+1} - p\| = \|\beta_n(T^n w_n - p) + \gamma_n(z_n - p) + \delta_n(Sz_n - p) + \gamma_n(u_n - z_n)\|$$
$$\leq \beta_n(1+\theta_n)\|w_n - p\| + (1-\beta_n)\|z_n - p\| + \gamma_n\alpha_n\|u_n - f(x_n)\|$$
$$\leq [1 - \frac{\alpha_n(1-\beta_n)(1-\delta)}{2} + \beta_n\theta_n(2+\theta_n)]\|x_n - p\| + \beta_n(1+\theta_n)\alpha_n M_1$$
$$+ \alpha_n(1-\beta_n)(M_0 + M_1 + \|(f-I)p\|)$$
$$\leq [1 - \frac{\alpha_n(1-\beta_n)(1-\delta)}{2} + \beta_n\frac{\alpha_n(1-\beta_n)(1-\delta)}{2}]\|x_n - p\| + \alpha_n(1-\beta_n)(M_0 + M_1\frac{1+\theta_n}{1-\beta_n} + \|(f-I)p\|)$$
$$= [1 - \frac{\alpha_n(1-\beta_n)^2(1-\delta)}{2}]\|x_n - p\| + \frac{\alpha_n(1-\beta_n)^2(1-\delta)}{2} \cdot \frac{2(M_0 + M_1\frac{1+\theta_n}{1-\beta_n} + \|(f-I)p\|)}{(1-\delta)(1-\beta_n)}.$$

Hence,

$$\|x_n - p\| \leq \max\{\|x_{n_0} - p\|, \frac{2(M_0 + M_1\frac{2}{1-b} + \|(f-I)p\|)}{(1-\delta)(1-b)}\}, \quad \forall n \geq n_0.$$

Thus, sequence $\{x_n\}$ is bounded.

Step 2. We claim that for all $n \geq n_0$,

$$\|x_n - p\|^2 - \|x_{n+1} - p\|^2 + \alpha_n M_4 \geq (1-\alpha_n)(1-\beta_n)(1-\mu)[\|w_n - y_n\|^2 + \|u_n - y_n\|^2],$$

with constant $M_4 > 0$. Indeed, utilizing Lemmas 4 and 7 and the convexity of $\|\cdot\|^2$, one reaches

$$\|x_{n+1} - p\|^2 = \|\beta_n(T^n w_n - p) + \gamma_n(z_n - p) + \delta_n(Sz_n - p) + \gamma_n(u_n - z_n)\|^2$$
$$\leq \beta_n\|T^n w_n - p\|^2 + (1-\beta_n)\|\frac{1}{1-\beta_n}[\gamma_n(z_n - p) + \delta_n(Sz_n - p)]\|^2$$
$$+ 2(1-\beta_n)\alpha_n\|u_n - f(x_n)\|\|x_{n+1} - p\|$$
$$\leq \beta_n(1+\theta_n)^2\|w_n - p\|^2 + (1-\beta_n)\{[1 - \alpha_n(1-\delta)]\|x_n - p\|^2 + (1-\alpha_n)\Lambda_n$$
$$- (1-\alpha_n)(1-\mu)[\|w_n - y_n\|^2 + \|u_n - y_n\|^2] + 2\alpha_n\langle(f-I)p, z_n - p\rangle\}$$
$$+ 2(1-\beta_n)\alpha_n\|u_n - f(x_n)\|\|x_{n+1} - p\|$$
$$\leq \beta_n(1+\theta_n)^2(\|x_n - p\|^2 + \Lambda_n) + (1-\beta_n)\{[1 - \alpha_n(1-\delta)]\|x_n - p\|^2 + (1-\alpha_n)\Lambda_n$$
$$- (1-\alpha_n)(1-\mu)[\|w_n - y_n\|^2 + \|u_n - y_n\|^2] + \alpha_n M_2\},$$

(27)

where $\Lambda_n := \theta_n(2+\theta_n)\|x_n-p\|^2 + \sigma_n\|x_n-x_{n-1}\|[2(1+\theta_n)\|x_n-p\| + \sigma_n\|x_n-x_{n-1}\|]$, and $\sup_{n\geq 1} 2(\|(f-I)p\|\|z_n-p\| + \|u_n-f(x_n)\|\|x_{n+1}-p\|) \leq M_2$ for some $M_2 > 0$. Also, from Equation (16) we have

$$\begin{aligned}\Lambda_n &= \theta_n(2+\theta_n)\|x_n-p\|^2 + \sigma_n\|x_n-x_{n-1}\|[2(1+\theta_n)\|x_n-p\| + \sigma_n\|x_n-x_{n-1}\|]\\ &\leq \alpha_n\{\tfrac{\theta_n}{\alpha_n}(2+\theta_n)\|x_n-p\|^2 + M_1\|[2(1+\theta_n)\|x_n-p\| + \alpha_n M_1]\} \leq \alpha_n M_3,\end{aligned} \quad (28)$$

where $\sup_{n\geq 1}\{\tfrac{\theta_n}{\alpha_n}(2+\theta_n)\|x_n-p\|^2 + M_1\|[2(1+\theta_n)\|x_n-p\| + \alpha_n M_1]\} \leq M_3$ for some $M_3 > 0$. Note that $\theta_n(2+\theta_n) \leq \tfrac{\alpha_n(1-\beta_n)(1-\delta)}{2}$, $\forall n \geq n_0$. Substituting Equation (28) for Equation (27), we obtain that for all $n \geq n_0$,

$$\begin{aligned}\|x_{n+1}-p\|^2 &\leq [1-\alpha_n(1-\beta_n)(1-\delta) + \beta_n\theta_n(2+\theta_n)]\|x_n-p\|^2 + \beta_n(1+\theta_n)^2\alpha_n M_3\\ &\quad + (1-\beta_n)(1-\alpha_n)\alpha_n M_3 - (1-\alpha_n)(1-\beta_n)(1-\mu)[\|w_n-y_n\|^2\\ &\quad + \|u_n-y_n\|^2] + (1-\beta_n)\alpha_n M_2\\ &\leq \|x_n-p\|^2 - (1-\alpha_n)(1-\beta_n)(1-\mu)[\|w_n-y_n\|^2 + \|u_n-y_n\|^2] + \alpha_n M_4,\end{aligned}$$

where $M_4 := M_2 + 4M_3$. This immediately implies that for all $n \geq n_0$,

$$(1-\alpha_n)(1-\beta_n)(1-\mu)[\|w_n-y_n\|^2 + \|u_n-y_n\|^2] \leq \|x_n-p\|^2 - \|x_{n+1}-p\|^2 + \alpha_n M_4.$$

Step 3. We claim that $\exists M > 0$ s.t.

$$\begin{aligned}\|x_{n+1}&-p\|^2\\ &\leq [1-\tfrac{(1-2\delta)\delta_n-\gamma_n}{1-\alpha_n\gamma_n}\alpha_n]\|x_n-p\|^2 + \tfrac{[(1-2\delta)\delta_n-\gamma_n]\alpha_n}{1-\alpha_n\gamma_n} \cdot \{\tfrac{2\gamma_n}{(1-2\delta)\delta_n-\gamma_n}\|f(x_n)-p\|\|z_n-x_{n+1}\|\\ &\quad + \tfrac{2\delta_n}{(1-2\delta)\delta_n-\gamma_n}\|f(x_n)-p\|\|z_n-x_n\| + \tfrac{2\delta_n}{(1-2\delta)\delta_n-\gamma_n}\langle f(p)-p, x_n-p\rangle\\ &\quad + \tfrac{\gamma_n+\delta_n}{(1-2\delta)\delta_n-\gamma_n}(\tfrac{\theta_n}{\alpha_n}\cdot\tfrac{2M^2(1+b(1+\theta_n)^2)}{1-b} + \tfrac{\sigma_n}{\alpha_n}\|x_n-x_{n-1}\|\tfrac{3M(1+b\theta_n(2+\theta_n))}{1-b})\}.\end{aligned} \quad (29)$$

In fact, we get

$$\|w_n-p\|^2 \leq [(1+\theta_n)\|x_n-p\| + \sigma_n\|x_n-x_{n-1}\|]^2 \leq \|x_n-p\|^2 + \theta_n 2M^2 + \sigma_n\|x_n-x_{n-1}\|3M, \quad (30)$$

where $\exists M > 0$ s.t. $\sup_{n\geq 1}\{(1+\theta_n)\|x_n-p\|, \sigma_n\|x_n-x_{n-1}\|\} \leq M$. From Algorithm 4 and the convexity of $\|\cdot\|^2$, we have

$$\begin{aligned}\|x_{n+1}-p\|^2 &= \|\beta_n(T^n w_n-p) + \gamma_n(z_n-p) + \delta_n(Sz_n-p) + \gamma_n(u_n-z_n)\|^2\\ &\leq \beta_n\|T^n w_n-p\|^2 + (1-\beta_n)\|\tfrac{1}{1-\beta_n}[\gamma_n(z_n-p) + \delta_n(Sz_n-p)]\|^2\\ &\quad + 2\gamma_n\alpha_n\langle u_n-p, x_{n+1}-p\rangle + 2\gamma_n\alpha_n\langle p-f(x_n), x_{n+1}-p\rangle,\end{aligned}$$

which together with Lemma 4, leads to

$$\begin{aligned}\|x_{n+1}-p\|^2 &\leq \beta_n(1+\theta_n)^2\|w_n-p\|^2 + (1-\beta_n)\|z_n-p\|^2 + 2\gamma_n\alpha_n\|u_n-p\|\|x_{n+1}-p\|\\ &\quad + 2\gamma_n\alpha_n\langle p-f(x_n), x_{n+1}-p\rangle\\ &\leq \beta_n(1+\theta_n)^2(\|x_n-p\|^2 + \theta_n 2M^2 + \sigma_n\|x_n-x_{n-1}\|3M) + (1-\beta_n)[(1-\alpha_n)\|u_n-p\|^2\\ &\quad + 2\alpha_n\langle f(x_n)-p, z_n-p\rangle] + \gamma_n\alpha_n(\|u_n-p\|^2 + \|x_{n+1}-p\|^2)\\ &\quad + 2\gamma_n\alpha_n\langle p-f(x_n), x_{n+1}-p\rangle.\end{aligned}$$

By Step 3 of Algorithm 4, and from Equation (30) we know that $\|u_n - p\|^2 \leq \|x_n - p\|^2 + \theta_n 2M^2 + \sigma_n \|x_n - x_{n-1}\| 3M$. Hence, we have

$$\begin{aligned}
\|x_{n+1} - p\|^2 &\leq [1 - \alpha_n(1 - \beta_n)]\|x_n - p\|^2 + \beta_n \theta_n 2M^2 + (1 - \beta_n)(1 - \alpha_n)(\theta_n 2M^2 \\
&\quad + \sigma_n \|x_n - x_{n-1}\| 3M) + 2\alpha_n \delta_n \langle f(x_n) - p, z_n - p \rangle + \gamma_n \alpha_n (\|x_n - p\|^2 \\
&\quad + \|x_{n+1} - p\|^2) + (1 - \beta_n) \alpha_n (\theta_n 2M^2 + \sigma_n \|x_n - x_{n-1}\| 3M) \\
&\quad + 2\gamma_n \alpha_n \langle f(x_n) - p, z_n - x_{n+1} \rangle + \beta_n (1 + \theta_n)^2 (\theta_n 2M^2 + \sigma_n \|x_n - x_{n-1}\| 3M) \\
&\leq [1 - \alpha_n(1 - \beta_n)]\|x_n - p\|^2 + 2\gamma_n \alpha_n \|f(x_n) - p\| \|z_n - x_{n+1}\| \\
&\quad + 2\alpha_n \delta_n \langle f(x_n) - p, x_n - p \rangle + 2\alpha_n \delta_n \langle f(x_n) - p, z_n - x_n \rangle \\
&\quad + \gamma_n \alpha_n (\|x_n - p\|^2 + \|x_{n+1} - p\|^2) + (1 - \beta_n)[\theta_n \tfrac{2M^2(1+\beta_n(1+\theta_n)^2)}{1-\beta_n} \\
&\quad + \sigma_n \|x_n - x_{n-1}\| \tfrac{3M(1+\beta_n \theta_n(2+\theta_n))}{1-\beta_n}] \\
&\leq [1 - \alpha_n(1 - \beta_n)]\|x_n - p\|^2 + 2\gamma_n \alpha_n \|f(x_n) - p\| \|z_n - x_{n+1}\| \\
&\quad + 2\alpha_n \delta_n \delta \|x_n - p\|^2 + 2\alpha_n \delta_n \langle f(p) - p, x_n - p \rangle + 2\alpha_n \delta_n \|f(x_n) - p\| \|z_n - x_n\| \\
&\quad + \gamma_n \alpha_n (\|x_n - p\|^2 + \|x_{n+1} - p\|^2) + (1 - \beta_n)[\theta_n \tfrac{2M^2(1+b(1+\theta_n)^2)}{1-b} \\
&\quad + \sigma_n \|x_n - x_{n-1}\| \tfrac{3M(1+b\theta_n(2+\theta_n))}{1-b}],
\end{aligned}$$

which immediately yields Equation (29).

Step 4. We claim the strong convergence of $\{x_n\}$ to a unique solution $x^* \in \Omega$ of HVI Equation (12). In fact, using the same reasoning as in Step 4 of the proof of Theorem 1, we derive the desired conclusion. This completes the proof. □

Next, we shall show how to solve the VIP and CFPP in the following illustrating example.

The initial point $x_0 = x_1$ is randomly chosen in $\mathbf{R} = (-\infty, \infty)$. Take $f(x) = \frac{1}{4}\sin x$, $\gamma = l = \mu = \frac{1}{2}$, $\sigma_n = \alpha_n = \frac{1}{n+1}$, $\beta_n = \frac{1}{3}$, $\gamma_n = \frac{1}{6}$, and $\delta_n = \frac{1}{2}$. Then we know that $\delta = \frac{1}{4}$ and $f(\mathbf{R}) \subset [-\frac{1}{4}, \frac{1}{4}]$.

We first provide an example of Lipschitz continuous and pseudomonotone mapping A, asymptotically nonexpansive mapping T and strictly pseudocontractive mapping S with $\Omega = \text{Fix}(T) \cap \text{Fix}(S) \cap \text{VI}(C, A) \neq \emptyset$. Let $C = [-1.5, 1]$ and $H = \mathbf{R}$ with the inner product $\langle a, b \rangle = ab$ and induced norm $\|\cdot\| = |\cdot|$. Let $A, T, S : H \to H$ be defined as $Ax := \frac{1}{1+|\sin x|} - \frac{1}{1+|x|}$, $Tx := \frac{4}{5}\sin x$ and $Sx := \frac{1}{3}x + \frac{1}{2}\sin x$ for all $x \in H$. Now, we first show that A is pseudomonotone and Lipschitz continuous with $L = 2$ such that $A(C)$ is bounded. Indeed, it is clear that $A(C)$ is bounded. Moreover, for all $x, y \in H$ we have

$$\begin{aligned}
\|Ax - Ay\| &= \left|\tfrac{1}{1+\|\sin x\|} - \tfrac{1}{1+\|x\|} - \tfrac{1}{1+\|\sin y\|} + \tfrac{1}{1+\|y\|}\right| \\
&\leq \left|\tfrac{\|\sin y\| - \|\sin x\|}{(1+\|\sin x\|)(1+\|\sin y\|)}\right| + \left|\tfrac{\|y\| - \|x\|}{(1+\|x\|)(1+\|y\|)}\right| \\
&\leq \|\sin x - \sin y\| + \|x - y\| \leq 2\|x - y\|.
\end{aligned}$$

This implies that A is Lipschitz continuous with $L = 2$. Next, we show that A is pseudomonotone. For any given $x, y \in H$, it is clear that the relation holds:

$$\langle Ax, y - x \rangle = \left(\frac{1}{1+|\sin x|} - \frac{1}{1+|x|}\right)(y - x) \geq 0 \Rightarrow \langle Ay, y - x \rangle = \left(\frac{1}{1+|\sin y|} - \frac{1}{1+|y|}\right)(y - x) \geq 0.$$

Furthermore, it is easy to see that T is asymptotically nonexpansive with $\theta_n = (\frac{4}{5})^n$, $\forall n \geq 1$, such that $\|T^{n+1}x_n - T^n x_n\| \to 0$ as $n \to \infty$. Indeed, we observe that

$$\|T^n x - T^n y\| \leq \frac{4}{5}\|T^{n-1}x - T^{n-1}y\| \leq \cdots \leq \left(\frac{4}{5}\right)^n \|x - y\| \leq (1 + \theta_n)\|x - y\|,$$

and

$$\|T^{n+1}x_n - T^n x_n\| \leq (\tfrac{4}{5})^{n-1}\|T^2 x_n - T x_n\| = (\tfrac{4}{5})^{n-1}\|\tfrac{4}{5}\sin(Tx_n) - \tfrac{4}{5}\sin x_n\| \leq 2(\tfrac{4}{5})^n \to 0, \ (n \to \infty).$$

It is clear that Fix$(T) = \{0\}$ and

$$\lim_{n\to\infty} \frac{\theta_n}{\alpha_n} = \lim_{n\to\infty} \frac{(4/5)^n}{1/(n+1)} = 0.$$

Moreover, it is readily seen that $\sup_{n\geq 1} |(T^n - f)x_n| = \sup_{n\geq 1} |\tfrac{4}{5}\sin(T^{n-1}x_n) - \tfrac{1}{4}\sin x_n| \leq \tfrac{21}{20} < \infty$. In addition, it is clear that S is strictly pseudocontractive with constant $\zeta = \tfrac{1}{4}$. Indeed, we observe that for all $x, y \in H$,

$$\|Sx - Sy\|^2 \leq [\tfrac{1}{3}\|x - y\| + \tfrac{1}{2}\|\sin x - \sin y\|]^2 \leq \|x - y\|^2 + \tfrac{1}{4}\|(I - S)x - (I - S)y\|^2.$$

It is clear that $(\gamma_n + \delta_n)\zeta = (\tfrac{1}{6} + \tfrac{1}{2}) \cdot \tfrac{1}{4} \leq \tfrac{1}{6} = \gamma_n < (1 - 2\delta)\delta_n = (1 - 2 \cdot \tfrac{1}{4}) \cdot \tfrac{1}{2} = \tfrac{1}{4}$ for all $n \geq 1$. Therefore, $\Omega = \text{Fix}(T) \cap \text{Fix}(S) \cap \text{VI}(C, A) = \{0\} \neq \emptyset$. In this case, Algorithm 3 can be rewritten as follows:

$$\begin{cases} w_n = T^n x_n + \tfrac{1}{n+1}(x_n - x_{n-1}), \\ y_n = P_C(w_n - \tau_n A w_n), \\ z_n = \tfrac{1}{n+1} f(x_n) + \tfrac{n}{n+1} P_{C_n}(w_n - \tau_n A y_n), \\ x_{n+1} = \tfrac{1}{3} T^n x_n + \tfrac{1}{6} P_{C_n}(w_n - \tau_n A y_n) + \tfrac{1}{2} S z_n, \quad \forall n \geq 1, \end{cases}$$

where C_n and τ_n are picked up as in Algorithm 3. Thus, by Theorem 1, we know that $\{x_n\}$ converges to $0 \in \Omega$ if and only if $|x_n - x_{n+1}| + |x_n - T^n x_n| \to 0, \ (n \to \infty)$.

On the other hand, Algorithm 4 can be rewritten as follows:

$$\begin{cases} w_n = T^n x_n + \tfrac{1}{n+1}(x_n - x_{n-1}), \\ y_n = P_C(w_n - \tau_n A w_n), \\ z_n = \tfrac{1}{n+1} f(x_n) + \tfrac{n}{n+1} P_{C_n}(w_n - \tau_n A y_n), \\ x_{n+1} = \tfrac{1}{3} T^n w_n + \tfrac{1}{6} P_{C_n}(w_n - \tau_n A y_n) + \tfrac{1}{2} S z_n, \quad \forall n \geq 1, \end{cases}$$

where C_n and τ_n are picked up as in Algorithm 4. Thus, by Theorem 2, we know that $\{x_n\}$ converges to $0 \in \Omega$ if and only if $|x_n - x_{n+1}| + |x_n - T^n x_n| \to 0, \ (n \to \infty)$.

Author Contributions: The authors made equal contributions to this paper. Conceptualization, methodology, formal analysis and investigation: L.-C.C., A.P., C.-F.W. and J.-C.Y.; writing—original draft preparation: L.-C.C. and A.P.; writing—review and editing: C.-F.W. and J.-C.Y.

Funding: This research was partially supported by the Innovation Program of Shanghai Municipal Education Commission (15ZZ068), Ph.D. Program Foundation of Ministry of Education of China (20123127110002) and Program for Outstanding Academic Leaders in Shanghai City (15XD1503100). This research was also supported by the Ministry of Science and Technology, Taiwan [grant number: 107-2115-M-037-001].

Conflicts of Interest: The authors declare no conflict of interest.

References

1. Korpelevich, G.M. The extragradient method for finding saddle points and other problems. *Ekon. Mat. Metod.* **1976**, *12*, 747–756.
2. Bin Dehaish, B.A. Weak and strong convergence of algorithms for the sum of two accretive operators with applications. *J. Nonlinear Convex Anal.* **2015**, *16*, 1321–1336.

3. Bin Dehaish, B.A. A regularization projection algorithm for various problems with nonlinear mappings in Hilbert spaces. *J. Inequal. Appl.* **2015**, *2015*, 51. [CrossRef]
4. Ceng, L.C.; Ansari, Q.H.; Yao, J.C. Some iterative methods for finding fixed points and for solving constrained convex minimization problems. *Nonlinear Aanl.* **2011**, *74*, 5286–5302. [CrossRef]
5. Ceng, L.C.; Guu, S.M.; Yao, J.C. Finding common solutions of a variational inequality, a general system of variational inequalities, and a fixed-point problem via a hybrid extragradient method. *Fixed Point Theory Appl.* **2011**, *2011*, 626159. [CrossRef]
6. Ceng, L.C.; Ansari, Q.H.; Wong, N.C.; Yao, J.C. An extragradient-like approximation method for variational inequalities and fixed point problems. *Fixed Point Theory Appl.* **2011**, *2011*, 22. [CrossRef]
7. Liu, L.; Qin, X. Iterative methods for fixed points and zero points of nonlinear mappings with applications. *Optimization* **2019**. [CrossRef]
8. Nguyen, L.V.; Qin, X. Some results on strongly pseudomonotone quasi-variational inequalities. *Set-Valued Var. Anal.* **2019**. [CrossRef]
9. Ansari, Q.H.; Babu, F.; Yao, J.C. Regularization of proximal point algorithms in Hadamard manifolds. *J. Fixed Point Theory Appl.* **2019**, *21*, 25. [CrossRef]
10. Ceng, L.C.; Guu, S.M.; Yao, J.C. Hybrid iterative method for finding common solutions of generalized mixed equilibrium and fixed point problems. *Fixed Point Theory Appl.* **2012**, *2012*, 92. [CrossRef]
11. Takahashi, W.; Wen, C.F.; Yao, J.C. The shrinking projection method for a finite family of demimetric mappings with variational inequality problems in a Hilbert space. *Fixed Point Theory* **2018**, *19*, 407–419. [CrossRef]
12. Chang, S.S.; Wen, C.F.; Yao, J.C. Common zero point for a finite family of inclusion problems of accretive mappings in Banach spaces. *Optimization* **2018**, *67*, 1183–1196. [CrossRef]
13. Chang, S.S.; Wen, C.F.; Yao, J.C. Zero point problem of accretive operators in Banach spaces. *Bull. Malays. Math. Sci. Soc.* **2019**, *42*, 105–118. [CrossRef]
14. Zhao, X.; Ng, K.F.; Li, C.; Yao, J.C. Linear regularity and linear convergence of projection-based methods for solving convex feasibility problems. *Appl. Math. Optim.* **2018** *78*, 613–641. [CrossRef]
15. Latif, A.; Ceng, L.C.; Ansari, Q.H. Multi-step hybrid viscosity method for systems of variational inequalities defined over sets of solutions of an equilibrium problem and fixed point problems. *Fixed Point Theory Appl.* **2012**, *2012*, 186. [CrossRef]
16. Ceng, L.C.; Ansari, Q.H.; Yao, J.C. An extragradient method for solving split feasibility and fixed point problems. *Comput. Math. Appl.* **2012**, *64*, 633–642. [CrossRef]
17. Ceng, L.C.; Ansari, Q.H.; Yao, J.C. Relaxed extragradient methods for finding minimum-norm solutions of the split feasibility problem. *Nonlinear Anal.* **2012**, *75*, 2116–2125. [CrossRef]
18. Qin, X.; Cho, S.Y.; Wang, L. Strong convergence of an iterative algorithm involving nonlinear mappings of nonexpansive and accretive type. *Optimization* **2018**, *67*, 1377–1388. [CrossRef]
19. Cai, G.; Shehu, Y.; Iyiola, O.S. Strong convergence results for variational inequalities and fixed point problems using modified viscosity implicit rules. *Numer. Algorithms* **2018**, *77*, 535–558. [CrossRef]
20. Censor, Y.; Gibali, A.; Reich, S. The subgradient extragradient method for solving variational inequalities in Hilbert space. *J. Optim. Theory Appl.* **2011**, *148*, 318–335. [CrossRef]
21. Kraikaew, R.; Saejung, S. Strong convergence of the Halpern subgradient extragradient method for solving variational inequalities in Hilbert spaces. *J. Optim. Theory Appl.* **2014**, *163*, 399–412. [CrossRef]
22. Thong, D.V.; Hieu, D.V. Modified subgradient extragradient method for variational inequality problems. *Numer. Algorithms* **2018**, *79*, 597–610. [CrossRef]
23. Thong, D.V.; Hieu, D.V. Inertial subgradient extragradient algorithms with line-search process for solving variational inequality problems and fixed point problems. *Numer. Algorithms* **2019**, *80*, 1283–1307. [CrossRef]
24. Cho, S.Y.; Qin, X. On the strong convergence of an iterative process for asymptotically strict pseudocontractions and equilibrium problems. *Appl. Math. Comput.* **2014** *235*, 430–438. [CrossRef]
25. Ceng, L.C.; Yao, J.C. Relaxed and hybrid viscosity methods for general system of variational inequalities with split feasibility problem constraint. *Fixed Point Theory Appl.* **2013**, *2013*, 43. [CrossRef]

26. Ceng, L.C.; Petrusel, A.; Yao, J.C.; Yao, Y. Hybrid viscosity extragradient method for systems of variational inequalities, fixed points of nonexpansive mappings, zero points of accretive operators in Banach spaces. *Fixed Point Theory* **2018**, *19*, 487–502. [CrossRef]
27. Ceng, L.C.; Yuan, Q. Hybrid Mann viscosity implicit iteration methods for triple hierarchical variational inequalities, systems of variational inequalities and fixed point problems. *Mathematics* **2019**, *7*, 338. [CrossRef]
28. Ceng, L.C.; Petrusel, A.; Yao, J.C.; Yao, Y. Systems of variational inequalities with hierarchical variational inequality constraints for Lipschitzian pseudocontractions. *Fixed Point Theory* **2019**, *20*, 113–134. [CrossRef]
29. Ceng, L.C.; Latif, A.; Ansari, Q.H.; Yao, J.C. Hybrid extragradient method for hierarchical variational inequalities. *Fixed Point Theory Appl.* **2014**, *2014*, 222. [CrossRef]
30. Takahahsi, W.; Yao, J.C. The split common fixed point problem for two finite families of nonlinear mappings in Hilbert spaces. *J. Nonlinear Convex Anal.* **2019**, *20*, 173–195.
31. Ceng, L.C.; Latif, A.; Yao, J.C. On solutions of a system of variational inequalities and fixed point problems in Banach spaces. *Fixed Point Theory Appl.* **2013**, *2013*, 176. [CrossRef]
32. Ceng, L.C.; Shang, M. Hybrid inertial subgradient extragradient methods for variational inequalities and fixed point problems involving asymptotically nonexpansive mappings. *Optimization* **2019**. [CrossRef]
33. Yao, Y.; Liou, Y.C.; Kang, S.M. Approach to common elements of variational inequality problems and fixed point problems via a relaxed extragradient method. *Comput. Math. Appl.* **2010**, *59*, 3472–3480. [CrossRef]
34. Ceng, L.C.; Petruşel, A.; Yao, J.C. Composite viscosity approximation methods for equilibrium problem, variational inequality and common fixed points. *J. Nonlinear Convex Anal.* **2014**, *15*, 219–240.
35. Ceng, L.C.; Kong, Z.R.; Wen, C.F. On general systems of variational inequalities. *Comput. Math. Appl.* **2013**, *66*, 1514–1532. [CrossRef]
36. Ceng, L.C.; Ansari, Q.H.; Yao, J.C. Relaxed extragradient iterative methods for variational inequalities. *Appl. Math. Comput.* **2011**, *218*, 1112–1123. [CrossRef]
37. Ceng, L.C.; Wen, C.F.; Yao, Y. Iterative approaches to hierarchical variational inequalities for infinite nonexpansive mappings and finding zero points of m-accretive operators. *J. Nonlinear Var. Anal.* **2017**, *1*, 213–235.
38. Zaslavski, A.J. *Numerical Optimization with Computational Errors*; Springer: Cham, Switzerland, 2016.
39. Cottle, R.W.; Yao, J.C. Pseudo-monotone complementarity problems in Hilbert space. *J. Optim. Theory Appl.* **1992**, *75*, 281–295. [CrossRef]
40. Xu, H.K; Kim, T.H. Convergence of hybrid steepest-descent methods for variational inequalities. *J. Optim. Theory Appl.* **2003**, *119*, 185–201. [CrossRef]
41. Ceng, L.C.; Xu, H.K.; Yao, J.C. The viscosity approximation method for asymptotically nonexpansive mappings in Banach spaces. *Nonlinear Anal.* **2008**, *69*, 1402–1412. [CrossRef]

© 2019 by the authors. Licensee MDPI, Basel, Switzerland. This article is an open access article distributed under the terms and conditions of the Creative Commons Attribution (CC BY) license (http://creativecommons.org/licenses/by/4.0/).

Article

On Mann Viscosity Subgradient Extragradient Algorithms for Fixed Point Problems of Finitely Many Strict Pseudocontractions and Variational Inequalities

Lu-Chuan Ceng [1], Adrian Petruşel [2,3] and Jen-Chih Yao [4,*]

1. Department of Mathematics, Shanghai Normal University, Shanghai 200234, China; zenglc@shnu.edu.cn
2. Department of Mathematics, Babeş-Bolyai University, Cluj-Napoca 400084; petrusel@math.ubbcluj.ro
3. Academy of Romanian Scientists, Bucharest 050044, Romania
4. Research Center for Interneural Computing, China Medical University Hospital, Taichung 40447, Taiwan
* Correspondence: yaojc@mail.cmu.edu.tw

Received: 21 August 2019; Accepted: 26 September 2019; Published: 4 October 2019

Abstract: In a real Hilbert space, we denote CFPP and VIP as common fixed point problem of finitely many strict pseudocontractions and a variational inequality problem for Lipschitzian, pseudomonotone operator, respectively. This paper is devoted to explore how to find a common solution of the CFPP and VIP. To this end, we propose Mann viscosity algorithms with line-search process by virtue of subgradient extragradient techniques. The designed algorithms fully assimilate Mann approximation approach, viscosity iteration algorithm and inertial subgradient extragradient technique with line-search process. Under suitable assumptions, it is proven that the sequences generated by the designed algorithms converge strongly to a common solution of the CFPP and VIP, which is the unique solution to a hierarchical variational inequality (HVI).

Keywords: method with line-search process; pseudomonotone variational inequality; strictly pseudocontractive mappings; common fixed point; sequentially weak continuity

MSC: 47H05; 47H09; 47H10; 90C52

1. Introduction and Preliminaries

Throughout this article, we suppose that the real vector space H is a Hilbert one and the nonempty subset C of H is a convex and closed one. An operator $S : C \to H$ is called:

(i) L-Lipschitzian if there exists $L > 0$ such that $\|Su - Sv\| \leq L\|u - v\| \; \forall u, v \in C$;

(ii) sequentially weakly continuous if for any $\{w_n\} \subset C$, the following implication holds: $w_n \rightharpoonup w \Rightarrow Sw_n \rightharpoonup Sw$;

(iii) pseudomonotone if $\langle Su, u - v \rangle \leq 0 \Rightarrow \langle Sv, u - v \rangle \leq 0 \; \forall u, v \in C$;

(iv) monotone if $\langle Su - Sv, v - u \rangle \leq 0 \; \forall u, v \in C$;

(v) γ-strongly monotone if $\exists \gamma > 0$ s.t. $\langle Su - Sw, u - w \rangle \geq \gamma \|u - w\|^2 \; \forall u, w \in C$.

It is not difficult to observe that monotonicity ensures the pseudomonotonicity. A self-mapping $S : C \to C$ is called a η-strict pseudocontraction if the relation holds: $\langle Su - Sv, u - v \rangle \leq \|u - v\|^2 - \frac{1-\eta}{2}\|(I - S)u - (I - S)v\|^2 \; \forall u, v \in C$ for some $\eta \in [0, 1)$. By [1] we know that, in the case where S is η-strictly pseudocontractive, S is Lipschitzian, i.e., $\|Su - Sv\| \leq \frac{1+\eta}{1-\eta}\|u - v\| \; \forall u, v \in C$. It is clear that the class of strict pseudocontractions includes the class of nonexpansive operators, i.e., $\|Su - Sv\| \leq \|u - v\| \; \forall u, v \in C$. Both classes of nonlinear operators received much attention and many numerical algorithms were designed for calculating their fixed points in Hilbert or Banach spaces; see e.g., [2–11].

Let A be a self-mapping on H. The classical variational inequality problem (VIP) is to find $z \in C$ such that $\langle Az, y - z \rangle \geq 0 \ \forall y \in C$. The solution set of such a VIP is indicated by VI(C, A). To the best of our knowledge, one of the most effective methods for solving the VIP is the gradient-projection method. Recently, many authors numerically investigated the VIP in finite dimensional spaces, Hilbert spaces or Banach spaces; see e.g., [12–20].

In 2014, Kraikaew and Saejung [21] suggested a Halpern-type gradient-like algorithm to deal with the VIP

$$\begin{cases} v_k = P_C(u_k - \ell A u_k), \\ C_k = \{v \in H : \langle u_k - \ell A u_k - v_k, v_k - v \rangle \geq 0\}, \\ w_k = P_{C_k}(u_n - \ell A v_k), \\ u_{k+1} = \varrho_k u_0 + (1 - \varrho_k) w_k \quad \forall k \geq 0, \end{cases}$$

where $\ell \in (0, \frac{1}{L})$, $\{\varrho_k\} \subset (0,1)$, $\lim_{k \to \infty} \varrho_k = 0$, $\sum_{k=1}^{\infty} \varrho_k = +\infty$, and established strong convergence theorems for approximation solutions in Hilbert spaces. Later, Thong and Hieu [22] designed an inertial algorithm, i.e., for arbitrarily given $u_0, u_1 \in H$, the sequence $\{u_k\}$ is constructed by

$$\begin{cases} z_k = u_k + \varrho_k(u_k - u_{k-1}), \\ v_k = P_C(z_k - \ell A z_k), \\ C_k = \{v \in H : \langle z_k - \ell A z_k - v_k, v_k - v \rangle \geq 0\}, \\ u_{k+1} = P_{C_k}(z_n - \ell A v_k) \quad \forall k \geq 1, \end{cases}$$

with $\ell \in (0, \frac{1}{L})$. Under mild assumptions, they proved that $\{u_k\}$ converge weakly to a point of VI(C, A). Very recently, Thong and Hieu [23] suggested two inertial algorithms with linear-search process, to solve the VIP for Lipschitzian, monotone operator A and the FPP for a quasi-nonexpansive operator S satisfying a demiclosedness property in H. Under appropriate assumptions, they proved that the sequences constructed by the suggested algorithms converge weakly to a point of Fix(S) $\cap$ VI(C, A). Further research on common solutions problems, we refer the readers to [24–38].

In this paper, we first introduce Mann viscosity algorithms via subgradient extragradient techniques, and then establish some strong convergence theorems in Hilbert spaces. It is remarkable that our algorithms involve line-search process.

The following lemmas are useful for the convergence analysis of our algorithms in the sequel.

Lemma 1. [39] *Let the operator A be pseudomonotone and continuous on C. Given a point $w \in C$. Then the relation holds: $\langle Aw, w - y \rangle \leq 0 \ \forall y \in C \Leftrightarrow \langle Ay, w - y \rangle \leq 0 \ \forall y \in C$.*

Lemma 2. [40] *Suppose that $\{s_k\}$ is a sequence in $[0, +\infty)$ such that $s_{k+1} \leq t_k b_k + (1 - t_k) s_k \ \forall k \geq 1$, where $\{t_k\}$ and $\{b_k\}$ lie in real line $\mathbf{R} := (-\infty, \infty)$, such that:*
(a) $\{t_k\} \subset [0,1]$ and $\sum_{k=1}^{\infty} t_k = \infty$;
(b) $\limsup_{k \to \infty} b_k \leq 0$ or $\sum_{k=1}^{\infty} |t_k b_k| < \infty$. *Then $s_k \to 0$ as $k \to \infty$.*

From Ceng et al. [2] it is not difficult to find that the following lemmas hold.

Lemma 3. *Let Γ be an η-strictly pseudocontractive self-mapping on C. Then $I - \Gamma$ is demiclosed at zero.*

Lemma 4. *For $l = 1, ..., N$, let Γ_l be an η_l-strictly pseudocontractive self-mapping on C. Then for $l = 1, ..., N$, the mapping Γ_l is an η-strict pseudocontraction with $\eta = \max\{\eta_l : 1 \leq l \leq N\}$, such that*

$$\|\Gamma_l u - \Gamma_l v\| \leq \frac{1+\eta}{1-\eta} \|u - v\| \quad \forall u, v \in C.$$

Lemma 5. *Let Γ be an η-strictly pseudocontractive self-mapping on C. Given two reals $\gamma, \beta \in [0, +\infty)$. If $(\gamma + \beta)\eta \leq \gamma$, then $\|\gamma(u - v) + \beta(\Gamma u - \Gamma v)\| \leq (\gamma + \beta)\|u - v\| \ \forall u, v \in C$.*

2. Main Results

Our first algorithm is specified below.

Algorithm 1

Initial Step: Given $x_0, x_1 \in H$ arbitrarily. Let $\gamma > 0$, $l \in (0,1)$, $\mu \in (0,1)$.
Iteration Steps: Compute x_{n+1} below:
Step 1. Put $v_n = x_n - \sigma_n(x_{n-1} - x_n)$ and calculate $u_n = P_C(v_n - \ell_n A v_n)$, where ℓ_n is picked to be the largest $\ell \in \{\gamma, \gamma l, \gamma l^2, ...\}$ s.t.
$$\ell \|Av_n - Au_n\| \leq \mu \|v_n - u_n\|. \tag{1}$$
Step 2. Calculate $z_n = (1 - \alpha_n)P_{C_n}(v_n - \ell_n A u_n) + \alpha_n f(x_n)$ with $C_n := \{v \in H : \langle v_n - \ell_n A v_n - u_n, u_n - v \rangle \geq 0\}$.
Step 3. Calculate
$$x_{n+1} = \gamma_n P_{C_n}(v_n - \ell_n A u_n) + \delta_n T_n z_n + \beta_n x_n. \tag{2}$$
Update $n := n+1$ and return to Step 1.
In this section, we always suppose that the following hypotheses hold:
T_k is a ζ_k-strictly pseudocontractive self-mapping on H for $k = 1, ..., N$ s.t. $\zeta \in [0,1)$ with $\zeta = \max\{\zeta_k : 1 \leq k \leq N\}$.
A is L-Lipschitzian, pseudomonotone self-mapping on H, and sequentially weakly continuous on C, such that $\Omega := \cap_{k=1}^N \text{Fix}(T_k) \cap \text{VI}(C, A) \neq \emptyset$.
$f : H \to C$ is a δ-contraction with $\delta \in [0, \frac{1}{2})$.
$\{\sigma_n\} \subset [0,1]$ and $\{\alpha_n\}, \{\beta_n\}, \{\gamma_n\}, \{\delta_n\} \subset (0,1)$ are such that:
(i) $\beta_n + \gamma_n + \delta_n = 1$ and $\sup_{n \geq 1} \frac{\sigma_n}{\alpha_n} < \infty$;
(ii) $(1 - 2\delta)\delta_n > \gamma_n \geq (\gamma_n + \delta_n)\zeta$ $\forall n \geq 1$ and $\liminf_{n \to \infty}((1 - 2\delta)\delta_n - \gamma_n) > 0$;
(iii) $\lim_{n \to \infty} \alpha_n = 0$ and $\sum_{n=1}^\infty \alpha_n = \infty$;
(iv) $\liminf_{n \to \infty} \beta_n > 0$, $\liminf_{n \to \infty} \delta_n > 0$ and $\limsup_{n \to \infty} \beta_n < 1$.
Following Xu and Kim [40], we denote $T_n := T_{n \bmod N}$, $\forall n \geq 1$, where the mod function takes values in $\{1, 2, ..., N\}$, i.e., whenever $n = jN + q$ for some $j \geq 0$ and $0 \leq q < N$, we obtain that $T_n = T_N$ in the case of $q = 0$ and $T_n = T_q$ in the case of $0 < q < N$.

Lemma 6. *The Armijo-like search rule (1) is well defined, and $\min\{\gamma, \frac{\mu l}{L}\} \leq \ell_n \leq \gamma$.*

Proof. Obviously, (1) holds for all $\gamma l^m \leq \frac{\mu}{L}$. So, ℓ_n is well defined and $\ell_n \leq \gamma$. In the case of $\ell_n = \gamma$, the inequality is true. In the case of $\ell_n < \gamma$, (1) ensures $\|Av_n - AP_C(v_n - \frac{\ell_n}{l} Av_n)\| > \frac{\mu}{\ell_n/l} \|v_n - P_C(v_n - \frac{\ell_n}{l} Av_n)\|$. The L-Lipschitzian property of A yields $\ell_n > \frac{\mu l}{L}$. □

Lemma 7. *Let $\{v_n\}, \{u_n\}$ and $\{z_n\}$ be the sequences constructed by Algorithm 1. Then*
$$\begin{aligned}\|z_n - \omega\|^2 &\leq (1 - \alpha_n)\|v_n - \omega\|^2 + \alpha_n \delta \|x_n - \omega\|^2 - (1 - \alpha_n)(1 - \mu)[\|v_n - u_n\|^2 \\ &\quad + \|h_n - u_n\|^2] + 2\alpha_n \langle f\omega - \omega, z_n - \omega \rangle \quad \forall \omega \in \Omega,\end{aligned} \tag{3}$$
where $h_n := P_{C_n}(v_n - \ell_n A u_n)$ $\forall n \geq 1$.

Proof. First, taking an arbitrary $p \in \Omega \subset C \subset C_n$, we observe that
$$\begin{aligned} 2\|h_n - p\|^2 &\leq 2\langle h_n - p, v_n - \ell_n A u_n - p \rangle \\ &= \|h_n - p\|^2 + \|v_n - p\|^2 - \|h_n - v_n\|^2 - 2\langle \ell_n A u_n, h_n - p \rangle. \end{aligned}$$

So, it follows that $\|v_n - p\|^2 - 2\langle h_n - p, \ell_n A u_n \rangle - \|h_n - v_n\|^2 \geq \|h_n - p\|^2$, which together with (1), we deduce that $0 \geq \langle p - u_n, A u_n \rangle$ and
$$\begin{aligned} \|h_n - p\|^2 &\leq \|v_n - p\|^2 - \|h_n - v_n\|^2 + 2\ell_n(\langle A u_n, p - u_n \rangle + \langle A u_n, u_n - h_n \rangle) \\ &\leq \|v_n - p\|^2 - \|u_n - h_n\|^2 - \|v_n - u_n\|^2 + 2\langle u_n - v_n + \ell_n A u_n, u_n - h_n \rangle. \end{aligned} \tag{4}$$

Since $h_n = P_{C_n}(v_n - \ell_n A u_n)$ with $C_n := \{v \in H : \langle u_n - v_n + \ell_n A v_n, u_n - v\rangle \leq 0\}$, we have $\langle u_n - v_n + \ell_n A v_n, u_n - h_n\rangle \leq 0$, which together with (1), implies that

$$\begin{aligned} 2\langle u_n - v_n + \ell_n A u_n, u_n - h_n\rangle &= 2\langle u_n - v_n + \ell_n A v_n, u_n - h_n\rangle + 2\ell_n\langle A v_n - A u_n, h_n - u_n\rangle \\ &\leq 2\mu\|u_n - v_n\|\|u_n - h_n\| \leq \mu(\|v_n - u_n\|^2 + \|h_n - u_n\|^2). \end{aligned}$$

Therefore, substituting the last inequality for (4), we infer that

$$\|h_n - p\|^2 \leq \|v_n - p\|^2 - (1 - \mu)\|v_n - u_n\|^2 - (1 - \mu)\|h_n - u_n\|^2 \quad \forall p \in \Omega. \tag{5}$$

In addition, we have

$$z_n - p = (1 - \alpha_n)(h_n - p) + \alpha_n(f - I)p + \alpha_n(f(x_n) - f(p)).$$

Using the convexity of the function $h(t) = t^2 \; \forall t \in \mathbf{R}$, from (5) we get

$$\begin{aligned} \|z_n - p\|^2 &\leq [\alpha_n \delta \|x_n - p\| + (1 - \alpha_n)\|h_n - p\|]^2 + 2\alpha_n\langle(f - I)p, z_n - p\rangle \\ &\leq \alpha_n \delta \|x_n - p\|^2 + (1 - \alpha_n)\|h_n - p\|^2 + 2\alpha_n\langle(f - I)p, z_n - p\rangle \\ &\leq \alpha_n \delta \|x_n - p\|^2 + (1 - \alpha_n)\|v_n - p\|^2 - (1 - \alpha_n)(1 - \mu)[\|v_n - u_n\|^2 \\ &\quad + \|h_n - u_n\|^2] + 2\alpha_n\langle(f - I)p, z_n - p\rangle. \end{aligned}$$

□

Lemma 8. *Let $\{x_n\}, \{u_n\}$, and $\{z_n\}$ be bounded sequences constructed by Algorithm 1. If $x_n - x_{n+1} \to 0$, $v_n - u_n \to 0$, $v_n - z_n \to 0$ and $\exists \{v_{n_i}\} \subset \{v_n\}$ s.t. $v_{n_i} \rightharpoonup z \in H$, then $z \in \Omega$.*

Proof. According to Algorithm 1, we get $\sigma_n(x_n - x_{n-1}) = v_n - x_n \; \forall n \geq 1$, and hence $\|x_n - x_{n-1}\| \geq \|v_n - x_n\|$. Using the assumption $x_n - x_{n+1} \to 0$, we have

$$\lim_{n \to \infty} \|v_n - x_n\| = 0. \tag{6}$$

So,

$$\|z_n - x_n\| \leq \|v_n - z_n\| + \|v_n - x_n\| \to 0.$$

Since $\{x_n\}$ is bounded, from $v_n = x_n - \sigma_n(x_{n-1} - x_n)$ we know that $\{v_n\}$ is a bounded vector sequence. According to (5), we obtain that $h_n := P_{C_n}(v_n - \ell_n A u_n)$ is a bounded vector sequence. Also, by Algorithm 1 we get $\alpha_n f(x_n) + h_n - x_n - \alpha_n h_n = z_n - x_n$. So, the boundedness of $\{x_n\}, \{h_n\}$ guarantees that as $n \to \infty$,

$$\|h_n - x_n\| = \|z_n - x_n - \alpha_n f(x_n) + \alpha_n h_n\| \leq \|z_n - x_n\| + \alpha_n(\|f(x_n)\| + \|h_n\|) \to 0.$$

It follows that

$$x_{n+1} - z_n = \gamma_n(h_n - x_n) + \delta_n(T_n z_n - z_n) + (1 - \delta_n)(x_n - z_n),$$

which immediately yields

$$\begin{aligned} \delta_n\|T_n z_n - z_n\| &= \|x_{n+1} - x_n + x_n - z_n - (1 - \delta_n)(x_n - z_n) - \gamma_n(h_n - x_n)\| \\ &= \|x_{n+1} - x_n + \delta_n(x_n - z_n) - \gamma_n(h_n - x_n)\| \\ &\leq \|x_{n+1} - x_n\| + \|x_n - z_n\| + \|h_n - x_n\|. \end{aligned}$$

Since $x_n - x_{n+1} \to 0$, $z_n - x_n \to 0$, $h_n - x_n \to 0$ and $\liminf_{n\to\infty} \delta_n > 0$, we obtain $\|z_n - T_n z_n\| \to 0$ as $n \to \infty$. This further implies that

$$\begin{aligned}\|x_n - T_n x_n\| &\leq \|x_n - z_n\| + \|z_n - T_n z_n\| + \tfrac{1+\zeta}{1-\zeta}\|z_n - x_n\| \\ &\leq \tfrac{2}{1-\zeta}\|x_n - z_n\| + \|z_n - T_n z_n\| \to 0 \quad (n \to \infty).\end{aligned} \quad (7)$$

We have $\langle v_n - \ell_n A v_n - u_n, v - u_n\rangle \leq 0 \ \forall v \in C$, and

$$\langle v_n - u_n, v - u_n\rangle + \ell_n \langle A v_n, u_n - v_n\rangle \leq \ell_n \langle A v_n, v - v_n\rangle \quad \forall v \in C. \quad (8)$$

Note that $\ell_n \geq \min\{\gamma, \tfrac{\mu l}{L}\}$. So, $\liminf_{i\to\infty} \langle A v_{n_i}, v - v_{n_i}\rangle \geq 0 \ \forall v \in C$. This yields $\liminf_{i\to\infty}\langle A u_{n_i}, v - u_{n_i}\rangle \geq 0 \ \forall v \in C$. Since $v_n - x_n \to 0$ and $v_{n_i} \rightharpoonup z$, we get $x_{n_i} \rightharpoonup z$. We may assume $k = n_i \bmod N$ for all i. By the assumption $x_n - x_{n+k} \to 0$, we have $x_{n_i+j} \rightharpoonup z$ for all $j \geq 1$. Hence, $\|x_{n_i+j} - T_{k+j} x_{n_i+j}\| = \|x_{n_i+j} - T_{n_i+j} x_{n_i+j}\| \to 0$. Then the demiclosedness principle implies that $z \in \text{Fix}(T_{k+j})$ for all j. This ensures that

$$z \in \bigcap_{k=1}^{N} \text{Fix}(T_k). \quad (9)$$

We now take a sequence $\{\varsigma_i\} \subset (0,1)$ satisfying $\varsigma_i \downarrow 0$ as $i \to \infty$. For all $i \geq 1$, we denote by m_i the smallest natural number satisfying

$$\langle A u_{n_j}, v - u_{n_j}\rangle + \varsigma_i \geq 0 \quad \forall j \geq m_i. \quad (10)$$

Since $\{\varsigma_i\}$ is decreasing, it is clear that $\{m_i\}$ is increasing. Noticing that $\{u_{m_i}\} \subset C$ ensures $A u_{m_i} \neq 0 \ \forall i \geq 1$, we set $e_{m_i} = \frac{A u_{m_i}}{\|A u_{m_i}\|^2}$, we get $\langle A u_{m_i}, e_{m_i}\rangle = 1 \ \forall i \geq 1$. So, from (10) we get $\langle A u_{m_i}, v + \varsigma_i e_{m_i} - u_{m_i}\rangle \geq 0 \ \forall i \geq 1$. Also, the pseudomonotonicity of A implies $\langle A(v + \varsigma_i e_{m_i}), v + \varsigma_i e_{m_i} - u_{m_i}\rangle \geq 0 \ \forall i \geq 1$. This immediately leads to

$$\langle Av - A(v + \varsigma_i h_{m_i}), v + \varsigma_i e_{m_i} - u_{m_i}\rangle - \varsigma_i \langle Av, h_{m_i}\rangle \leq \langle Av, v - u_{m_i}\rangle \quad \forall i \geq 1. \quad (11)$$

We claim $\lim_{i\to\infty} \varsigma_i e_{m_i} = 0$. Indeed, from $v_{n_i} \rightharpoonup z$ and $v_n - u_n \to 0$, we obtain $u_{n_i} \rightharpoonup z$. So, $\{u_n\} \subset C$ ensures $z \in C$. Also, the sequentially weak continuity of A guarantees that $A u_{n_i} \rightharpoonup Az$. Thus, we have $Az \neq 0$ (otherwise, z is a solution). Moreover, the sequentially weak lower semicontinuity of $\|\cdot\|$ ensures $0 < \|Az\| \leq \liminf_{i\to\infty} \|A u_{n_i}\|$. Since $\{u_{m_i}\} \subset \{u_{n_i}\}$ and $\varsigma_i \downarrow 0$ as $i \to \infty$, we deduce that $0 \leq \limsup_{i\to\infty}\|\varsigma_i e_{m_i}\| = \limsup_{i\to\infty} \frac{\varsigma_i}{\|A u_{m_i}\|} \leq \frac{\limsup_{i\to\infty} \varsigma_i}{\liminf_{i\to\infty} \|A u_{n_i}\|} = 0$. Hence we get $\varsigma_i e_{m_i} \to 0$.

Finally we claim $z \in \Omega$. In fact, letting $i \to \infty$, we conclude that the right hand side of (11) tends to zero by the Lipschitzian property of A, the boundedness of $\{u_{m_i}\}$, $\{h_{m_i}\}$ and the limit $\lim_{i\to\infty} \varsigma_i e_{m_i} = 0$. Thus, we get $\langle Av, v - z\rangle = \liminf_{i\to\infty} \langle Av, v - u_{m_i}\rangle \geq 0 \ \forall v \in C$. So, $z \in \text{VI}(C, A)$. Therefore, from (9) we have $z \in \bigcap_{k=1}^{N} \text{Fix}(T_k) \cap \text{VI}(C, A) = \Omega$. $\square$

Theorem 1. *Assume $A(C)$ is bounded. Let $\{x_n\}$ be constructed by Algorithm 1. Then*

$$x_n \to x^* \in \Omega \iff \begin{cases} x_n - x_{n+1} \to 0, \\ \sup_{n \geq 1} \|x_n - f x_n\| < \infty \end{cases}$$

where $x^* \in \Omega$ is the unique solution to the hierarchical variational inequality (HVI): $\langle (I - f)x^*, x^* - \omega\rangle \leq 0, \ \forall \omega \in \Omega$.

Proof. Taking into account condition (iv) on $\{\gamma_n\}$, we may suppose that $\{\beta_n\} \subset [a,b] \subset (0,1)$. Applying Banach's Contraction Principle, we obtain existence and uniqueness of a fixed point $x^* \in H$ for the mapping $P_\Omega \circ f$, which means that $x^* = P_\Omega f(x^*)$. Hence, the HVI

$$\langle (I-f)x^*, x^* - \omega \rangle \leq 0, \quad \forall \omega \in \Omega \tag{12}$$

has a unique solution $x^* \in \Omega := \cap_{k=1}^N \text{Fix}(T_k) \cap \text{VI}(C,A)$.

It is now obvious that the necessity of the theorem is true. In fact, if $x_n \to x^* \in \Omega$, then we get $\sup_{n \geq 1} \|x_n - f(x_n)\| \leq \sup_{n \geq 1} (\|x_n - x^*\| + \|x^* - f(x^*)\| + \|f(x^*) - f(x_n)\|) < \infty$ and

$$\|x_n - x_{n+1}\| \leq \|x_n - x^*\| + \|x_{n+1} - x^*\| \to 0 \quad (n \to \infty).$$

For the sufficient condition, let us suppose $x_n - x_{n+1} \to 0$ and $\sup_{n \geq 1} \|(I-f)x_n\| < \infty$. The sufficiency of our conclusion is proved in the following steps. □

Step 1. We show the boundedness of $\{x_n\}$. In fact, let p be an arbitrary point in Ω. Then $T_n p = p \ \forall n \geq 1$, and

$$\|v_n - p\|^2 - (1-\mu)\|h_n - u_n\|^2 - (1-\mu)\|v_n - u_n\|^2 \geq \|h_n - p\|^2, \tag{13}$$

which hence leads to

$$\|v_n - p\| \geq \|h_n - p\| \quad \forall n \geq 1. \tag{14}$$

By the definition of v_n, we have

$$\|v_n - p\| \leq \|x_n - p\| + \sigma_n \|x_n - x_{n-1}\| \leq \|x_n - p\| + \alpha_n \cdot \frac{\sigma_n}{\alpha_n} \|x_n - x_{n-1}\|. \tag{15}$$

Noticing $\sup_{n \geq 1} \frac{\sigma_n}{\alpha_n} < \infty$ and $\sup_{n \geq 1} \|x_n - x_{n-1}\| < \infty$, we obtain that $\sup_{n \geq 1} \frac{\sigma_n}{\alpha_n} \|x_n - x_{n-1}\| < \infty$. This ensures that $\exists M_1 > 0$ s.t.

$$\frac{\sigma_n}{\alpha_n} \|x_n - x_{n-1}\| \leq M_1 \quad \forall n \geq 1. \tag{16}$$

Combining (14)–(16), we get

$$\|h_n - p\| \leq \|v_n - p\| \leq \|x_n - p\| + \alpha_n M_1 \quad \forall n \geq 1. \tag{17}$$

Note that $A(C)$ is bounded, $u_n = P_C(v_n - \ell_n A v_n)$, $f(H) \subset C \subset C_n$ and $h_n = P_{C_n}(v_n - \ell_n A u_n)$. Hence we know that $\{Au_n\}$ is bounded. So, from $\sup_{n \geq 1} \|(I-f)x_n\| < \infty$, it follows that

$$\begin{aligned}\|h_n - f(x_n)\| &\leq \|v_n - \ell_n A u_n - f(x_n)\| \\ &\leq \|x_n - x_{n-1}\| + \|x_n - f(x_n)\| + \gamma \|Au_n\| \leq M_0,\end{aligned}$$

where $\exists M_0 > 0$ s.t. $M_0 \geq \sup_{n \geq 1}(\|x_n - x_{n-1}\| + \|x_n - f(x_n)\| + \gamma \|Au_n\|)$ (due to the assumption $x_n - x_{n+1} \to 0$). Consequently,

$$\begin{aligned}\|z_n - p\| &\leq \alpha_n \delta \|x_n - p\| + (1-\alpha_n)\|h_n - p\| + \alpha_n \|(f-I)p\| \\ &\leq (1-\alpha_n(1-\delta))\|x_n - p\| + \alpha_n(M_1 + \|(f-I)p\|),\end{aligned}$$

which together with $(\gamma_n + \delta_n)\zeta \leq \gamma_n$, yields

$$\begin{aligned}\|x_{n+1} - p\| &\leq \beta_n \|x_n - p\| + (1-\beta_n)\|\tfrac{1}{1-\beta_n}[\gamma_n(z_n - p) + \delta_n(T_n z_n - p)]\| + \gamma_n \|h_n - z_n\| \\ &\leq \beta_n \|x_n - p\| + (1-\beta_n)[(1-\alpha_n(1-\delta))\|x_n - p\| + \alpha_n(M_0 + M_1 + \|(f-I)p\|)] \\ &= [1 - \alpha_n(1-\beta_n)(1-\delta)]\|x_n - p\| + \alpha_n(1-\beta_n)(1-\delta)\tfrac{M_0 + M_1 + \|(f-I)p\|}{1-\delta}.\end{aligned}$$

This shows that $\|x_n - p\| \leq \max\{\|x_1 - p\|, \frac{M_0 + M_1 + \|(I-f)p\|}{1-\delta}\}$ $\forall n \geq 1$. Thus, $\{x_n\}$ is bounded, and so are the sequences $\{h_n\}, \{v_n\}, \{u_n\}, \{z_n\}, \{T_n z_n\}$.

Step 2. We show that $\exists M_4 > 0$ s.t.

$$(1-\alpha_n)(1-\beta_n)(1-\mu)[\|w_n - y_n\|^2 + \|u_n - y_n\|^2] \leq \|x_n - p\|^2 - \|x_{n+1} - p\|^2 + \alpha_n M_4.$$

In fact, using Lemma 7 and the convexity of $\|\cdot\|^2$, we get

$$\begin{aligned}
\|x_{n+1} - p\|^2 &\leq \|\beta_n(x_n - p) + \gamma_n(z_n - p) + \delta_n(T_n z_n - p)\|^2 + 2\gamma_n \alpha_n \langle h_n - f(x_n), x_{n+1} - p\rangle \\
&\leq \beta_n \|x_n - p\|^2 + (1-\beta_n)\|z_n - p\|^2 + 2(1-\beta_n)\alpha_n \|h_n - f(x_n)\|\|x_{n+1} - p\| \\
&\leq \beta_n \|x_n - p\|^2 + (1-\beta_n)\{\alpha_n \delta \|x_n - p\|^2 + (1-\alpha_n)\|v_n - p\|^2 \\
&\quad - (1-\alpha_n)(1-\mu)[\|v_n - u_n\|^2 + \|h_n - u_n\|^2] + \alpha_n M_2\},
\end{aligned} \tag{18}$$

where $\exists M_2 > 0$ s.t. $M_2 \geq \sup_{n\geq 1} 2(\|(f-I)p\|\|z_n - p\| + \|u_n - f(x_n)\|\|x_{n+1} - p\|)$. Also,

$$\begin{aligned}
\|v_n - p\|^2 &\leq \|x_n - p\|^2 + \alpha_n(2M_1 \|x_n - p\| + \alpha_n M_1^2) \\
&\leq \|x_n - p\|^2 + \alpha_n M_3,
\end{aligned} \tag{19}$$

where $\exists M_3 > 0$ s.t. $M_3 \geq \sup_{n\geq 1}(2M_1 \|x_n - p\| + \beta_n M_1^2)$. Substituting (19) for (18), we have

$$\begin{aligned}
\|x_{n+1} - p\|^2 &\leq \beta_n \|x_n - p\|^2 + (1-\beta_n)\{(1-\alpha_n(1-\delta))\|x_n - p\|^2 + (1-\alpha_n)\alpha_n M_3 \\
&\quad - (1-\alpha_n)(1-\mu)[\|v_n - u_n\|^2 + \|h_n - u_n\|^2] + \alpha_n M_2\} \\
&\leq \|x_n - p\|^2 - (1-\alpha_n)(1-\beta_n)(1-\mu)[\|v_n - u_n\|^2 + \|h_n - u_n\|^2] + \alpha_n M_4,
\end{aligned} \tag{20}$$

where $M_4 := M_2 + M_3$. This immediately implies that

$$(1-\alpha_n)(1-\beta_n)(1-\mu)[\|v_n - u_n\|^2 + \|h_n - u_n\|^2] \leq \|x_n - p\|^2 - \|x_{n+1} - p\|^2 + \alpha_n M_4. \tag{21}$$

Step 3. We show that $\exists M > 0$ s.t.

$$\begin{aligned}
\|x_{n+1} - p\|^2 &\leq [1 - \frac{(1-2\delta)\delta_n - \gamma_n}{1-\alpha_n \gamma_n}\alpha_n]\|x_n - p\|^2 + \frac{[(1-2\delta)\delta_n - \gamma_n]\alpha_n}{1-\alpha_n \gamma_n} \cdot \{\frac{2\gamma_n}{(1-2\delta)\delta_n - \gamma_n}\|f(x_n) - p\|\|z_n - x_{n+1}\| \\
&\quad + \frac{2\delta_n}{(1-2\delta)\delta_n - \gamma_n}\|f(x_n) - p\|\|z_n - x_n\| + \frac{2\delta_n}{(1-2\delta)\delta_n - \gamma_n}\langle f(p) - p, x_n - p\rangle \\
&\quad + \frac{\gamma_n + \delta_n}{(1-2\delta)\delta_n - \gamma_n} \cdot \frac{\sigma_n}{\alpha_n}\|x_n - x_{n-1}\|3M\}.
\end{aligned}$$

In fact, we get

$$\begin{aligned}
\|v_n - p\|^2 &\leq \|x_n - p\|^2 + \sigma_n \|x_n - x_{n-1}\|(2\|x_n - p\| + \sigma_n \|x_n - x_{n-1}\|) \\
&\leq \|x_n - p\|^2 + \sigma_n \|x_n - x_{n-1}\|3M,
\end{aligned} \tag{22}$$

where $\exists M > 0$ s.t. $M \geq \sup_{n\geq 1}\{\|x_n - p\|, \sigma_n \|x_n - x_{n-1}\|\}$. By Algorithm 1 and the convexity of $\|\cdot\|^2$, we have

$$\begin{aligned}
\|x_{n+1} - p\|^2 &\leq \|\beta_n(x_n - p) + \gamma_n(z_n - p) + \delta_n(T_n z_n - p)\|^2 + 2\gamma_n \alpha_n \langle h_n - f(x_n), x_{n+1} - p\rangle \\
&\leq \beta_n \|x_n - p\|^2 + (1-\beta_n)\|\frac{1}{1-\beta_n}[\gamma_n(z_n - p) + \delta_n(T_n z_n - p)]\|^2 \\
&\quad + 2\gamma_n \alpha_n \langle h_n - p, x_{n+1} - p\rangle + 2\gamma_n \alpha_n \langle p - f(x_n), x_{n+1} - p\rangle,
\end{aligned}$$

which leads to

$$\begin{aligned}
\|x_{n+1} - p\|^2 &\leq \beta_n \|x_n - p\|^2 + (1-\beta_n)[(1-\alpha_n)\|h_n - p\|^2 + 2\alpha_n \langle f(x_n) - p, z_n - p\rangle] \\
&\quad + \gamma_n \alpha_n (\|h_n - p\|^2 + \|x_{n+1} - p\|^2) + 2\gamma_n \alpha_n \langle p - f(x_n), x_{n+1} - p\rangle.
\end{aligned}$$

Using (17) and (22) we obtain that $\|h_n - p\|^2 \le \|x_n - p\|^2 + \sigma_n \|x_n - x_{n-1}\| 3M$. Hence,

$$\begin{aligned}
\|x_{n+1} - p\|^2 &\le [1 - \alpha_n(1-\beta_n)]\|x_n - p\|^2 + (1-\beta_n)(1-\alpha_n)\sigma_n\|x_n - x_{n-1}\|3M \\
&\quad + 2\alpha_n\delta_n\langle f(x_n) - p, z_n - p\rangle + \gamma_n\alpha_n(\|x_n - p\|^2 + \|x_{n+1} - p\|^2) \\
&\quad + (1-\beta_n)\alpha_n\sigma_n\|x_n - x_{n-1}\|3M + 2\gamma_n\alpha_n\langle f(x_n) - p, z_n - x_{n+1}\rangle \\
&\le [1 - \alpha_n(1-\beta_n)]\|x_n - p\|^2 + 2\gamma_n\alpha_n\|f(x_n) - p\|\|z_n - x_{n+1}\| \\
&\quad + 2\alpha_n\delta_n\langle f(x_n) - p, x_n - p\rangle + 2\alpha_n\delta_n\langle f(x_n) - p, z_n - x_n\rangle \\
&\quad + \gamma_n\alpha_n(\|x_n - p\|^2 + \|x_{n+1} - p\|^2) + (1-\beta_n)\sigma_n\|x_n - x_{n-1}\|3M \\
&\le [1 - \alpha_n(1-\beta_n)]\|x_n - p\|^2 + 2\gamma_n\alpha_n\|f(x_n) - p\|\|z_n - x_{n+1}\| \\
&\quad + 2\alpha_n\delta_n\delta\|x_n - p\|^2 + 2\alpha_n\delta_n\langle f(p) - p, x_n - p\rangle + 2\alpha_n\delta_n\|f(x_n) - p\|\|z_n - x_n\| \\
&\quad + \gamma_n\alpha_n(\|x_n - p\|^2 + \|x_{n+1} - p\|^2) + (1-\beta_n)\sigma_n\|x_n - x_{n-1}\|3M,
\end{aligned}$$

which immediately yields

$$\begin{aligned}
\|x_{n+1} - p\|^2 &\le [1 - \tfrac{(1-2\delta)\delta_n - \gamma_n}{1 - \alpha_n\gamma_n}\alpha_n]\|x_n - p\|^2 + \tfrac{[(1-2\delta)\delta_n - \gamma_n]\alpha_n}{1 - \alpha_n\gamma_n} \cdot \{\tfrac{2\gamma_n}{(1-2\delta)\delta_n - \gamma_n}\|f(x_n) - p\|\|z_n - x_{n+1}\| \\
&\quad + \tfrac{2\delta_n}{(1-2\delta)\delta_n - \gamma_n}\|f(x_n) - p\|\|z_n - x_n\| + \tfrac{2\delta_n}{(1-2\delta)\delta_n - \gamma_n}\langle f(p) - p, x_n - p\rangle \\
&\quad + \tfrac{\gamma_n + \delta_n}{(1-2\delta)\delta_n - \gamma_n} \cdot \tfrac{\sigma_n}{\alpha_n}\|x_n - x_{n-1}\|3M\}.
\end{aligned} \tag{23}$$

Step 4. We show that $x_n \to x^* \in \Omega$, where x^* is the unique solution of (12). Indeed, putting $p = x^*$, we infer from (23) that

$$\begin{aligned}
\|x_{n+1} - x^*\|^2 &\le [1 - \tfrac{(1-2\delta)\delta_n - \gamma_n}{1 - \alpha_n\gamma_n}\alpha_n]\|x_n - x^*\|^2 + \tfrac{[(1-2\delta)\delta_n - \gamma_n]\alpha_n}{1 - \alpha_n\gamma_n} \cdot \{\tfrac{2\gamma_n}{(1-2\delta)\delta_n - \gamma_n}\|f(x_n) - x^*\|\|z_n - x_{n+1}\| \\
&\quad + \tfrac{2\delta_n}{(1-2\delta)\delta_n - \gamma_n}\|f(x_n) - x^*\|\|z_n - x_n\| + \tfrac{2\delta_n}{(1-2\delta)\delta_n - \gamma_n}\langle f(x^*) - x^*, x_n - x^*\rangle \\
&\quad + \tfrac{\gamma_n + \delta_n}{(1-2\delta)\delta_n - \gamma_n} \cdot \tfrac{\sigma_n}{\alpha_n}\|x_n - x_{n-1}\|3M\}.
\end{aligned} \tag{24}$$

It is sufficient to show that $\limsup_{n\to\infty}\langle (f - I)x^*, x_n - x^*\rangle \le 0$. From (21), $x_n - x_{n+1} \to 0$, $\alpha_n \to 0$ and $\{\beta_n\} \subset [a, b] \subset (0, 1)$, we get

$$\begin{aligned}
&\limsup_{n\to\infty}(1 - \alpha_n)(1 - b)(1 - \mu)[\|v_n - u_n\|^2 + \|h_n - u_n\|^2] \\
&\le \limsup_{n\to\infty}[(\|x_n - p\| + \|x_{n+1} - p\|)\|x_n - x_{n+1}\| + \alpha_n M_4] = 0.
\end{aligned}$$

This ensures that

$$\lim_{n\to\infty}\|v_n - u_n\| = 0 \quad \text{and} \quad \lim_{n\to\infty}\|h_n - u_n\| = 0. \tag{25}$$

Consequently,

$$\|x_n - u_n\| \le \|x_n - v_n\| + \|v_n - u_n\| \to 0 \quad (n \to \infty).$$

Since $z_n = \alpha_n f(x_n) + (1 - \alpha_n)h_n$ with $h_n := P_{C_n}(v_n - \ell_n A u_n)$, we get

$$\begin{aligned}
\|z_n - u_n\| &= \|\alpha_n f(x_n) - \alpha_n h_n + h_n - u_n\| \\
&\le \alpha_n(\|f(x_n)\| + \|h_n\|) + \|h_n - u_n\| \to 0 \quad (n \to \infty),
\end{aligned} \tag{26}$$

and hence

$$\|z_n - x_n\| \le \|z_n - u_n\| + \|u_n - x_n\| \to 0 \quad (n \to \infty). \tag{27}$$

Obviously, combining (25) and (26), guarantees that

$$\|v_n - z_n\| \le \|v_n - u_n\| + \|u_n - z_n\| \to 0 \quad (n \to \infty).$$

From the boundedness of $\{x_n\}$, it follows that $\exists\{x_{n_i}\} \subset \{x_n\}$ s.t.

$$\limsup_{n\to\infty}\langle(f-I)x^*, x_n - x^*\rangle = \lim_{i\to\infty}\langle(f-I)x^*, x_{n_i} - x^*\rangle. \tag{28}$$

Since $\{x_n\}$ is bounded, we may suppose that $x_{n_i} \rightharpoonup \tilde{x}$. Hence from (28) we get

$$\limsup_{n\to\infty}\langle(f-I)x^*, x_n - x^*\rangle = \lim_{i\to\infty}\langle(f-I)x^*, x_{n_i} - x^*\rangle = \langle(f-I)x^*, \tilde{x} - x^*\rangle. \tag{29}$$

It is easy to see from $v_n - x_n \to 0$ and $x_{n_i} \rightharpoonup \tilde{x}$ that $v_{n_i} \rightharpoonup \tilde{x}$. Since $x_n - x_{n+1} \to 0$, $v_n - u_n \to 0$, $v_n - z_n \to 0$ and $v_{n_i} \rightharpoonup \tilde{x}$, by Lemma 8 we infer that $\tilde{x} \in \Omega$. Therefore, from (12) and (29) we conclude that

$$\limsup_{n\to\infty}\langle(f-I)x^*, x_n - x^*\rangle = \langle(f-I)x^*, \tilde{x} - x^*\rangle \leq 0. \tag{30}$$

Note that $\liminf_{n\to\infty} \frac{(1-2\delta)\delta_n - \gamma_n}{1-\alpha_n\gamma_n} > 0$. It follows that $\sum_{n=0}^{\infty} \frac{(1-2\delta)\delta_n - \gamma_n}{1-\alpha_n\gamma_n}\alpha_n = \infty$. It is clear that

$$\limsup_{n\to\infty}\{\frac{2\gamma_n}{(1-2\delta)\delta_n - \gamma_n}\|f(x_n) - x^*\|\|z_n - x_{n+1}\| + \frac{2\delta_n}{(1-2\delta)\delta_n - \gamma_n}\|f(x_n) - x^*\|\|z_n - x_n\|$$
$$+ \frac{2\delta_n}{(1-2\delta)\delta_n - \gamma_n}\langle f(x^*) - x^*, x_n - x^*\rangle + \frac{\gamma_n + \delta_n}{(1-2\delta)\delta_n - \gamma_n} \cdot \frac{\sigma_n}{\alpha_n}\|x_n - x_{n-1}\|3M\} \leq 0. \tag{31}$$

Therefore, by Lemma 2 we immediately deduce that $x_n \to x^*$.

Next, we introduce another Mann viscosity algorithm with line-search process by the subgradient extragradient technique.

Algorithm 2

Initial Step: Given $x_0, x_1 \in H$ arbitrarily. Let $\gamma > 0$, $l \in (0,1)$, $\mu \in (0,1)$.
Iteration Steps: Compute x_{n+1} below:
Step 1. Put $v_n = x_n - \sigma_n(x_{n-1} - x_n)$ and calculate $u_n = P_C(v_n - \ell_n A v_n)$, where ℓ_n is picked to be the largest $\ell \in \{\gamma, \gamma l, \gamma l^2, ...\}$ s.t.

$$\ell\|Av_n - Au_n\| \leq \mu\|v_n - u_n\|. \tag{32}$$

Step 2. Calculate $z_n = (1-\alpha_n)P_{C_n}(v_n - \ell_n Au_n) + \alpha_n f(x_n)$ with $C_n := \{v \in H : \langle v_n - \ell_n Av_n - u_n, u_n - v\rangle \geq 0\}$.
Step 3. Calculate

$$x_{n+1} = \gamma_n P_{C_n}(v_n - \ell_n Au_n) + \delta_n T_n z_n + \beta_n v_n. \tag{33}$$

Update $n := n+1$ and return to Step 1.

It is remarkable that Lemmas 6, 7 and 8 remain true for Algorithm 2.

Theorem 2. *Assume $A(C)$ is bounded. Let $\{x_n\}$ be constructed by Algorithm 2. Then*

$$x_n \to x^* \in \Omega \iff \begin{cases} x_n - x_{n+1} \to 0, \\ \sup_{n\geq 1}\|(I-f)x_n\| < \infty \end{cases}$$

where $x^* \in \Omega$ is the unique solution of the HVI: $\langle(I-f)x^*, x^* - \omega\rangle \leq 0, \forall \omega \in \Omega$.

Proof. For the necessity of our proof, we can observe that, by a similar approach to that in the proof of Theorem 1, we obtain that there is a unique solution $x^* \in \Omega$ of (12).

We show the sufficiency below. To this aim, we suppose $x_n - x_{n+1} \to 0$ and $\sup_{n\geq 1}\|(I-f)x_n\| < \infty$, and prove the sufficiency by the following steps. □

Step 1. We show the boundedness of $\{x_n\}$. In fact, by the similar inference to that in Step 1 for the proof of Theorem 1, we obtain that (13)–(17) hold. So, using Algorithm 2 and (17) we obtain

$$\|z_n - p\| \leq (1 - \alpha_n(1-\delta))\|x_n - p\| + \alpha_n(M_1 + \|(f-I)p\|),$$

which together with $(\gamma_n + \delta_n)\zeta \leq \gamma_n$, yields

$$\begin{aligned}
\|x_{n+1} - p\| &\leq \beta_n\|v_n - p\| + (1-\beta_n)\|\tfrac{1}{1-\beta_n}[\gamma_n(z_n - p) + \delta_n(T_n z_n - p)]\| + \gamma_n\|h_n - z_n\| \\
&\leq \beta_n(\|x_n - p\| + \alpha_n M_1) + (1-\beta_n)[(1-\alpha_n(1-\delta))\|x_n - p\| \\
&\quad + \alpha_n(M_0 + M_1 + \|(f-I)p\|)] \\
&= [1 - \alpha_n(1-\beta_n)(1-\delta)]\|x_n - p\| + \alpha_n(1-\beta_n)(1-\delta)\tfrac{M_0 + \tfrac{1}{1-\beta_n}M_1 + \|(f-I)p\|}{1-\delta}.
\end{aligned}$$

Therefore, we get the boundedness of $\{x_n\}$ and hence the one of sequences $\{h_n\}, \{v_n\}, \{u_n\}, \{z_n\}, \{T_n z_n\}$.

Step 2. We show that $\exists M_4 > 0$ s.t.

$$(1-\alpha_n)(1-\beta_n)(1-\mu)[\|w_n - y_n\|^2 + \|u_n - y_n\|^2] \leq \|x_n - p\|^2 - \|x_{n+1} - p\|^2 + \alpha_n M_4.$$

In fact, by Lemma 7 and the convexity of $\|\cdot\|^2$, we get

$$\begin{aligned}
\|x_{n+1} - p\|^2 &\leq \|\beta_n(v_n - p) + \gamma_n(z_n - p) + \delta_n(T_n z_n - p)\|^2 + 2\gamma_n \alpha_n\langle h_n - f(x_n), x_{n+1} - p\rangle \\
&\leq \beta_n\|v_n - p\|^2 + (1-\beta_n)\|z_n - p\|^2 + 2(1-\beta_n)\alpha_n\|h_n - f(x_n)\|\|x_{n+1} - p\| \\
&\leq \beta_n\|v_n - p\|^2 + (1-\beta_n)\{\alpha_n \delta\|x_n - p\|^2 + (1-\alpha_n)\|v_n - p\|^2 \\
&\quad - (1-\alpha_n)(1-\mu)[\|v_n - u_n\|^2 + \|h_n - u_n\|^2] + \alpha_n M_2\},
\end{aligned} \quad (34)$$

where $\exists M_2 > 0$ s.t. $M_2 \geq \sup_{n\geq 1} 2(\|(f-I)p\|\|z_n - p\| + \|u_n - f(x_n)\|\|x_{n+1} - p\|)$. Also,

$$\begin{aligned}
\|v_n - p\|^2 &\leq \|x_n - p\|^2 + \alpha_n(2M_1\|x_n - p\| + \alpha_n M_1^2) \\
&\leq \|x_n - p\|^2 + \alpha_n M_3,
\end{aligned} \quad (35)$$

where $\exists M_3 > 0$ s.t. $M_3 \geq \sup_{n\geq 1}(2M_1\|x_n - p\| + \beta_n M_1^2)$. Substituting (35) for (34), we have

$$\begin{aligned}
\|x_{n+1} - p\|^2 &\leq \beta_n\|x_n - p\|^2 + (1-\beta_n)\{(1-\alpha_n(1-\delta))\|x_n - p\|^2 + (1-\alpha_n)\alpha_n M_3 \\
&\quad - (1-\alpha_n)(1-\mu)[\|v_n - u_n\|^2 + \|h_n - u_n\|^2] + \alpha_n M_2\} + \beta_n \alpha_n M_3 \\
&= \|x_n - p\|^2 - (1-\alpha_n)(1-\beta_n)(1-\mu)[\|v_n - u_n\|^2 + \|h_n - u_n\|^2] + \alpha_n M_4,
\end{aligned} \quad (36)$$

where $M_4 := M_2 + M_3$. This ensures that

$$(1-\alpha_n)(1-\beta_n)(1-\mu)[\|v_n - u_n\|^2 + \|h_n - u_n\|^2] \leq \|x_n - p\|^2 - \|x_{n+1} - p\|^2 + \alpha_n M_4. \quad (37)$$

Step 3. We show that $\exists M > 0$ s.t.

$$\begin{aligned}
\|x_{n+1} - p\|^2 &\leq [1 - \tfrac{(1-2\delta)\delta_n - \gamma_n}{1-\alpha_n\gamma_n}\alpha_n]\|x_n - p\|^2 + \tfrac{[(1-2\delta)\delta_n - \gamma_n]\alpha_n}{1-\alpha_n\gamma_n} \cdot \{\tfrac{2\gamma_n}{(1-2\delta)\delta_n - \gamma_n}\|f(x_n) - p\|\|z_n - x_{n+1}\| \\
&\quad + \tfrac{2\delta_n}{(1-2\delta)\delta_n - \gamma_n}\|f(x_n) - p\|\|z_n - x_n\| + \tfrac{2\delta_n}{(1-2\delta)\delta_n - \gamma_n}\langle f(p) - p, x_n - p\rangle \\
&\quad + \tfrac{1}{(1-2\delta)\delta_n - \gamma_n} \cdot \tfrac{\sigma_n}{\alpha_n}\|x_n - x_{n-1}\|3M\}.
\end{aligned}$$

In fact, we get

$$\begin{aligned}
\|v_n - p\|^2 &\leq \|x_n - p\|^2 + \sigma_n\|x_n - x_{n-1}\|(2\|x_n - p\| + \sigma_n\|x_n - x_{n-1}\|) \\
&\leq \|x_n - p\|^2 + \sigma_n\|x_n - x_{n-1}\|3M,
\end{aligned} \quad (38)$$

where $\exists M > 0$ s.t. $M \geq \sup_{n\geq 1}\{\|x_n - p\|, \sigma_n\|x_n - x_{n-1}\|\}$. Using Algorithm 1 and the convexity of $\|\cdot\|^2$, we get

$$\begin{aligned}
\|x_{n+1} - p\|^2 &\leq \|\beta_n(v_n - p) + \gamma_n(z_n - p) + \delta_n(T_n z_n - p)\|^2 + 2\gamma_n\alpha_n\langle h_n - f(x_n), x_{n+1} - p\rangle \\
&\leq \beta_n\|v_n - p\|^2 + (1 - \beta_n)\|\tfrac{1}{1-\beta_n}[\gamma_n(z_n - p) + \delta_n(T_n z_n - p)]\|^2 \\
&\quad + 2\gamma_n\alpha_n\langle h_n - p, x_{n+1} - p\rangle + 2\gamma_n\alpha_n\langle p - f(x_n), x_{n+1} - p\rangle,
\end{aligned}$$

which leads to

$$\begin{aligned}
\|x_{n+1} - p\|^2 &\leq \beta_n\|v_n - p\|^2 + (1 - \beta_n)[(1 - \alpha_n)\|h_n - p\|^2 + 2\alpha_n\langle f(x_n) - p, z_n - p\rangle] \\
&\quad + \gamma_n\alpha_n(\|h_n - p\|^2 + \|x_{n+1} - p\|^2) + 2\gamma_n\alpha_n\langle p - f(x_n), x_{n+1} - p\rangle.
\end{aligned}$$

Using (17) and (38) we deduce that $\|h_n - p\|^2 \leq \|v_n - p\|^2 \leq \|x_n - p\|^2 + \sigma_n\|x_n - x_{n-1}\|3M$. Hence,

$$\begin{aligned}
\|x_{n+1} - p\|^2 &\leq [1 - \alpha_n(1 - \beta_n)]\|x_n - p\|^2 + [1 - \alpha_n(1 - \beta_n)]\sigma_n\|x_n - x_{n-1}\|3M \\
&\quad + 2\alpha_n\delta_n\langle f(x_n) - p, z_n - p\rangle + \gamma_n\alpha_n(\|x_n - p\|^2 + \|x_{n+1} - p\|^2) \\
&\quad + (1 - \beta_n)\alpha_n\sigma_n\|x_n - x_{n-1}\|3M + 2\gamma_n\alpha_n\langle f(x_n) - p, z_n - x_{n+1}\rangle \\
&\leq [1 - \alpha_n(1 - \beta_n)]\|x_n - p\|^2 + 2\gamma_n\alpha_n\|f(x_n) - p\|\|z_n - x_{n+1}\| \\
&\quad + 2\alpha_n\delta_n\langle f(x_n) - p, x_n - p\rangle + 2\alpha_n\delta_n\langle f(x_n) - p, z_n - x_n\rangle \\
&\quad + \gamma_n\alpha_n(\|x_n - p\|^2 + \|x_{n+1} - p\|^2) + \sigma_n\|x_n - x_{n-1}\|3M \\
&\leq [1 - \alpha_n(1 - \beta_n)]\|x_n - p\|^2 + 2\gamma_n\alpha_n\|f(x_n) - p\|\|z_n - x_{n+1}\| \\
&\quad + 2\alpha_n\delta_n\delta\|x_n - p\|^2 + 2\alpha_n\delta_n\langle f(p) - p, x_n - p\rangle + 2\alpha_n\delta_n\|f(x_n) - p\|\|z_n - x_n\| \\
&\quad + \gamma_n\alpha_n(\|x_n - p\|^2 + \|x_{n+1} - p\|^2) + \sigma_n\|x_n - x_{n-1}\|3M,
\end{aligned}$$

which immediately yields

$$\begin{aligned}
\|x_{n+1} - p\|^2 &\leq [1 - \tfrac{(1-2\delta)\delta_n - \gamma_n}{1 - \alpha_n\gamma_n}\alpha_n]\|x_n - p\|^2 + \tfrac{[(1-2\delta)\delta_n - \gamma_n]\alpha_n}{1 - \alpha_n\gamma_n} \cdot \{\tfrac{2\gamma_n}{(1-2\delta)\delta_n - \gamma_n}\|f(x_n) - p\|\|z_n - x_{n+1}\| \\
&\quad + \tfrac{2\delta_n}{(1-2\delta)\delta_n - \gamma_n}\|f(x_n) - p\|\|z_n - x_n\| + \tfrac{2\delta_n}{(1-2\delta)\delta_n - \gamma_n}\langle f(p) - p, x_n - p\rangle \\
&\quad + \tfrac{1}{(1-2\delta)\delta_n - \gamma_n} \cdot \tfrac{\sigma_n}{\alpha_n}\|x_n - x_{n-1}\|3M\}.
\end{aligned}$$
(39)

Step 4. In order to show that $x_n \to x^* \in \Omega$, which is the unique solution of (12), we can follow a similar method to that in Step 4 for the proof of Theorem 1.

Finally, we apply our main results to solve the VIP and common fixed point problem (CFPP) in the following illustrating example.

The starting point $x_0 = x_1$ is randomly picked in the real line. Put $f(u) = \frac{1}{8}\sin u$, $\gamma = l = \mu = \frac{1}{2}$, $\sigma_n = \alpha_n = \frac{1}{n+1}$, $\beta_n = \frac{1}{3}$, $\gamma_n = \frac{1}{6}$ and $\delta_n = \frac{1}{2}$.

We first provide an example of Lipschitzian, pseudomonotone self-mapping A satisfying the boundedness of $A(C)$ and strictly pseudocontractive self-mapping T_1 with $\Omega = \text{Fix}(T_1) \cap \text{VI}(C, A) \neq \emptyset$. Let $C = [-1, 2]$ and H be the real line with the inner product $\langle a, b\rangle = ab$ and induced norm $\|\cdot\| = |\cdot|$. Then f is a δ-contractive map with $\delta = \frac{1}{8} \in [0, \frac{1}{2})$ and $f(H) \subset C$ because $\|f(u) - f(v)\| = \frac{1}{8}\|\sin u - \sin v\| \leq \frac{1}{8}\|u - v\|$ for all $u, v \in H$.

Let $A : H \to H$ and $T_1 : H \to H$ be defined as $Au := \frac{1}{1+|\sin u|} - \frac{1}{1+|u|}$, and $T_1 u := \frac{1}{2}u - \frac{3}{8}\sin u$ for all $u \in H$. Now, we first show that A is L-Lipschitzian, pseudomonotone operator with $L = 2$, such that $A(C)$ is bounded. In fact, for all $u, v \in H$ we get

$$\begin{aligned}
\|Au - Av\| &\leq |\tfrac{1}{1+\|u\|} - \tfrac{1}{1+\|v\|}| + |\tfrac{1}{1+\|\sin u\|} - \tfrac{1}{1+\|\sin v\|}| \\
&= |\tfrac{\|v\| - \|u\|}{(1+\|u\|)(1+\|v\|)}| + |\tfrac{\|\sin v\| - \|\sin u\|}{(1+\|\sin u\|)(1+\|\sin v\|)}| \\
&\leq \tfrac{\|u - v\|}{(1+\|u\|)(1+\|v\|)} + \tfrac{\|\sin u - \sin v\|}{(1+\|\sin u\|)(1+\|\sin v\|)} \\
&\leq 2\|u - v\|.
\end{aligned}$$

This implies that A is 2-Lipschitzian. Next, we show that A is pseudomonotone. For any given $u, v \in H$, it is clear that the relation holds:

$$\langle Au, u - v \rangle = \left(\frac{1}{1 + |\sin u|} - \frac{1}{1 + |u|}\right)(u - v) \leq 0 \Rightarrow \langle Av, u - v \rangle = \left(\frac{1}{1 + |\sin v|} - \frac{1}{1 + |v|}\right)(u - v) \leq 0.$$

Furthermore, it is easy to see that T_1 is strictly pseudocontractive with constant $\zeta_1 = \frac{1}{4}$. In fact, we observe that for all $u, v \in H$,

$$\|T_1 u - T_1 v\| \leq \frac{1}{2}\|u - v\| + \frac{3}{8}\|\sin u - \sin v\| \leq \|u - v\| + \frac{1}{4}\|(I - T_1)u - (I - T_1)v\|.$$

It is clear that $(\gamma_n + \delta_n)\zeta_1 = (\frac{1}{6} + \frac{1}{2}) \cdot \frac{1}{4} \leq \frac{1}{6} = \gamma_n < (1 - 2\delta)\delta_n = (1 - 2 \cdot \frac{1}{8})\frac{1}{2} = \frac{3}{8}$ for all $n \geq 1$. In addition, it is clear that $\text{Fix}(T_1) = \{0\}$ and $A0 = 0$ because the derivative $d(T_1 u)/du = \frac{1}{2} - \frac{3}{8}\cos u > 0$. Therefore, $\Omega = \{0\} \neq \emptyset$. In this case, Algorithm 1 can be rewritten below:

$$\begin{cases} v_n = x_n - \frac{1}{n+1}(x_{n-1} - x_n), \\ u_n = P_C(v_n - \ell_n A v_n), \\ z_n = \frac{1}{n+1}f(x_n) + \frac{n}{n+1}P_{C_n}(v_n - \ell_n A u_n), \\ x_{n+1} = \frac{1}{3}x_n + \frac{1}{6}P_{C_n}(v_n - \ell_n A u_n) + \frac{1}{2}T_1 z_n \quad \forall n \geq 1, \end{cases}$$

with $\{C_n\}$ and $\{\ell_n\}$, selected as in Algorithm 1. Then, by Theorem 1, we know that $x_n \to 0 \in \Omega$ iff $x_n - x_{n+1} \to 0$ $(n \to \infty)$ and $\sup_{n \geq 1} |x_n - \frac{1}{8}\sin x_n| < \infty$.

On the other hand, Algorithm 2 can be rewritten below:

$$\begin{cases} v_n = x_n - \frac{1}{n+1}(x_{n-1} - x_n), \\ u_n = P_C(v_n - \ell_n A v_n), \\ z_n = \frac{1}{n+1}f(x_n) + \frac{n}{n+1}P_{C_n}(v_n - \ell_n A u_n), \\ x_{n+1} = \frac{1}{3}v_n + \frac{1}{6}P_{C_n}(v_n - \ell_n A u_n) + \frac{1}{2}T_1 z_n \quad \forall n \geq 1, \end{cases}$$

with $\{C_n\}$ and $\{\ell_n\}$, selected as in Algorithm 2. Then, by Theorem 2, we know that $x_n \to 0 \in \Omega$ iff $x_n - x_{n+1} \to 0$ $(n \to \infty)$ and $\sup_{n \geq 1} |x_n - \frac{1}{8}\sin x_n| < \infty$.

Author Contributions: All authors contributed equally to this manuscript.

Funding: This research was partially supported by the Innovation Program of Shanghai Municipal Education Commission (15ZZ068), Ph.D. Program Foundation of Ministry of Education of China (20123127110002) and Program for Outstanding Academic Leaders in Shanghai City (15XD1503100).

Conflicts of Interest: The authors certify that they have no affiliations with or involvement in any organization or entity with any financial or non-financial interest in the subject matter discussed in this manuscript.

References

1. Browder, F.E.; Petryshyn, W.V. Construction of fixed points of nonlinear mappings in Hilbert space. *J. Math. Anal. Appl.* 1967, *1967*, 197–228. [CrossRef]
2. Ceng, L.C.; Kong, Z.R.; Wen, C.F. On general systems of variational inequalities. *Comput. Math. Appl.* 2013, *66*, 1514–1532. [CrossRef]
3. Nguyen, L.V. Some results on strongly pseudomonotone quasi-variational inequalities. *Set-Valued Var. Anal.* 2019. [CrossRef]
4. Bin Dehaish, B.A. Weak and strong convergence of algorithms for the sum of two accretive operators with applications. *J. Nonlinear Convex Anal.* 2015, *16*, 1321–1336.
5. Qin, X.; Cho, S.Y.; Wang, L. Strong convergence of an iterative algorithm involving nonlinear mappings of nonexpansive and accretive type. *Optimization* 2018, *67*, 1377–1388. [CrossRef]
6. Liu, L. A hybrid steepest method for solving split feasibility problems inovling nonexpansive mappings. *J. Nonlinear Convex Anal.* 2019, *20*, 471–488.

7. Ceng, L.C.; Ansari, Q.H.; Yao, J.C. Some iterative methods for finding fixed points and for solving constrained convex minimization problems. *Nonlinear Anal.* **2011** *74*, 5286–5302. [CrossRef]
8. Ceng, L.C.; Ansari, Q.H.; Yao, J.C. An extragradient method for solving split feasibility and fixed point problems. *Comput. Math. Appl.* **2012**, *64*, 633–642. [CrossRef]
9. Ceng, L.C.; Ansari, Q.H.; Yao, J.C. Relaxed extragradient methods for finding minimum-norm solutions of the split feasibility problem. *Nonlinear Anal.* **2012**, *75*, 2116–2125. [CrossRef]
10. Qin, X.; Cho, S.Y.; Yao, J.C. Weak and strong convergence of splitting algorithms in Banach spaces. *Optimization* **2019**. [CrossRef]
11. Cho, S.Y.; Li, W.; Kang, S.M. Convergence analysis of an iterative algorithm for monotone operators. *J. Inequal. Appl.* **2013**, *2013*, 199. [CrossRef]
12. Cho, S.Y. Strong convergence analysis of a hybrid algorithm for nonlinear operators in a Banach space. *J. Appl. Anal. Comput.* **2018**, *8*, 19–31.
13. Ceng, L.C.; Petrusel, A.; Yao, J.C.; Yao, Y. Hybrid viscosity extragradient method for systems of variational inequalities, fixed points of nonexpansive mappings, zero points of accretive operators in Banach spaces. *Fixed Point Theory* **2018**, *19*, 487–501. [CrossRef]
14. Ceng, L.C.; Petrusel, A.; Yao, J.C.; Yao, Y. Systems of variational inequalities with hierarchical variational inequality constraints for Lipschitzian pseudocontractions. *Fixed Point Theory* **2019**, *20*, 113–134. [CrossRef]
15. Cho, S.Y.; Kang, S.M. Approximation of fixed points of pseudocontraction semigroups based on a viscosity iterative process. *Appl. Math. Lett.* **2011**, *24*, 224–228. [CrossRef]
16. Cho, S.Y.; Kang, S.M. Approximation of common solutions of variational inequalities via strict pseudocontractions. *Acta Math. Sci.* **2012**, *32*, 1607–1618. [CrossRef]
17. Ceng, L.C.; Yuan, Q. Hybrid Mann viscosity implicit iteration methods for triple hierarchical variational inequalities, systems of variational inequalities and fixed point problems. *Mathematics* **2019**, *7*, 142. [CrossRef]
18. Qin, X.; Yao, J.C. Projection splitting algorithms for nonself operators. *J. Nonlinear Convex Anal.* **2017**, *18*, 925–935.
19. Qin, X.; Yao, J.C. Weak convergence of a Mann-like algorithm for nonexpansive and accretive operators. *J. Inequal. Appl.* **2016**, *2016*, 232. [CrossRef]
20. Ceng, L.C.; Wong, M.M.; Yao, J.C. A hybrid extragradient-like approximation method with regularization for solving split feasibility and fixed point problems. *J. Nonlinear Convex Anal.* **2013**, *14*, 163–182.
21. Kraikaew, R.; Saejung, S. Strong convergence of the Halpern subgradient extragradient method for solving variational inequalities in Hilbert spaces. *J. Optim. Theory Appl.* **2014**, *163*, 399–412. [CrossRef]
22. Thong, D.V.; Hieu, D.V. Modified subgradient extragradient method for variational inequality problems. *Numer. Algorithms* **2018**, *79*, 597–610. [CrossRef]
23. Thong, D.V.; Hieu, D.V. Inertial subgradient extragradient algorithms with line-search process for solving variational inequality problems and fixed point problems. *Numer. Algorithms* **2019**, *80*, 1283–1307. [CrossRef]
24. Takahahsi, W.; Yao, J.C. The split common fixed point problem for two finite families of nonlinear mappings in Hilbert spaces. *J. Nonlinear Convex Anal.* **2019**, *20*, 173–195.
25. Ansari, Q.H.; Babu, F.; Yao, J.C. Regularization of proximal point algorithms in Hadamard manifolds. *J. Fixed Point Theory Appl.* **2019**, *21*, 25. [CrossRef]
26. Qin, X.; Cho, S.Y.; Wang, L. Iterative algorithms with errors for zero points of m-accretive operators. *Fixed Point Theory Appl.* **2013**, *2013*, 148. [CrossRef]
27. Takahashi, W.; Wen, C.F.; Yao, J.C. The shrinking projection method for a finite family of demimetric mappings with variational inequality problems in a Hilbert space. *Fixed Point Theory* **2018**, *19*, 407–419. [CrossRef]
28. Zhao, X.; Ng, K.F.; Li, C.; Yao, J.C. Linear regularity and linear convergence of projection-based methods for solving convex feasibility problems. *Appl. Math. Optim.* **2018**, *78*, 613–641. [CrossRef]
29. Cho, S.Y.; Bin Dehaish, B.A. Weak convergence of a splitting algorithm in Hilbert spaces. *J. Appl. Anal. Comput.* **2017**, *7*, 427–438.
30. Cho, S.Y.; Qin, X. On the strong convergence of an iterative process for asymptotically strict pseudocontractions and equilibrium problems. *Appl. Math. Comput.* **2014**, *235*, 430–438. [CrossRef]
31. Chang, S.S.; Wen, C.F.; Yao, J.C. Common zero point for a finite family of inclusion problems of accretive mappings in Banach spaces. *Optimization* **2018**, *67*, 1183–1196. [CrossRef]

32. Chang, S.S.; Wen, C.F.; Yao, J.C. Zero point problem of accretive operators in Banach spaces. *Bull. Malays. Math. Sci. Soc.* **2019**, *42*, 105–118. [CrossRef]
33. Qin, X.; Cho, S.Y.; Wang, L. A regularization method for treating zero points of the sum of two monotone operators. *Fixed Point Theory Appl.* **2014**, *2014*, 75. [CrossRef]
34. Ceng, L.C.; Ansari, Q.H.; Wong, N.C.; Yao, J.C. An extragradient-like approximation method for variational inequalities and fixed point problems. *Fixed Point Theory Appl.* **2011**, *2011*, 18. [CrossRef]
35. Ceng, L.C.; Petrusel, A.; Yao, J.C. Composite viscosity approximation methods for equilibrium problem, variational inequality and common fixed points. *J. Nonlinear Convex Anal.* **2014**, *15*, 219–240.
36. Ceng, L.C.; Plubtieng, S.; Wong, M.M.; Yao, J.C. System of variational inequalities with constraints of mixed equilibria, variational inequalities, and convex minimization and fixed point problems. *J. Nonlinear Convex Anal.* **2015**, *16*, 385–421.
37. Ceng, L.C.; Gupta, H.; Ansari, Q.H. Implicit and explicit algorithms for a system of nonlinear variational inequalities in Banach spaces. *J. Nonlinear Convex Anal.* **2015**, *16*, 965–984.
38. Ceng, L.C.; Guu, S.M.; Yao, J.C. Hybrid iterative method for finding common solutions of generalized mixed equilibrium and fixed point problems. *Fixed Point Theory Appl.* **2012**, *2012*, 92. [CrossRef]
39. Cottle, R.W.; Yao, J.C. Pseudo-monotone complementarity problems in Hilbert space. *J. Optim. Theory Appl.* **1992**, *75*, 281–295. [CrossRef]
40. Xu, H.K.; Kim, T.H. Convergence of hybrid steepest-descent methods for variational inequalities. *J. Optim. Theory Appl.* **2003**, *119*, 185–201. [CrossRef]

© 2019 by the authors. Licensee MDPI, Basel, Switzerland. This article is an open access article distributed under the terms and conditions of the Creative Commons Attribution (CC BY) license (http://creativecommons.org/licenses/by/4.0/).

Article

Generalized Nonsmooth Exponential-Type Vector Variational-Like Inequalities and Nonsmooth Vector Optimization Problems in Asplund Spaces

Syed Shakaib Irfan [1,†], Mijanur Rahaman [2,†], Iqbal Ahmad [1,*] and Rais Ahmad [2,†] and Saddam Husain [2,†]

1. College of Engineering, Qassim University, Buraidah 51452, Al-Qassim, Saudi Arabia; shakaib@qec.edu.sa
2. Department of Mathematics, Aligarh Muslim University, Aligarh-202002, India; mrahman96@yahoo.com (M.R.); raisain_123@rediffmail.com (R.A.); saddamhusainamu26@gmail.com (S.H.)
* Correspondence: iqbal@qec.edu.sa; Tel.: +966-16-13762
† These authors contributed equally to this work.

Received: 27 February 2019; Accepted: 4 April 2019; Published: 10 April 2019

Abstract: The aim of this article is to study new types of generalized nonsmooth exponential type vector variational-like inequality problems involving Mordukhovich limiting subdifferential operator. We establish some relationships between generalized nonsmooth exponential type vector variational-like inequality problems and vector optimization problems under some invexity assumptions. The celebrated Fan-KKM theorem is used to obtain the existence of solution of generalized nonsmooth exponential-type vector variational like inequality problems. In support of our main result, some examples are given. Our results presented in this article improve, extend, and generalize some known results offer in the literature.

Keywords: vector variational-like inequalities; vector optimization problems; limiting (p,r)-α-(η,θ)-invexity; Lipschitz continuity; Fan-KKM theorem

1. Introduction

The vector variational inequality has been introduced and studied in [1] in finite-dimensional Euclidean spaces. Vector variational inequalities have emerged as an efficient tool to provide imperative requirements for the solution of vector optimization problems. Vector variational-like inequalities for nonsmooth mappings are useful generalizations of vector variational inequalities. For more details on vector variational inequalities and their generalizations, see the references [2–8]. In 1998, Giannessi [9] proved a necessary and sufficient condition for the existence of an efficient solution of a vector optimization problem for differentiable and convex mappings by using a Minty type vector variational inequality problem. Under different assumptions, many researchers have studied vector optimization problems by using different types of Minty type vector variational inequality problems. Yang et al. [8] generalized the result of Giannessi [9] for differentiable but pseudoconvex mappings.

On the other hand, Yang and Yang [10] considered vector variational-like inequality problem and showed relationships between vector variational-like inequality and vector optimization problem under the assumptions of pseudoinvexity or invariant pseudomonotonicity. Later, some researchers extended above problems in the direction of nonsmooth mappings. Rezaie and Zafarani [11] established a correspondence between a solution of the generalized vector variational-like inequality problem and the nonsmooth vector optimization problem under the same assumptions of Yang and Yang [10] in the setting of Clarke's subdifferentiability. Due to the fact that Clarke's subdifferentiability is bigger class than Mordukhovich limiting subdifferentiability, many authors studied the vector variational-like inequality problems and vector optimization problems by means of Mordukhovich limiting subdifferential. Later, Long et al. [12] and Oveisiha and Zafarani [13] studied generalized vector variational-like inequality problem and discussed the relationships between generalized vector variational-like inequality problem and nonsmooth vector optimization problem for pseudoinvex mappings, whereas Chen and Huang [14] obtained similar results for invex mappings by means of Mordukhovich limiting subdifferential.

Due to several applications of invex sets and exponential mappings in engineering, economics, population growth, mathematical modelling problems, Antczak [15] introduced exponential (p,r)-invex sets and mappings. After that, Mandal and Nahak [16] introduced (p,r)-ρ-(η,θ)-invexity mapping which is the generalization of the result of Antczak [15]. By using (p,r)-invexity, Jayaswal and Choudhury [17] introduced exponential type vector variational-like inequality problem involving locally Lipschitz mappings.

In this paper, we introduce generalized nonsmooth exponential-type vector variational like inequality problems involving Mordukhovich limiting subdifferential in Asplund spaces. We obtain some relationships between an efficient solution of nonsmooth vector optimization problems and this generalized nonsmooth exponential-type vector variational like inequality problems using limiting (p,r)-α-(η,θ)-invexity mapping. Employing the Fan-KKM theorem, we establish an existence result for our problem in Asplund spaces.

2. Preliminaries

Suppose that X is a real Banach space with dual space X^* and $\langle \cdot, \cdot \rangle$ is duality pairing between them. Assume that $K \subseteq X$ is a nonempty subset, $C \subset \mathbb{R}^n$ is a pointed, closed, convex cone with nonempty interior, i.e., $int C \neq \emptyset$ and $f : K \longrightarrow \mathbb{R}$ is a non-differentiable mapping. When the mappings are non-differentiable, many authors used the concept of subdifferential such as Fréchet subdifferential, Mordukhovich limiting subdifferential, and Clarke subdifferential operators. Now, we mention some notions and results already known in the literature.

Definition 1. *Suppose that $f : X \longrightarrow \mathbb{R}$ is a proper lower semicontinuous mapping on Banach space X. Then, the mapping f is said to be Fréchet subdifferentiable and ξ^* is Fréchet subderivative of f at x (i.e., $\xi^* \in \partial_F f(x)$) if, $x \in dom f$ and*

$$\liminf_{\|h\| \to 0} \frac{f(x+h) - f(x) - \langle \xi^*, h \rangle}{\|h\|} \geq 0.$$

Definition 2 ([18]). *Suppose that Ω is a nonempty subset of a normed vector space X. Then, for any $x \in X$ and $\varepsilon \geq 0$, the set of ε-normals to Ω at x is defined as*

$$\widehat{N}_\varepsilon(x; \Omega) = \left\{ x^* \in X^* : \limsup_{u \xrightarrow{\Omega} x} \frac{\langle x^*, u - x \rangle}{\|u - x\|} \leq \varepsilon \right\}.$$

For $\tilde{x} \in \Omega$, the limiting normal cone to Ω at $\tilde{x}$ is

$$N(\tilde{x}; \Omega) = \limsup_{x \xrightarrow{\Omega} \tilde{x}, \varepsilon \downarrow 0} \widehat{N}_\varepsilon(x; \Omega).$$

Consider a mapping $f : X \longrightarrow \mathbb{R} \cup \{\pm\infty\}$ and a finite point $\tilde{x} \in X$. Then, the limiting subdifferential of f at $\tilde{x}$ is the following set

$$\partial_L f(\tilde{x}) = \{ x^* \in X^* : (x^*, -1) \in N((\tilde{x}, f(\tilde{x})); epi f) \},$$

where $epi f$ is defined as $epi f = \{(x, a) \in X \times \mathbb{R} : f(x) \leq a\}$. If $|f(\tilde{x})| = \infty$, then we put $\partial_L f(\tilde{x}) = \emptyset$.

Remark 1 ([18]). *It is noted that the Clarke subdifferential is larger class than the Fréchet subdifferential and the limiting subdifferential with the relation $\partial_F f(x) \subseteq \partial_L f(x) \subseteq \partial_C f(x)$.*

Definition 3. *A Banach space X is said to be Asplund space if K is any open subset of X and $f : K \longrightarrow \mathbb{R}$ is continuous convex mapping, then f is Fréchet subdifferentiable at any point of a dense subset of K.*

Remark 2. *It is remarked that a Banach space X has the Asplundity property if every separable subspace of X has separable dual. The concept of Asplund space depicts the differentiability characteristics of continuous convex mappings on Euclidean space. All the spaces which are reflexive Banach spaces are Asplund. The space of convergent real sequences c_0 (whose limit is 0) is non-reflexive separable Banach space, but its is an Asplund space. For more details, we refer to [19].*

Definition 4. *A bi-mapping $\eta : K \times K \longrightarrow K$ is said to be affine with respect to the first argument if, for any $\lambda \in [0,1]$ and $u_1, u_2 \in K$ with $u = \lambda u_1 + (1-\lambda) u_2 \in K$ such that*

$$\eta(\lambda u_1 + (1-\lambda) u_2, v) = \lambda \eta(u_1, v) + (1-\lambda) \eta(u_2, v), \quad \forall v \in K.$$

Definition 5. *A bi-mapping $\eta : K \times K \longrightarrow X$ is said to be continuous in the first argument if,*

$$\|\eta(u,z) - \eta(v,z)\| \to 0 \text{ as } \|u-v\| \to 0, \quad \forall u, v \in K, \ z \text{ is fixed}.$$

Definition 6 ([20]). *Suppose that K is a subset of a topological vector space Y. A set-valued mapping $T : K \longrightarrow 2^Y$ is called a KKM-mapping if, for each nonempty finite subset $\{y_1, y_2, \cdots, y_n\} \subset K$, we have*

$$Co\{y_1, y_2, \cdots, y_n\} \subseteq \bigcup_{i=1}^n T(y_i),$$

where Co denotes the convex hull.

Theorem 1 (Fan-KKM Theorem [20]). *Suppose that K is a subset of a topological vector space Y and $T : K \longrightarrow 2^Y$ is a KKM-mapping. If, for each $y \in K, T(y)$ is closed and for at least one $y \in K, T(y)$ is compact, then*

$$\bigcap_{y \in K} T(y) \neq \emptyset.$$

Definition 7. *A mapping $f : X \longrightarrow \mathbb{R}^n$ is called locally Lipschitz continuous at x_0 if, there exists a $L > 0$ and a neighbourhood N of x_0 such that*

$$\|f(y) - f(z)\| \leq L \|y - z\|, \quad \forall y, z \in N(x_0).$$

If f is locally Lipschitz continuous for each x_0 in X, then f is locally Lipschitz continuous mapping on X.

Slightly changing the structure of definition of (p,r)-α-(η, θ)-invexity defined in [16], we have the following definition.

Definition 8. Suppose that $f : X \longrightarrow \mathbb{R}^n$ is a locally Lipschitz continuous mapping, $e = (1, 1, \cdots, 1) \in \mathbb{R}^n$ and p, r are arbitrary real numbers. If there exist the mappings $\eta, \theta : X \times X \longrightarrow X$ and a constant $\alpha \in \mathbb{R}$ such that one of the following relations

$$\frac{1}{r}\left\{\exp^{r(f(x)-f(u))} - 1\right\} \geq \frac{1}{p}\left\langle \xi; \left(\exp^{p\eta(x,u)} - e\right)\right\rangle + \alpha\|\theta(x,u)\|^2 e \ (> \text{ if } x \neq u) \text{ for } p \neq 0, r \neq 0,$$

$$\frac{1}{r}\left\{\exp^{r(f(x)-f(u))} - 1\right\} \geq \langle \xi; \eta(x,u)\rangle + \alpha\|\theta(x,u)\|^2 e \ (> \text{ if } x \neq u) \text{ for } p = 0, r \neq 0,$$

$$f(x) - f(u) \geq \frac{1}{p}\left\langle \xi; \left(\exp^{p\eta(x,u)} - e\right)\right\rangle + \alpha\|\theta(x,u)\|^2 e \ (> \text{ if } x \neq u) \text{ for } p \neq 0, r = 0,$$

$$f(x) - f(u) \geq \langle \xi; \eta(x,u)\rangle + \alpha\|\theta(x,u)\|^2 e \ (> \text{ if } x \neq u) \text{ for } p = 0, r = 0,$$

holds for each $\xi \in \partial_L f(u)$, then f is called limiting (p,r)-α-(η,θ)-invex (strictly limiting (p,r)-α-(η,θ)-invex) with respect to η and θ at the point u on X. If f is limiting (p,r)-α-(η,θ)-invex with respect to η and θ at each $u \in X$, then f is limiting (p,r)-α-(η,θ)-invex with respect to the same η and θ on X.

Remark 3. We only consider the case when $p \neq 0, r \neq 0$ to prove the results. We exclude other cases as it is straightforward in terms of altering inequality. Throughout the proof of the results, we assume that $r > 0$. Under other condition $r < 0$, the direction in the proof will be reversed.

Problem 1. Suppose that $f = (f_1, f_2, \cdots, f_n) : K \longrightarrow \mathbb{R}^n$ is a vector-valued mapping such that each $f_i : K \longrightarrow \mathbb{R}$ $(i = 1, 2, \cdots, n)$ is locally Lipschitz continuous mapping. The nonsmooth vector optimization problem is to

$$\underset{C}{\text{Maximize}} f(x) = (f_1(x), f_2(x), \cdots, f_n(x)) \tag{P_1}$$

subject to $x \in K$,

where $C \in \mathbb{R}^n$ is a pointed, closed and convex cone with $\text{int} C \neq \emptyset$.

Definition 9. Suppose that $f : K \longrightarrow \mathbb{R}^n$ is a vector-valued mapping. A point $\bar{x} \in K$ is called

(i) *an efficient solution (Pareto solution) of* (P_1) *if and only if*

$$f(y) - f(\bar{x}) \notin -C \setminus \{0\}, \ \forall y \in K;$$

(ii) *a weak efficient solution (weak Pareto solution) of* (P_1) *if and only if*

$$f(y) - f(\bar{x}) \notin -\text{int} C, \ \forall y \in K.$$

Now, we introduce following two kinds of generalized nonsmooth exponential-type vector variational-like inequality problems. Suppose that $K \neq \emptyset$ is a subset of an Asplund space X and $C \subset \mathbb{R}^n$ is a pointed, closed and convex cone such that $\text{int} C \neq \emptyset$. Assume that $f = (f_1, f_2, \cdots, f_n) : K \longrightarrow \mathbb{R}^n$ is a non-differentiable locally Lipschitz continuous mapping, $\eta, \theta : K \times K \longrightarrow X$ are the continuous mappings, β, p is an arbitrary real number and $e = (1, 1, \cdots, 1) \in \mathbb{R}^n$.

Problem 2. Generalized nonsmooth exponential-type strong vector variational like inequality problem is to find a vector $\bar{x} \in K$ such that

$$\left.\begin{array}{l} \frac{1}{p}\left\langle \xi; \left(\exp^{p\eta(y,\bar{x})} - e\right)\right\rangle + \beta\|\theta(y,\bar{x})\|^2 e \notin -C \setminus \{0\}, \text{ for } p \neq 0, \\ \langle \xi; \eta(y,\bar{x})\rangle + \beta\|\theta(y,\bar{x})\|^2 e \notin -C \setminus \{0\}, \text{ for } p = 0, \end{array}\right\} \forall \xi \in \partial_L f(\bar{x}), y \in K; \tag{P_2}$$

Problem 3. *Generalized nonsmooth exponential-type weak vector variational like inequality problem is to find a vector $\bar{x} \in K$ such that*

$$\left.\begin{array}{ll} \frac{1}{p}\left\langle \xi; \left(\exp^{p\eta(y,\bar{x})} - e\right)\right\rangle + \beta\|\theta(y,\bar{x})\|^2 e \not\in -\mathrm{int}C, & \text{for } p \neq 0, \\ \langle \xi; \eta(y,\bar{x})\rangle + \beta\|\theta(y,\bar{x})\|^2 e \not\in -\mathrm{int}C, & \text{for } p = 0, \end{array}\right\} \forall \xi \in \partial_L f(\bar{x}), y \in K. \quad (P_3)$$

Special Cases:

(i) If $\theta \equiv 0$ and $\partial_L f(\cdot) = \partial f(\cdot)$, i.e., the Clarke subdifferential operator, then (P_2) and (P_3) reduces to nonsmooth exponential-type vector variational like inequality problem and nonsmooth exponential-type weak vector variational like inequality problem considered and studied by Jayswal and Choudhury [17].

(ii) For $p = 0$, a similar analogue of problems (P_2) and (P_3) was introduced and studied by Oveisiha and Zafarani [13].

Apparently, it shows that the solution of (P_2) is also a solution of (P_3). We construct the following example in support of (P_2).

Example 1. *Let us consider $X = \mathbb{R}$, $K = [-1,1]$, $C = \mathbb{R}^2_+$, $p = 1$ and the mapping f be defined as $f = (f_1, f_2)$ by*

$$f_1(x) = \begin{cases} x, & \text{if } x \geq 0, \\ 0, & \text{if } x < 0, \end{cases} \quad \text{and} \quad f_2(x) = \begin{cases} x^2 + 2x, & \text{if } x \geq 0, \\ 0, & \text{if } x < 0. \end{cases}$$

Now, the limiting subdifferential of f is

$$\partial_L f(x) = \begin{cases} (1, 2x+2), & \text{if } x > 0, \\ \{(s,t) : s \in [0,1], t \in [0,2]\}, & \text{if } x = 0, \\ (0,0), & \text{if } x < 0. \end{cases}$$

Define the mappings $\eta, \theta : K \times K \longrightarrow X$ by

$$\eta(y,x) = \ln(|y-x|+1) \quad \text{and} \quad \theta(y,x) = \frac{y-x}{2}, \quad \forall y, x \in K.$$

Then, the problem (P_2) is to find a point $\bar{x} \in K$ such that

$$\left\langle \xi; \left(\exp^{\eta(y,\bar{x})} - e\right)\right\rangle + \beta\|\theta(y,\bar{x})\|^2 e \not\in -C \setminus \{0\}, \quad \forall \xi \in \partial_L f(x), y \in K,$$

which is equivalent to say that

$$\left\langle \partial_L f(\bar{x}); \left(\exp^{\eta(y,\bar{x})} - e\right)\right\rangle + \beta\|\theta(y,\bar{x})\|^2 e \not\subseteq -C \setminus \{0\}, \quad \forall \xi \in \partial_L f(x), y \in K.$$

For $\bar{x} = 0$ and $\beta \geq 4$, we can see that

$$\left\langle \partial_L f(\bar{x}); \left(\exp^{\eta(y,\bar{x})} - e\right)\right\rangle + \beta\|\theta(y,\bar{x})\|^2 e$$

$$= \left\{\left(s\left(\exp^{\ln(|y-x|+1)} - e\right), t\left(\exp^{\ln(|y-x|+1)} - e\right)\right) : s \in [0,1], t \in [0,2]\right\} + \beta\left\|\frac{y-\bar{x}}{2}\right\|^2 e$$

$$= \{(s(|y-x|), t(|y-x|)) : s \in [0,1], t \in [0,2]\} + \frac{\beta}{4}\|y-\bar{x}\|^2 e$$

$$= \{(s|y|, t|y|) : s \in [0,1], t \in [0,2]\} + \frac{\beta}{4}|y|^2 e$$

$$\not\subseteq -C \setminus \{0\}.$$

Hence, $\bar{x} = 0$ is the solution of the problem (P_2).

3. Main Results

Now, we prove a result which ensures that the solution of (P_2) is an efficient solution of (P_1).

Theorem 2. *Suppose that $K \neq \emptyset$ is a subset of Asplund space X, $C = \mathbb{R}^n_+$ and $f = (f_1, f_2, \cdots, f_n) : K \longrightarrow \mathbb{R}^n$ is a locally Lipschitz continous mapping on K. Let $\eta, \theta : K \times K \longrightarrow X$ be the mappings such that each f_i ($i = 1, 2, \cdots, n$) is limiting (p, r)-α_i-(η, θ)-invex mapping with respect to η and θ. If $\bar{x} \in K$ is a solution of (P_2), then $\bar{x}$ is an efficient solution of (P_1).*

Proof. Assume that $\bar{x} \in K$ is a solution of (P_2). We will prove that $\bar{x} \in K$ is an efficient solution of (P_1). Indeed, let us assume that $\bar{x} \in K$ is not an efficient solution of (P_1). Then, $\exists y \in K$ such that

$$(f_1(y) - f_1(\bar{x}), f_2(y) - f_2(\bar{x}), \cdots, f_n(y) - f_n(\bar{x})) = f_i(y) - f_i(\bar{x}) \in -C \setminus \{0\},$$

which implies that
$$f_i(y) - f_i(\bar{x}) \leq 0, \quad \forall i = 1, 2, \cdots, n, \tag{1}$$

and strict inequality holds for some $1 \leq k \leq n$.

Since $C = \mathbb{R}^n_+$, exponential mapping is monotonic and $r > 0$, then from (1), we have

$$\frac{1}{r}\left(\exp^{r(f_i(y) - f_i(\bar{x}))} - 1\right) \leq 0, \quad \forall i = 1, 2, \cdots, n. \tag{2}$$

Since each f_i is limiting (p, r)-α_i-(η, θ)-invex mapping with respect to η and θ at $\bar{x}$, therefore for all $\zeta_i \in \partial_L f_i(\bar{x})$, we have

$$\frac{1}{r}\left(\exp^{r(f_i(y) - f_i(\bar{x}))} - 1\right) \geq \frac{1}{p}\left\langle \zeta_i; \left(\exp^{p\eta(y,\bar{x})} - e\right)\right\rangle + \alpha_i \|\theta(y, \bar{x})\|^2 e. \tag{3}$$

Set $\beta = \min\{\alpha_1, \alpha_2, \cdots, \alpha_n\}$, therefore from (3), we have

$$\frac{1}{r}\left(\exp^{r(f_i(y) - f_i(\bar{x}))} - 1\right) \geq \frac{1}{p}\left\langle \zeta_i; \left(\exp^{p\eta(y,\bar{x})} - e\right)\right\rangle + \beta \|\theta(y, \bar{x})\|^2 e. \tag{4}$$

Now by using (2) and (4), we get

$$\frac{1}{p}\left\langle \zeta_i; \left(\exp^{p\eta(y,\bar{x})} - e\right)\right\rangle + \beta \|\theta(y, \bar{x})\|^2 e \leq 0,$$

which implies that for all $\zeta_i \in \partial_L f_i(\bar{x})$

$$\frac{1}{p}\left\langle \zeta_i; \left(\exp^{p\eta(y,\bar{x})} - e\right)\right\rangle + \beta \|\theta(y, \bar{x})\|^2 e \in -C \setminus \{0\},$$

which counteracts the hypothesis that $\bar{x}$ is a solution of (P_2). Hence, $\bar{x}$ is an efficient solution of (P_1). This completes the proof. □

Next, we show the converse of the above conclusion.

Theorem 3. *Suppose that $f = (f_1, f_2, \cdots, f_n) : K \longrightarrow \mathbb{R}^n$ is a locally Lipschitz continuous mapping on K. If each $-f_i$ is limiting (p, r)-α_i-(η, θ)-invex mapping with respect to η and θ, and $\bar{x}$ is an efficient solution of (P_1), then $\bar{x}$ is a solution of (P_2).*

Proof. Assume that $\bar{x}$ is an efficient solution of (P_1). On contrary suppose that $\bar{x}$ is not a solution of (P_2). Then, each β ensures the existence of x_β satisfying

$$\frac{1}{p}\left\langle \xi_i; \left(\exp^{p\eta(x_\beta,\bar{x})} - e\right)\right\rangle + \beta\|\theta(x_\beta,\bar{x})\|^2 e \in -C \setminus \{0\},$$

for all $\xi_i \in \partial_L f_i(x_\beta)$. Since $C = \mathbb{R}^n_+$, from above relation, we have

$$\frac{1}{p}\left\langle \xi_i; \left(\exp^{p\eta(x_\beta,\bar{x})} - e\right)\right\rangle + \beta\|\theta(x_\beta,\bar{x})\|^2 e \leq 0, \tag{5}$$

and strict inequality holds for some $1 \leq k \leq n$.

As each $-f_i$ is limiting (p,r)-α_i-(η,θ)-invex mapping with respect to η and θ with constants α_i, therefore for any $y \in K$, $\exists \xi_i \in \partial_L f_i(y)$ such that

$$\frac{1}{r}\left(\exp^{r(-f_i(y)+f_i(\bar{x}))} - 1\right) \geq \frac{1}{p}\left\langle (-\xi_i); \left(\exp^{p\eta(y,\bar{x})} - e\right)\right\rangle + \alpha_i\|\theta(y,\bar{x})\|^2 e,$$

which implies that

$$\frac{1}{r}\left(\exp^{r(-f_i(y)+f_i(\bar{x}))} - 1\right) \geq \frac{1}{p}\left\langle (-\xi_i); \left(\exp^{p\eta(y,\bar{x})} - e\right)\right\rangle + \beta\|\theta(y,\bar{x})\|^2 e, \tag{6}$$

where $\beta = \min\{\alpha_1, \alpha_2, \cdots, \alpha_n\}$.

Using (5), (6) and monotonic property of exponential mapping, it is easy to deduce that $\exists y \in K$ such that

$$f_i(\bar{x}) - f_i(y) \geq 0,$$

and strict inequality holds for $i = k$ and equivalently

$$f_i(\bar{x}) - f_i(y) \in C \setminus \{0\},$$

which counteracts the hypothesis that $\bar{x}$ is an efficient solution of (P_1). Therefore, $\bar{x}$ is a solution of (P_2). This completes the proof. □

Based on equivalent arguments as used in Theorems 2 and 3, we have the following theorem which associates the problems (P_1) and (P_3).

Theorem 4. *Suppose that $K \neq \emptyset$ is a subset of Asplund space X, $C = \mathbb{R}^n_+$ and $f = (f_1, f_2, \cdots, f_n) : K \longrightarrow \mathbb{R}^n$ a locally Lipschitz continuous mapping on K. If each $-f_i$ ($1 \leq i \leq n$) is strictly limiting (p,r)-α_i-(η,θ)-invex mapping with respect to η and θ and $\bar{x} \in K$ is a weak efficient solution of (P_1), then $\bar{x} \in K$ is also a solution of (P_3). Conversely, if each f_i ($1 \leq i \leq n$) is limiting (p,r)-α_i-(η,θ)-invex mapping with respect to η and θ and $\bar{x} \in K$ is the solution of (P_3), then $\bar{x} \in K$ is also a weak efficient solution of (P_1).*

We contrive the following example in support of Theorem 4.

Example 2. *Let us consider $X = \mathbb{R}$, $K = [0,1]$, $C = \mathbb{R}^2_+$ and $p = 1$. Define the nonsmooth vector optimization problem*

$$\min_{C} f(x) = (f_1(x), f_2(x)) \tag{7}$$

$$\text{subject to } x \in K,$$

where $f_1(x) = \ln\left(x^2 + \sqrt{x} + 1\right)$ and $f_2(x) = \ln\left(x^2 + \frac{\sqrt{x}}{2}\right)$. Clearly, f is locally Lipschitz mapping at $x = 0$. Now, the limiting subdifferential of f is as follows:

$$\partial_L f(x) = \begin{cases} \left(\frac{2x + \frac{1}{2\sqrt{x}}}{x^2 + \sqrt{x} + 1}, \frac{4x + \frac{1}{2\sqrt{x}}}{2x^2 + \sqrt{x}}\right), & \text{if } x > 0, \\ \{(s, t) : s, t \in [0, \infty)\}, & \text{if } x = 0. \end{cases}$$

Define the mappings $\theta, \eta : K \times K \longrightarrow X$ by

$$\eta(y, x) = \ln\left(-\frac{\sqrt{y}}{2} + x + 1\right) \text{ and } \theta(y, x) = y - x, \quad \forall y, x \in K.$$

For $r = 1$, we can see that for $\alpha = 1$ at $\bar{x} = 0$

$$\left(\exp^{f_1(y) - f_1(\bar{x})} - 1\right) - \left\langle \xi_1; \left(\exp^{\eta(y, \bar{x})} - e\right)\right\rangle - \alpha \|\theta(y, \bar{x})\|^2$$

$$= \left(\exp^{\ln\left(\frac{y^2 + \sqrt{y} + 1}{\bar{x}^2 + \sqrt{\bar{x}} + 1}\right)} - 1\right) - \left\langle \xi_1; \left(\exp^{\ln\left(-\frac{\sqrt{y}}{2} + \bar{x} + 1\right)} - e\right)\right\rangle - \|y - \bar{x}\|^2$$

$$= \left(\frac{y^2 + \sqrt{y} + 1}{\bar{x}^2 + \sqrt{\bar{x}} + 1} - 1\right) - \left\langle \xi_1; \left(-\frac{\sqrt{y}}{2} + \bar{x} + 1\right) - e\right\rangle - \|y - \bar{x}\|^2$$

$$= \left(y^2 + \sqrt{y}\right) + \xi_1\left(\frac{\sqrt{y}}{2}\right) - |y|^2$$

$$= y^2 + \sqrt{y}\left(1 + \frac{\xi_1}{2}\right) - |y|^2 \geq 0.$$

Similarly, we can show that

$$\left(\exp^{f_2(y) - f_2(\bar{x})} - 1\right) - \left\langle \xi_2; \left(\exp^{\eta(y, \bar{x})} - e\right)\right\rangle - \alpha \|\theta(y, \bar{x})\|^2 \geq 0.$$

Therefore, f is $(1, 1)$-1-(η, θ)-invex mapping at $\bar{x} = 0$.

Now, problem (P_3) is to find $\bar{x} \in [0, 1]$ such that

$$\frac{1}{p}\left\langle \xi; \left(\exp^{p\eta(y, \bar{x})} - e\right)\right\rangle + \alpha \|\theta(y, \bar{x})\|^2 e \notin -\text{int} C, \quad \forall \xi \in \partial_L f(x), y \in K,$$

which is analogous to the following problem

$$\frac{1}{p}\left\langle \partial_L f(\bar{x}); \left(\exp^{p\eta(y, \bar{x})} - e\right)\right\rangle + \alpha \|\theta(y, \bar{x})\|^2 e \not\subseteq -\text{int} C, \quad \forall \xi \in \partial_L f(x), y \in K.$$

Now, for $\alpha = p = 1$, we deduce that

$$\left\langle \partial_L f(\bar{x}); \left(\exp^{\eta(y, \bar{x})} - e\right)\right\rangle + \alpha \|\theta(y, \bar{x})\|^2 e$$

$$= \left\{\left(s\left(\exp^{\ln(-\sqrt{y} - \bar{x} + 1)} - e\right), t\left(\exp^{\ln(-\sqrt{y} - \bar{x} + 1)} - e\right)\right) : s, t \in [0, \infty)\right\} + \|y - \bar{x}\|^2 e$$

$$= \{(s(-\sqrt{y} - \bar{x}), t(-\sqrt{y} - \bar{x})) : s, t \in [0, \infty)\} + \|y\|^2 e$$

$$\not\subseteq -\text{int} C.$$

Therefore, $\bar{x} = 0$ is the solution of the problem (P_3). One can easily show that $\bar{x} = 0$ is a weakly efficient solution of vector optimization problem (7) by using Theorem 4.

Following is the existence theorem for the solution of generalized nonsmooth exponential-type weak vector variational like inequality problem (P_3) by employing the Fan-KKM Theorem.

Theorem 5. *Suppose that $K \neq \emptyset$ is a convex subset of Asplund space X, C is a pointed, closed and convex cone, and $f = (f_1, f_2, \cdots, f_n) : K \longrightarrow \mathbb{R}^n$ is a locally Lipschitz mapping such that each f_i $(1 \leq i \leq n)$ is limiting (p,r)-α_i-(η, θ)-invex mapping with respect to η and θ with constants α_i. Suppose that $\eta, \theta : K \times K \longrightarrow X$ are the continuous mappings which are affine in the first argument, respectively and $\eta(x,x) = 0 = \theta(x,x)$, for all $x \in K$. For any compact subset $B \neq \emptyset$ of K and $y_0 \in B$ with the property*

$$\frac{1}{p} \left\langle \zeta; \left(\exp^{p\eta(y_0, x)} - e \right) \right\rangle + \beta \|\theta(y_0, x)\|^2 e \in -intC, \quad \forall x \in K \setminus B, \; \zeta \in \partial_L f(x), \tag{8}$$

where $\beta = \min\{\alpha_1, \alpha_2, \cdots, \alpha_n\}$, then generalized nonsmooth exponential-type weak vector variational like inequality problem (P_3) admits a solution.

Proof. For any $y \in K$, consider the mapping $F : K \longrightarrow 2^K$ define by

$$F(y) = \left\{ x \in K : \frac{1}{p} \left\langle \zeta; \left(\exp^{p\eta(y, x)} - e \right) \right\rangle + \beta \|\theta(y, x)\|^2 e \notin -intC, \; \forall \zeta \in \partial_L f(x) \right\}.$$

Since $y \in F(y)$, therefore F is nonempty.

Now, we will prove that F is a KKM-mapping on K. On contrary, assume that F is not a KKM-mapping. Therefore, we can find a finite set $\{x_1, x_2, \cdots, x_n\}$ and $t_i \geq 0, i = 1, 2, \cdots, n$ with $\sum_{i=1}^{n} t_i = 1$ such that

$$x_0 = \sum_{i=1}^{n} t_i x_i \notin \bigcup_{i=1}^{n} F(x_i),$$

which implies that $x_0 \notin F(x_i), \forall i = 1, 2, \cdots, n$, i.e.,

$$\frac{1}{p} \left\langle \zeta; \left(\exp^{p\eta(x_i, x_0)} - e \right) \right\rangle + \beta \|\theta(x_i, x_0)\|^2 e \in -intC, \; \forall i = 1, 2, \cdots, n.$$

In view of convexity of $(\exp^{\lambda x} - e)$, for all $x \in \mathbb{R}$ and for any $\lambda > 0$, and affinity of η and θ in the first argument with the property $\eta(x, x) = 0 = \theta(x, x)$, we obtain

$$\begin{aligned}
0 &= \frac{1}{p} \left\langle \zeta; \left(\exp^{p\eta(x_0, x_0)} - e \right) \right\rangle + \left(\frac{\beta}{\sum_{i=1}^{n} t_i} \right) \|\theta(x_0, x_0)\|^2 e \\
&= \frac{1}{p} \left\langle \zeta; \left(\exp^{p\eta\left(\sum_{i=1}^{n} t_i x_i, x_0 \right)} - e \right) \right\rangle + \left(\frac{\beta}{\sum_{i=1}^{n} t_i} \right) \left\| \theta \left(\sum_{i=1}^{n} t_i x_i, x_0 \right) \right\|^2 e \\
&= \frac{1}{p} \left\langle \zeta; \left(\exp^{p \sum_{i=1}^{n} t_i \eta(x_i, x_0)} - e \right) \right\rangle + \left(\frac{\beta}{\sum_{i=1}^{n} t_i} \right) \left\| \sum_{i=1}^{n} t_i \theta(x_i, x_0) \right\|^2 e \\
&\leq_C \frac{1}{p} \left\langle \zeta; \sum_{i=1}^{n} t_i \left(\exp^{p\eta(x_i, x_0)} - e \right) \right\rangle + \beta \left(\sum_{i=1}^{n} t_i \right) \|\theta(x_i, x_0)\|^2 e \\
&= \frac{1}{p} \sum_{i=1}^{n} t_i \left\langle \zeta; \left(\exp^{p\eta(x_i, x_0)} - e \right) \right\rangle + \beta \left(\sum_{i=1}^{n} t_i \right) \|\theta(x_i, x_0)\|^2 e \\
&\in -intC,
\end{aligned}$$

which implies that $0 \in -intC$ and hence, a contradiction. Therefore, F is a KKM-mapping.

Next, to show that $F(y)$ is closed set, for each $y \in K$, consider any sequence $\{x_n\}$ in $F(y)$ which converges to $\bar{x}$. This implies that

$$z_n = \frac{1}{p}\left\langle \xi_n; \left(\exp^{p\eta(y,x_n)} - e\right)\right\rangle + \beta\|\theta(y,x_n)\|^2 e \notin -intC, \forall \xi_n \in \partial_L f(x_n). \tag{9}$$

Using locally Lipschitz continuity property of f, we have

$$\|f(x) - f(y)\| \leq L\|x - y\|, \quad \forall x, y \in N(\bar{x}),$$

where $L > 0$ is a constant and $N(\bar{x})$ is the neighbourhood of $\bar{x}$. Then, we can find any $x \in N(\bar{x})$ and $\xi \in \partial_L f(x)$ such that

$$\|\xi\| \leq L.$$

Since $\partial_L f(x_n)$ is w^*-compact, then the sequence $\{\xi_n\}$ has a convergent subsequence, say $\{\xi_m\}$ in $\partial_L f(x_n)$ such that $\xi_m \to \bar{\xi} \in \partial_L f(\bar{x})$. Since η and θ are continuous mappings, we have

$$\bar{z} = \lim_m z_m = \frac{1}{p}\left\langle \bar{\xi}; \left(\exp^{p\eta(y,\bar{x})} - e\right)\right\rangle + \beta\|\theta(y,\bar{x})\|^2 e.$$

From (9), it follows that $\bar{z} \in intC$ and therefore, we have

$$\frac{1}{p}\left\langle \bar{\xi}; \left(\exp^{p\eta(y,\bar{x})} - e\right)\right\rangle + \beta\|\theta(y,\bar{x})\|^2 e \notin intC.$$

Hence $\bar{x} \in F(y)$, and thus $F(y)$ is closed set.

Using the hypothesis (8), for any compact subset $B \neq \emptyset$ of K and $y_0 \in B$, we have

$$\frac{1}{p}\left\langle \xi; \left(\exp^{p\eta(y_0,x)} - e\right)\right\rangle + \beta\|\theta(y_0,x)\|^2 e \in -intC, \quad \forall x \in K \setminus B, \xi \in \partial_L f(x),$$

which shows that $F(y_0) \in B$. Due to compactness of B, we have $F(y_0)$ is also compact. Therefore, by applying the Fan-KKM Theorem 1, we obtain

$$\bigcap_{y \in K} F(y) \neq \emptyset.$$

Therefore, $\exists \bar{x} \in K$ such that

$$\frac{1}{p}\left\langle \bar{\xi}; \left(\exp^{p\eta(y,\bar{x})} - e\right)\right\rangle + \beta\|\theta(y,\bar{x})\|^2 e \notin -intC, \quad \forall \bar{\xi} \in \partial_L f(\bar{x}).$$

Thus, generalized nonsmooth exponential-type weak vector variational like inequality problem (P_3) has a solution. This completes the proof. $\square$

4. Conclusions

We have introduced and studied a new type of generalized nonsmooth exponential type vector variational-like inequality problem involving Mordukhovich limiting subdifferential operator in Asplund spaces. We proved the relationships between our considered problems with vector optimization problems using the generalized concept of invexity, which we called limiting (p,r)-α-(η,θ)-invexity of mappings. We also derived the existence of a result for our considered problem using the Fan-KKM theorem. It is remarked that our problems and related results are more general than the previously known results.

Author Contributions: The authors S.S.I., M.R., I.A., R.A. and S.H. carried out this work and drafted the manuscript together. All the authors studied and validated the article.

Funding: The research was supported by the Deanship of Scientific Research, Qassim University, Saudi Arabia grant number 3611-qec-2018-1-14-S.

Acknowledgments: We are grateful for the comments and suggestions of the reviewers and Editor, which improve the paper a lot. The first and third authors are thankful to Deanship of Scientific Research, Qassim University, Saudi Arabia for technical and financial support of the research project 3611-qec-2018-1-14-S.

Conflicts of Interest: The authors declare no conflict of interest.

References

1. Giannessi, F. Theorems of alternative, quadratic programs and complementarity problems. In *Variational Inequalities and Complementarity Problems*; John Wiley & Sons: New York, NY, USA, 1980; pp. 151–186.
2. Ansari, Q.H.; Lee, G.M. Nonsmooth vector optimization problems and Minty vector variational inequalities. *J. Optim. Theory Appl.* **2010**, *145*, 1–16. [CrossRef]
3. Barbagallo, A.; Ragusa, M.A.; Scapellato, A. Preface of the "5th Symposium on Variational Inequalities and Equilibrium Problems". *AIP Conf. Proc.* **2017**, *1863*, 510001. [CrossRef]
4. Farajzadeh, A.; Noor, M.A.; Noor, K.I. Vector nonsmmoth variational-like inequalities and oplimization problems. *Nonlinear Anal.* **2009**, *71*, 3471–3476. [CrossRef]
5. Jabarootian, T.; Zafarani, J. Generalized invarient monotonicity and invexity of non-differentiable functions. *J. Glob. Optim.* **2006**, *36*, 537–564. [CrossRef]
6. Liu, Z.B.; Kim, J.K.; Huang, N.J. Existence of solutions for nonconvex and nonsmooth vector optimization problems. *J. Inequal. Appl.* **2008**, *2008*, 678014. [CrossRef]
7. Mishra, S.K.; Wang, S.Y.; Lai, K.K. On nonsmooth α-invex functions and vector variational-like inequality. *Optim. Lett.* **2011**, *8*, 91–98.
8. Yang, X.M.; Yang, X.Q.; Teo, K.L. Some remarks on the Minty vector variational inequality. *J. Optim. Theory Appl.* **2004**, *121*, 193–201. [CrossRef]
9. Giannessi, F. *On Minty Variational Principle*; New Trends in Mathematical Programming; Kluwer Academic: Dordrecht, The Netherland, 1997.
10. Yang, X.M.; Yang, X.Q. Vector variational-like inequality with pseudoinvexity. *Optimization* **2004**, *55*, 157–170. [CrossRef]
11. Rezaie, M.; Zafarani, J. Vector optimization and variational-like inequalities. *J. Glob. Optim.* **2009**, *43*, 47–66. [CrossRef]
12. Long, X.J.; Peng, J.W.; Wu, S.Y. Generelaized vector variational-like inequalities and nonsmooth vector optimization problems. *Optimization* **2012**, *61*, 1075–1086. [CrossRef]
13. Oveisiha, M.; Zafarani, J. Generalized Minty vector variational-like inequalities and vector optimization problems in Asplund spaces. *Optim. Lett.* **2013**, *7*, 709–721. [CrossRef]
14. Chen, B.; Huang, N.J. Vector variational-like inequalities and vector optimization problems in Asplund spaces. *Optim. Lett.* **2012**, *6*, 1513–1525. [CrossRef]
15. Antczak, T. (p, r)-invexity in multiobjective programming. *Eur. J. Oper. Res.* **2004**, *152*, 72–87. [CrossRef]
16. Mandal, P.; Nahak, C. Symmetric duality with (p, r)-ρ-(η, θ)-invexity. *Appl. Math. Comput.* **2011**, *217*, 8141–8148. [CrossRef]
17. Jayswal, A.; Choudhury, S. Exponential type vector variational-like inequalities ans nonsmooth vector optimization problem. *J. Appl. Math. Comput.* **2015**, *49*, 127–143. [CrossRef]
18. Mordukhovich, B.S. *Variational Analysis and Generalized Differential I, Basic Theory, Grundlehren Ser. (Fundamental Principles of Mathematical Sciences)*; Springer: New York, NY, USA, 2006; p. 330.
19. Mordukhovich, B.S.; Shao, Y. Extremal characterizations of Asplund spaces. *Proc. Am. Math. Soc.* **1996**, *124*, 197–205. [CrossRef]
20. Fan, K. A generalization of Tychonoff's fixed point theorem. *Math. Ann.* **1961**, *142*, 305–310. [CrossRef]

© 2019 by the authors. Licensee MDPI, Basel, Switzerland. This article is an open access article distributed under the terms and conditions of the Creative Commons Attribution (CC BY) license (http://creativecommons.org/licenses/by/4.0/).

Article

A Kind of New Higher-Order Mond-Weir Type Duality for Set-Valued Optimization Problems

Liu He [1,†], Qi-Lin Wang [1,*], Ching-Feng Wen [2,3,*], Xiao-Yan Zhang [1] and Xiao-Bing Li [1]

1. College of Mathematics and Statistics, Chongqing Jiaotong University, Chongqing 400074, China; heliu20170906@163.com(L.H.); zhangxy4732@163.com(X.-Y.Z.); xiaobinglicq@126.com(X.-B.L.)
2. Center for Fundamental Science; and Research Center for Nonlinear Analysis and Optimization, Kaohsiung Medical University, Kaohsiung 80708, Taiwan
3. Department of Medical Research, Kaohsiung Medical University Hospital, Kaohsiung 80708, Taiwan
* Correspondence: wangql97@126.com(Q.-L.W.); cfwen@kmu.edu.tw (C.-F.W.)
† Current address: No.66 Xuefu Rd., Nan'an Dist., Chongqing 400074, China.

Received: 9 March 2019; Accepted: 17 April 2019; Published: 24 April 2019

Abstract: In this paper, we introduce the notion of higher-order weak adjacent epiderivative for a set-valued map without lower-order approximating directions and obtain existence theorem and some properties of the epiderivative. Then by virtue of the epiderivative and Benson proper efficiency, we establish the higher-order Mond-Weir type dual problem for a set-valued optimization problem and obtain the corresponding weak duality, strong duality and converse duality theorems, respectively.

Keywords: set-valued optimization problems; higher-order weak adjacent epiderivatives; higher-order mond-weir type dual; benson proper efficiency

1. Introduction

The theory of duality and optimality conditions for optimization problems has received considerable attention (see [1–10]). The derivative (epiderivative) plays an important role in studying duality and optimality conditions for set-valued optimization problems. The contingent derivatives [1], the contingent epiderivatives [11] and the generalized contingent epiderivatives [12] for set-valued maps are employed by different authors to investigate necessary or/and sufficient optimality conditions for set-valued optimization problems. Later, the second-order epiderivatives [13], higher-order generalized contingent (adjacent) epiderivatives [14] and generalized higher-order contingent (adjacent) derivatives [15] for set-valued maps are used to study the second (or high) order necessary or/and sufficient optimality conditions for set-valued optimization problems. Chen et al. [2] utilized the weak efficiency to introduce higher-order weak adjacent (contingent) epiderivative for a set-valued map, they then investigate higher-order Mond-Weir (Wolfe) type duality and higher-order Kuhn-Tucker type optimality conditions for constrained set-valued optimization problems. Li et al. [3] used the higher-order contingent derivatives to discuss the weak duality, strong duality and converse duality of a higher-order Mond-Weir type dual for a set-valued optimization problem. Wang et al. [4] used the higher-order generalized adjacent derivative to extend the main results of [3] from convexity to non-convexity. Anh [6] used the higher-order radial derivatives [16] to discuss mixed duality of set-valued optimization problems.

It is well known that the lower-order approximating directions are very important to define the higher-order derivatives (epiderivatives) in [2–4,6,14,15]. This limits their practical applications when the lower-order approximating directions are unknown. So, it is necessary to introduce some higher-order derivatives (epiderivatives) without lower-order approximating directions. As we know, a few paper are devoted to this topic. Motivated by [17], Li et al. [7] proposed the higher-order upper and lower Studniarski derivatives of a set-valued map to establish necessary and sufficient conditions for a strict local minimizer of a constrained set-valued optimization problem. Anh [8] introduced the higher-order radial epiderivative to establish mixed type duality in constrained set-valued optimization problems. Anh [18] proposed the higher-order upper and lower Studniarski derivatives of a set-valued map to establish Fritz John type and Kuhn-Tucker type conditions, and discussed the higher-order Mond-Weir type dual for constrained set-valued optimization problems. Anh [19] further defined the notion of higher-order Studniarski epiderivative and established higher-order optimality conditions for a generalized set-valued optimization problems. Anh [20] noted that the epiderivatives in [8,19] is singleton, they proposed a notion of the higher-order generalized Studniarski epiderivative which is set-valued, and discussed its applications in optimality conditions and duality of set-valued optimization problems.

As we know that the existence conditions of weak efficient point are weaker than ones of efficient point for a set. Inspired by [2,8,18–20], we introduce the higher-order weak adjacent set without the lower-order approximating directions for set-valued maps. Furthermore, we use the higher-order weak adjacent set and weak efficiency to introduce the higher-order weak adjacent epiderivative for a set-valued map, we use it and Benson proper efficiency to discuss higher-order Mond-Weir type dual for a constrained set-valued optimization problem, and then obtain the corresponding weak duality, strong duality and converse duality, respectively.

The rest of the article is as follows. In Section 2, we recall some of definitions and notations to be needed in the paper, and so define the higher-order adjacent set of a set-valued map without lower-order approximating directions, which has some nice properties. In Section 3, we use the higher-order adjacent set of Section 2 to define the higher-order weak adjacent epiderivative for a set-valued map, and discuss its properties, such as existence and subdifferential. In Section 4, we introduce a higher-order Mond-Weir type dual for a constrained set-valued optimization problem and establish the corresponding weak duality, strong duality and converse duality, respectively.

2. Preliminaries

Throughout the paper, let X, Y and Z be three real normed linear spaces. The spaces Y and Z are partially ordered by nontrivial pointed closed convex cones $C \subseteq Y$ and $D \subseteq Z$ with nonempty interior, respectively. By 0_Y we denote the zero vector of Y. Y^* stands for the topological dual space of Y. The dual cone C^+ of C is defined as

$$C^+ := \{f \in Y^* | f(c) \geq 0, \forall c \in C\}.$$

Its quasi-interior C^{+i} is defined as

$$C^{+i} := \{f \in Y^* | f(c) > 0, \forall c \in C \setminus \{0_Y\}\}.$$

Let M be a nonempty subset of Y. We denote the closure, the interior and the cone hull of M by cl$M$, int$M$ and cone$M$, respectively. We denote by $B(c, r)$ the open ball of radius r centered at c. A nonempty subset B of C is called a base of C if and only if $C = $ cone$B$ and $0_Y \notin $ cl$B$.

Let $E \subseteq X$ be a nonempty subset and $F : E \to 2^Y$ be a set-valued map. The domain, graph and epigraph of F are, respectively, defined as

$$\text{dom}F := \{x \in E | F(x) \neq \emptyset\}, \quad \text{im}F := \{y \in Y | y \in F(x)\},$$

$$\text{gr}F := \{(x, y) \in E \times Y | y \in F(x), x \in E\}$$

and

$$\text{epi}F := \{(x, y) \in E \times Y | y \in F(x) + C, x \in E\}.$$

Definition 1. *[9] Let $M \subseteq Y$ and $y_0 \in M$.*
(i) y_0 is said to be a Pareto efficient point of M ($y_0 \in \text{Min}_C M$) if

$$(M - \{y_0\}) \cap (-C \setminus \{0_Y\}) = \emptyset.$$

(ii) Let $\text{int} C \neq \emptyset$. y_0 is said to be a weakly efficient point of M ($y_0 \in \text{WMin}_C M$) if

$$(M - \{y_0\}) \cap (-\text{int} C) = \emptyset.$$

Definition 2. *[10,21,22] (i) The cone C is called Daniell if any decreasing sequence in Y that has a lower bound converges to its infimum.*
(ii) A subset M of Y is said to be minorized if there is a $y \in Y$ such that

$$M \subseteq \{y\} + C.$$

(iii) The weak domination property is said to hold for a subset M of Y if

$$M \subseteq \text{WMin}_C M + \text{int} C \cup \{0_Y\}.$$

Definition 3. *Let $A \subseteq X \times Y$, $(x_0, y_0) \in \text{cl} A$ and $m \in \mathbb{N} \setminus \{0\}$.*
(i) [9] The mth-order adjacent set of A at $(x, v_1, \cdots, v_{m-1})$ is defined by

$$T_A^{\flat(m)}(x, v_1, \cdots, v_{m-1}) := \{y \in A | \forall t_n \to 0^+, \exists y_n \to y, s.t.$$
$$x_0 + t_n v_1 + \cdots + t_n^{m-1} v_{m-1} + t_n^m y_n \in A\},$$

where $v_i \in X (i = 1, \cdots, m-1)$.
(ii) [19] The mth-order Studniarski set of A at (x_0, y_0) is defined by

$$S_A^m(x_0, y_0) := \{(x, y) \in X \times Y | \exists t_n \to 0^+, \exists (x_n, y_n) \to (x, y),$$
$$s.t. (x_0 + t_n x_n, y_0 + t_n^m y_n) \in A\}.$$

Definition 4. *Let $K \subseteq X \times Y$, $(x_0, y_0) \in \text{cl} K$ and $m \in \mathbb{N} \setminus \{0\}$. The mth-order adjacent set of K at (x_0, y_0) is defined by*

$$T_K^{\flat(m)}(x_0, y_0) := \{(x, y) \in X \times Y | \forall t_n \to 0^+, \exists (x_n, y_n) \to (x, y),$$
$$s.t. (x_0 + t_n x_n, y_0 + t_n^m y_n) \in K\}.$$

We can obtain the equivalent characterization of $T_K^{\flat(m)}(x_0, y_0)$ in terms of sequences: $(x, y) \in T_K^{\flat(m)}(x_0, y_0)$ if and only if $\forall \{t_n\} \to 0^+, \exists \{(x'_n, y'_n)\} \subseteq K$ such that

$$\lim_{n \to \infty} \left(\frac{x'_n - x_0}{t_n}, \frac{y'_n - y_0}{t_n^m} \right) = (x, y).$$

Now, we establish a few properties of $T_K^{\flat(m)}(x_0, y_0)$.

Proposition 1. *Let $K \subseteq X \times Y$, $(x_0, y_0) \in K$ and $(x, y) \in T_K^{\flat(m)}(x_0, y_0)$. Then*

$$(\lambda x, \lambda^m y) \in T_K^{\flat(m)}(x_0, y_0), \forall \lambda \geq 0.$$

Proof. We divide λ into two cases to show the proposition.
Case 1: $\lambda = 0$. Note that $(x_0, y_0) \in K$; for any sequence $\{t_n\}$ with $t_n \to 0^+$, we choose $(x_n, y_n) = (0_X, 0_Y)$ such that $(x_0 + t_n x_n, y_0 + t_n^m y_n) \in K$. This means that $(0_X, 0_Y) \in T_K^{\flat(m)}(x_0, y_0)$.

Case 2: $\lambda > 0$. Let $(x,y) \in T_K^{\flat(m)}(x_0,y_0)$. Then for any sequence $\{t_n\}$ with $t_n \to 0^+$, there exists a sequence $\{(x_n, y_n)\} \subseteq K$ with $(x_n, y_n) \to (x,y)$ such that

$$K \ni (x_0 + t_n x_n, y_0 + t_n^m y_n) = (x_0 + (\frac{t_n}{\lambda})\lambda x_n, y_0 + (\frac{t_n}{\lambda})^m \lambda^m y_n).$$

Naturally, $\frac{t_n}{\lambda} \to 0^+$ and $(\lambda x_n, \lambda^m y_n) \to (\lambda x, \lambda^m y) \in T_K^{\flat(m)}(x_0, y_0)$. It completes the proof. □

Remark 1. Let $K \subseteq X \times Y$ and $(x_0, y_0) \in \text{cl}K$. The mth-order adjacent set $T_K^{\flat(m)}(x_0, y_0)$ of K at (x_0, y_0) may not be a cone; see Example 1.

Example 1. Let $K = \{(x,y) \in \mathbb{R}^2 | y \geq x^4, x \in \mathbb{R}\}$, $(x_0, y_0) = (0,0)$ and $m = 4$. A simple calculation shows that

$$T_K^{\flat(4)}(0,0) = \{(x,y) \in \mathbb{R}^2 | y \geq x^4\}.$$

Take $(x,y) = (1,1) \in T_K^{\flat(4)}(0,0)$ and $\lambda = 2$. Then $\lambda(x,y) = (2,2) \notin T_K^{\flat(4)}(0,0)$, i.e., $T_k^{\flat(4)}(0,0)$ is not a cone here.

Proposition 2. Let $F : E \to 2^Y$ be a set-valued map and $(x_0, y_0) \in \text{gr}F$. Then,
(i) $T_{\text{epi}F}^{\flat(m)}(x_0, y_0) = T_{\text{epi}F}^{\flat(m)}(x_0, y_0) + \{0_X\} \times C$;
(ii) $\{y \in Y | (x,y) \in T_{\text{epi}F}^{\flat(m)}(x_0, y_0)\} = \{y \in Y | (x,y) \in T_{\text{epi}F}^{\flat(m)}(x_0, y_0)\} + C, \forall x \in X$.

Proof. Since $0_Y \in C$, it is clearly that $T_{\text{epi}F}^{\flat(m)}(x_0, y_0) \subseteq T_{\text{epi}F}^{b(m)}(x_0, y_0) + \{0_X\} \times C$. Therefore we only need to prove $T_{\text{epi}F}^{\flat(m)}(x_0, y_0) + \{0_X\} \times C \subseteq T_{\text{epi}F}^{\flat(m)}(x_0, y_0)$.

Let $(u, v) \in T_{\text{epi}F}^{\flat(m)}(x_0, y_0)$ and $c \in C$. Then for any sequence $\{t_n\}$ with $t_n \to 0^+$, there exists a sequence $\{(u_n, v_n)\} \subseteq X \times Y$ with $(u_n, v_n) \to (u,v)$ such that

$$(x_0 + t_n u_n, y_0 + t_n^m v_n) \in \text{epi}F,$$

namely,

$$y_0 + t_n^m v_n \in F(x_0 + t_n u_n) + C.$$

Since $c \in C$, $t_n \to 0^+$ and $C + C \subseteq C$, one has

$$y_0 + t_n^m(v_n + c) \in F(x_0 + t_n u_n) + C + \{t_n^m c\} \subseteq F(x_0 + t_n u_n) + C.$$

Thus

$$(x_0 + t_n u_n, y_0 + t_n^m(v_n + c)) \in \text{epi}F.$$

This together with $(u_n, v_n + c) \to (u, v + c)$ implies $(u, v + c) \in T_{\text{epi}F}^{\flat(m)}(x_0, y_0)$, and so $T_{\text{epi}F}^{\flat(m)} + \{0_X\} \times C \subseteq T_{\text{epi}F}^{\flat(m)}$.

(ii) Obviously, (ii) follows from (i). The proof is complete. □

Proposition 3. Let $K \subseteq X \times Y$ and $(x_0, y_0) \in \text{cl}K$. If K is a convex set, then $T_K^{\flat(m)}(x_0, y_0)$ is a convex set.

Proof. Let $(x^i, y^i) \in T_K^{\flat(m)}(x_0, y_0)$ $(i = 1, 2)$ and $\lambda \in [0, 1]$. Then for any $t_n \to 0^+$, there exist $(x_n^i, y_n^i) \to (x^i, y^i)$ $(i = 1, 2)$ such that

$$(x_0 + t_n x_n^i, y_0 + t_n^m y_n^i) \in K \ (i = 1, 2).$$

From the convexity of K, we have

$$(x_0 + t_n[(1-\lambda)x_n^1 + \lambda x_n^2], y_0 + t_n^m[(1-\lambda)y_n^1 + \lambda y_n^2]) \in K.$$

It is obvious that

$$((1-\lambda)x_n^1 + \lambda x_n^2, (1-\lambda)y_n^1 + \lambda y_n^2) \to ((1-\lambda)x^1 + \lambda x^2, (1-\lambda)y^1 + \lambda y^2).$$

It follows from the definition of $T_K^{\flat(m)}(x_0, y_0)$ that

$$(1-\lambda)(x^1, y^1) + \lambda(x^2, y^2) = ((1-\lambda)x^1 + \lambda x^2, (1-\lambda)y^1 + \lambda y^2) \in T_K^{\flat(m)}(x_0, y_0).$$

Thus, $T_K^{\flat(m)}(x_0, y_0)$ is a convex set and the proof is complete. □

3. Higher-Order Weak Adjacent Epiderivatives

In this section, we introduce the notion of higher-order weak adjacent epiderivative of a set-valued map without lower-order approximating directions, and obtain some properties of the epiderivative.

Firstly, we recall the notions of mth-order weak adjacent epiderivative with lower-order approximating directions and generalized Studniarski epiderivative without lower-order approximating directions.

Definition 5. [2] *Let* $F : X \to 2^Y$, $(x_0, y_0) \in \operatorname{gr} F$ *and* $(u_i, v_i) \in X \times Y (i = 1, \cdots, m-1)$. *The mth-order weak adjacent epiderivative* $D_w^{\flat(m)} F(x_0, y_0, u_1, v_1 \cdots, u_{m-1}, v_{m-1})$ *of F at (x_0, y_0) for vectors* $(u_1, v_1), \cdots, (u_{m-1}, v_{m-1})$ *is the set-valued map from X to Y defined by*

$$D_w^{\flat(m)} F(x_0, y_0, u_1, v_1, \cdots, u_{m-1}, v_{m-1})(x)$$
$$:= \operatorname{WMin}_C \{y \in Y \mid (x, y) \in T_{\operatorname{epi} F}^{\flat(m)}(x_0, y_0, u_1, v_1, \cdots, u_{m-1}, v_{m-1})\}.$$

Definition 6. [20] *Let* $F : X \to 2^Y$ *and* $(x_0, y_0) \in \operatorname{gr} F$. *The mth-order generalized Studniarski epiderivative* $G\text{-}ED_S^m F(x_0, y_0)$ *of F at (x_0, y_0) is the set-valued map from X to Y defined by*

$$G\text{-}ED_S^m F(x_0, y_0)(x) := \operatorname{Min}_C \{y \in Y \mid (x, y) \in S_{\operatorname{epi} F}^m(x_0, y_0)\}.$$

Motivated by Definitions 5 and 6, we introduce the higher-order epiderivative without lower-order approximating directions.

Definition 7. *Let* $F : E \to 2^Y$ *and* $(x_0, y_0) \in \operatorname{gr} F$. *The mth-order weak adjacent epiderivative of F at (x_0, y_0) is a set-valued map* $ED_w^{\flat(m)} F(x_0, y_0) : E \to 2^Y$ *defined by*

$$ED_w^{\flat(m)} F(x_0, y_0)(x) := \operatorname{WMin}_C \{y \in Y \mid (x, y) \in T_{\operatorname{epi} F}^{\flat(m)}(x_0, y_0)\}.$$

Remark 2. *There are many examples show that* $ED_w^{\flat(m)} F(x_0, y_0)$ *possibly exists even if* $D_w^{\flat(m)} F(x_0, y_0, u_1, v_1, \cdots, u_{m-1}, v_{m-1})$ *and* $G\text{-}ED_S^m F(x_0, y_0)$ *do not; see Examples 2 and 3. Therefore it is interesting to study this derivative and employ it to investigate the Mond-Weir duality for set-valued optimization problems.*

Example 2. *Let* $E = X = Y = \mathbb{R}$, $C = \mathbb{R}_+$ *and* $F : E \to 2^Y$ *be defined by* $F(x) := \{y \in Y \mid y \geq x^2\}$. *Take* $(x_0, y_0) = (0, 0) \in \operatorname{gr} F$ *and* $(u, v) = (1, -1)$. *Then, simple calculations show that*

$$T_{\operatorname{epi} F}^{\flat(2)}((0, 0), (1, -1)) = \emptyset$$

and
$$T_{\text{epi}F}^{\flat(2)}(0,0) = \{(x,y) \in \mathbb{R} \times \mathbb{R} \mid x \in \mathbb{R}, y \geq x^2\}.$$

So, for any $x \in E$, $ED_w^{\flat(2)}F((0,0),(1,-1))(x) = \emptyset$, but $ED_w^{\flat(2)}F(0,0)(x) = \{x^2\}$.

Example 3. Let $E = X = \mathbb{R}$, $Y = \mathbb{R}^2$, $C = \mathbb{R}_+^2$ and $F : E \to 2^Y$ be defined by $F(x) := \{(y_1, y_2) \in Y \mid y_1 \in \mathbb{R}, y_2 \geq x^2\}$. Take $(x_0, y_0) = (0_X, 0_Y) \in \text{gr}F$. Then

$$S_{\text{epi}F}^2(0_X, 0_Y) = T_{\text{epi}F}^{\flat(2)}(0_X, 0_Y) = \{(x, (y_1, y_2)) \in \mathbb{R} \times \mathbb{R}^2 \mid x \in \mathbb{R}, y_1 \in \mathbb{R}, y_2 \geq x^2\}.$$

Therefore, for any $x \in E$, $ED_w^{\flat(2)}F(0_X, 0_Y)(x) = \{(y_1, y_2) \in \mathbb{R}^2 \mid y_1 \in \mathbb{R}, y_2 = x^2\}$, but $G\text{-}ED_S^2 F(0_X, 0_Y)(x) = \emptyset$.

Theorem 1. Let $F : E \to 2^Y$ and $(x_0, y_0) \in \text{gr}F$. Let C be a pointed closed convex cone and Daniell. If $P(x) := \{y \in Y \mid (x,y) \in T_{\text{epi}F}^{\flat(m)}(x_0, y_0)\}$ is minorized for all $x \in \text{dom}P$, then $ED_w^{\flat(m)}F(x_0, y_0)$ exists.

Proof. The proof is similar to that of Theorem 3.1 in [2]. □

Definition 8. [23] Let $M \subseteq \mathbb{R}^n$ be a nonempty set and $x_0 \in M$. M is called star-shaped at x_0, if for any point $x \in M$ with $x \neq x_0$, the segment

$$[x, x_0] := \{y \in M \mid y = (1-\lambda)x_0 + \lambda x, 0 \leq \lambda \leq 1\} \subseteq M.$$

Definition 9. [10] Let E be a nonempty convex set. The map F is said to be C-convex on E, if for any $x_1, x_2 \in E$ and $\lambda \in [0,1]$,

$$\lambda F(x_1) + (1-\lambda)F(x_2) \subseteq F(\lambda x_1 + (1-\lambda)x_2) + C.$$

Motivated by Definition 9, we introduce the following concept.

Definition 10. Let E be a star-shaped set at $x_0 \in E$. The map F is said to be generalized C-convex at x_0 on E, if for any $x \in E$ and $\lambda \in [0,1]$,

$$(1-\lambda)F(x_0) + \lambda F(x) \subseteq F((1-\lambda)x_0 + \lambda x) + C.$$

Remark 3. Let E be a convex set and $x_0 \in E$. If F is C-convex on E, then F is generalized C-convex at x_0 on E. However, the converse implication is not true.

To understand Remark 3, we give the following example.

Example 4. Let $E_1 = (-\infty, -1] \subseteq \mathbb{R}$, $E_2 = (-1, 1] \subseteq \mathbb{R}$, $E = X = E_1 \cup E_2 \subseteq \mathbb{R}$, $Y = \mathbb{R}$, $C = \mathbb{R}_+$ and $F : E \to 2^Y$ be defined by

$$F(x) = \begin{cases} \{y \in Y \mid y \geq 1\}, x \in E_1, \\ \{y \in Y \mid y \geq x^2\}, x \in E_2. \end{cases}$$

Take $x_0 = -1 \in E$. Then E is a convex set, and F is generalized C-convex at x_0 on E. Take $x_1 = -4 \in E_1 \subseteq E$, $x_2 = 0 \in E_2 \subseteq E$ and $\lambda = \frac{1}{2}$, then

$$\frac{1}{2}F(x_1) + \frac{1}{2}F(x_2) = \{y \mid y \geq \frac{1}{2}\}$$

and

$$F(\frac{1}{2}x_1 + \frac{1}{2}x_2) = \{y \mid y \geq 1\}.$$

Thus
$$\frac{1}{2}F(x_1) + (1 - \frac{1}{2})F(x_2) \not\subseteq F(\frac{1}{2}x_1 + (1 - \frac{1}{2})x_2) + C.$$

Therefore F is not C-convex on E.

Definition 11. [24] *Let $U \subseteq X$ be a star-shaped set at $x_0 \in U$. A set-valued map $F : U \to 2^Y$ is said to be decreasing-along-rays at x_0 if for any $x \in U$ and $0 \le t_1 \le t_2$ with $t_i x + (1 - t_i)x_0 \in U(i = 1, 2)$, one has*
$$F(t_1 x + (1 - t_1)x_0) \subseteq F(t_2 x + (1 - t_2)x_0) + C.$$

Next, we give an important property of the mth-order weak adjacent epiderivative.

Proposition 4. *Let E be a star-shaped set at $x_0 \in E$. Let $F : E \to 2^Y$ be a set-valued map and $(x_0, y_0) \in \mathrm{gr}F$. Suppose that the following conditions are satisfied:*
(i) F is decreasing-along-rays at x_0;
(ii) F is generalized C-convex at x_0 on E;
(iii) the set $P(x) := \{y \in Y \mid (x, y) \in T_{\mathrm{epi}F}^{\flat(m)}(x_0, y_0)\}$ fulfills the weak domination property for all $x \in \mathrm{dom}P$.
Then for all $x \in E$, one has $x - x_0 \in \Omega := \mathrm{dom}ED_w^{\flat(m)}F(x_0, y_0)$ and
$$F(x) - \{y_0\} \subseteq ED_w^{\flat(m)}F(x_0, y_0)(x - x_0) + C.$$

Proof. Let $x \in E$ and $y \in F(x)$. For any $\lambda_n \in (0, 1)$ with $\lambda_n \to 0^+$, $(\frac{\lambda_n}{2})^m \le \frac{\lambda_n}{2}$. Since E is a star-shaped set at x_0,
$$x_n := x_0 + \frac{\lambda_n}{2}(x - x_0) = (1 - \frac{\lambda_n}{2})x_0 + \frac{\lambda_n}{2}x \in E$$
and
$$x_0 + (\frac{\lambda_n}{2})^m(x - x_0) = (1 - (\frac{\lambda_n}{2})^m)x_0 + (\frac{\lambda_n}{2})^m x \in E.$$

Together this with conditions (i) and (ii) implies
$$y_n := y_0 + (\frac{\lambda_n}{2})^m(y - y_0) = (1 - (\frac{\lambda_n}{2})^m)y_0 + (\frac{\lambda_n}{2})^m y$$
$$\in (1 - (\frac{\lambda_n}{2})^m)F(x_0) + (\frac{\lambda_n}{2})^m F(x) \subseteq F((1 - (\frac{\lambda_n}{2})^m)x_0 + (\frac{\lambda_n}{2})^m x) + C$$
$$\subseteq F(x_n) + C + C \subseteq F(x_n) + C.$$

Hence, $(x_n, y_n) \in \mathrm{epi}F$. It follows from the definition of $T_K^{\flat(m)}(x_0, y_0)$ that $(x - x_0, y - y_0) \in T_{\mathrm{epi}F}^{\flat(m)}(x_0, y_0)$. Replacing $x - x_0 \in \mathrm{dom}P$ with x of condition (iii), from the definition of $ED_w^{\flat(m)}F(x_0, y_0)$, we have
$$P(x - x_0) \subseteq ED_w^{\flat(m)}F(x_0, y_0)(x - x_0) + \mathrm{int}C \cup \{0_Y\}$$
$$\subseteq ED_w^{\flat(m)}F(x_0, y_0)(x - x_0) + C.$$

Thus $x - x_0 \in \Omega$ and
$$F(x) - \{y_0\} \subseteq ED_w^{\flat(m)}F(x_0, y_0)(x - x_0) + C.$$

This completes the proof. □

We now give an example to explain Proposition 4.

Example 5. Let $E = [0, +\infty) \subseteq \mathbb{R}$, $Y = \mathbb{R}$, $C = \mathbb{R}_+$ and $F : E \to 2^Y$ be defined as $F(x) = \{y \in Y \mid y \geq 0\}$. Take $(x_0, y_0) = (0, 0) \in \text{gr}F$. Then, simple calculations show that $T_{\text{epi}F}^{\flat(2)}(0,0) = \mathbb{R}_+^2$ and

$$ED_w^{\flat(2)} F(0,0)(x - x_0) = \{0\}, \forall x \geq 0.$$

We can easily see that all conditions of Proposition 4 are satisfied. For any $x \in E$, one has $x - 0 \in \Omega := \text{dom}ED_w^{\flat(2)} F(0,0) = \{x \mid x \geq 0\}$ and

$$F(x) - \{y_0\} \subseteq ED_w^{\flat(2)} F(0,0)(x - x_0) + C.$$

Therefore Proposition 4 is applicable here.

The following examples show that every condition of Proposition 4 is necessary.

Example 6. Let $E = [0, +\infty) \subseteq \mathbb{R}$, $Y = \mathbb{R}$, $C = \mathbb{R}_+$ and $F : E \to 2^Y$ be a set-valued map satisfing $F(x) = \{y \in Y \mid y \geq x\}$. Take $(x_0, y_0) = (0, 0) \in \text{gr}F$. By a simple calculation, we obtain

$$T_{\text{epi}F}^{\flat(2)}(0,0) = \{(0, y) \in \mathbb{R} \times \mathbb{R} \mid y \geq 0\}$$

and

$$ED_w^{\flat(2)} F(0,0)(x) = \begin{cases} \{0\}, x = 0, \\ \varnothing, x \neq 0. \end{cases}$$

Thus $x - 0 \notin \Omega := \text{dom}ED_w^{\flat(2)} F(0,0) = \{0\}$, for any $x \in (0, +\infty)$.
Obviously, the conditions (ii) and (iii) of Proposition 4 are satisfied except condition (i), and

$$F(x) - \{y_0\} \nsubseteq ED_w^{\flat(2)} F(x_0, y_0)(x - x_0) + C, x \in (0, +\infty).$$

Thus Proposition 4 does not hold here and the condition (i) of Proposition 4 is essential.

Example 7. Let $E_1 = [0, 1] \subseteq \mathbb{R}$, $E_2 = (1, +\infty) \subseteq \mathbb{R}$, $E = X = E_1 \cup E_2 \subseteq \mathbb{R}$, $Y = \mathbb{R}$, $C = \mathbb{R}_+$ and $F : E \to 2^Y$ be given by

$$F(x) = \begin{cases} \{y \in \mathbb{R} \mid y \geq -x^2\}, x \in E_1, \\ \{y \in \mathbb{R} \mid y \geq -x^3\}, x \in E_2. \end{cases}$$

Take $(x_0, y_0) = (0, 0) \in \text{gr}F = \text{epi}F$. Then,

$$T_{\text{epi}F}^{\flat(2)}(0,0) = \{(x, y) \in \mathbb{R} \times \mathbb{R} \mid y \geq -x^2, x \geq 0\}$$

and

$$ED_w^{\flat(2)} F(0,0)(x - x_0) = \{y \in \mathbb{R} \mid y = -x^2\}, \forall x \geq 0.$$

Clearly, the conditions (i) and (iii) of Proposition 4 are satisfied except condition (ii), and for any $x \in E_2$,

$$F(x) - \{y_0\} \nsubseteq ED_w^{\flat(2)} F(x_0, y_0)(x - x_0) + C.$$

Therefore Proposition 4 does not hold here and the condition (ii) of Proposition 4 is essential.

Example 8. Let $E = X = \mathbb{R}$, $Y = \mathbb{R}^2$, $C = \mathbb{R}_+^2$ and $F : E \to 2^Y$ be defined by $F(x) := \{(y_1, y_2) \in Y \mid y_1 \in \mathbb{R}, y_2 \geq 0\}$. Take $(x_0, y_0) = (0, (0, 1)) \in \text{gr}F$. Then a simple calculation shows that

$$T_{\text{epi}F}^{\flat(2)}(0, (0,1)) = \mathbb{R} \times \mathbb{R}^2.$$

This means that: (i) dom$P = \mathbb{R}$ and $P(x) = \mathbb{R}^2, \forall x \in$ dom$P$; (ii) $ED_w^{\flat(2)}F(0,(0,1)) = \emptyset$ for each $x \in \mathbb{R}$. Obviously, $P(x) := \{y \in Y \mid (x,y) \in \mathbb{R} \times \mathbb{R}^2\}$ does not fulfill the weak domination property for each $x \in \mathbb{R}$ and $\Omega = \emptyset$. Thus Proposition 4 does not hold here and the condition (iii) of Proposition 4 is essential.

4. Higher-Order Mond-Weir Type Duality

In this section, by virtue of the higher-order weak adjacent epiderivative of a set-valued map, we establish Mond-Weir duality theorems for a constrained optimization problem under Benson proper efficiency.

Let $E \subseteq X$, $F : E \to 2^Y$ and $G : E \to 2^Z$ be two set-valued maps. We consider the following constrained set-valued optimization problem:

$$\text{(SOP)} \begin{cases} \text{Min}_C & F(x), \\ \text{s.t.} & x \in E, G(x) \cap (-D) \neq \emptyset. \end{cases}$$

Let $M := \{x \in E \mid G(x) \cap (-D) \neq \emptyset\}$ and $F(M) := \bigcup_{x \in M} F(x)$. We denote $F(x) \times G(x)$ by $(F,G)(x)$. The point $(x_0, y_0) \in E \times Y$ is said to be a feasible solution of (SOP) if $x_0 \in M$ and $y_0 \in F(x_0)$.

Definition 12. [25] *The feasible solution (x_0, y_0) is called a Benson proper efficient solution of (SOP) if*

$$\text{clcone}(F(M) + C - \{y_0\}) \cap (-C) = \{0_Y\}.$$

Let $(\tilde{x}, \tilde{y}, \tilde{z}) \in \text{gr}(F,G)$, $\nu \in Y^*$, $\omega \in Z^*$ and $x \in \Theta := \text{dom} ED_w^{\flat(m)}(F,G)(\tilde{x}, \tilde{y}, \tilde{z})$. Inspired by [2], We establish a new higher-order Mond-Weir type dual problem (DSOP) of (SOP) as follows:

$$\begin{align}
\max \quad & \tilde{y} \\
\text{s.t.} \quad & \nu(y) + \omega(z) \geq 0, \forall (y,z) \in ED_w^{\flat(m)}(F,G)(\tilde{x}, \tilde{y}, \tilde{z})(x), x \in \Theta, \tag{1}\\
& \omega(\tilde{z}) \geq 0, \tag{2}\\
& \nu \in C^{+i}, \tag{3}\\
& \omega \in D^+. \tag{4}
\end{align}$$

The point $(\tilde{x}, \tilde{y}, \tilde{z}, \nu, \omega)$ is called a feasible solution of (DSOP) if $(\tilde{x}, \tilde{y}, \tilde{z}, \nu, \omega)$ satisfies conditions (1), (2), (3) and (4) of (DSOP). A feasible solution $(x_0, y_0, z_0, \nu_0, \omega_0)$ is called a maximal solution of (DSOP) if for all $\tilde{y} \in M_D$, $(\{\tilde{y}\} - \{y_0\}) \cap (C \setminus \{0_Y\}) = \emptyset$, where $M_D := \{\tilde{y} \in F(\tilde{x}) \mid (\tilde{x}, \tilde{y}, \tilde{z}) \in \text{gr}(F,G), \nu \in C^{+i}, \omega \in D^+$, and $(\tilde{x}, \tilde{y}, \tilde{z}, \nu, \omega)$ is the feasible solution of (DSOP)$\}$.

Definition 13. [26] *Let $K \subseteq X$, the interior tangent cone of K at x_0 is defined by*

$$IT_K(x_0) := \{\mu \in X \mid \exists \lambda > 0, \forall t \in (0, \lambda), \forall \mu' \in B_X(\mu, \lambda), x_0 + t\mu' \in K\},$$

where $B_X(\mu, \lambda)$ stands for the closed ball centered at $\mu \in X$ and of radius λ.

Theorem 2. (Weak Duality) *Let E be a star-shaped set at $\tilde{x} \in E$ and $(\tilde{x}, \tilde{y}, \tilde{z}) \in \text{gr}(F,G)$. Let (x_0, y_0) and $(\tilde{x}, \tilde{y}, \tilde{z}, \nu, \omega)$ be the feasible solution of (SOP) and (DSOP), respectively. Then the weak duality: $\nu(y_0) \geq \nu(\tilde{y})$ holds if the following conditions are satisfied:*
  (i) (F,G) *is decreasing-along-rays at $\tilde{x}$;*
  (ii) (F,G) *is generalized $C \times D$-convex at $\tilde{x}$ on E;*
  (iii) *the set $P_{(F,G)}(x_0 - \tilde{x}) := \{y \in Y | (x_0 - \tilde{x}, y, z) \in T_{\text{epi}(F,G)}^{\flat(m)}(\tilde{x}, \tilde{y}, \tilde{z})\}$ fulfills the weak domination property.*

Proof. Since (x_0, y_0) is a feasible solution of (SOP), $G(x_0) \cap (-D) \neq \emptyset$. Take $z_0 \in G(x_0) \cap (-D)$. It follows from (2) and (4) that
$$w(z_0 - \tilde{z}) \leq 0. \tag{5}$$
From Proposition 4 it follows that $x_0 - \tilde{x} \in S := \text{dom} ED_w^{\flat(m)}(F,G)(\tilde{x},\tilde{y},\tilde{z})$ and
$$(y_0, z_0) - (\tilde{y}, \tilde{z}) \in ED_w^{\flat(m)}(F,G)(\tilde{x},\tilde{y},\tilde{z})(x_0 - \tilde{x}) + C \times D. \tag{6}$$

Noting that $v \in C^{+i}$ and $w \in D^+$, we have by (1) and (6) that $v(y_0 - \tilde{y}) + w(z_0 - \tilde{z}) \geq 0$. Combining this with (5), one has
$$v(y_0 - \tilde{y}) \geq 0.$$

Thus $v(y_0) \geq v(\tilde{y})$ and the proof is complete. □

Theorem 2 is an extension of [2], Theorem 4.1 from cone convexity to generalized cone convexity. Now, we give an example to illustrate that Theorem 2 can apply but [2], Theorem 4.1 dose not.

Example 9. Let $X = Y = Z = \mathbb{R}$, $C = D = \mathbb{R}_+$, $F : E \to 2^Y$ be given as $F(x) = \{y \in Y \mid y \geq 0\}$ and $G : E \to 2^Z$ be defined by
$$G(x) = \begin{cases} \{z \in Z \mid z \geq 0\}, & x \leq 0, \\ \mathbb{R}, & x > 0. \end{cases}$$
Then sets of the feasible solutions for (DSOP) and (SOP) are $\{(\tilde{x},\tilde{y},\tilde{z},v,w) \mid \tilde{x} = 0, \tilde{y} = 0, \tilde{z} \geq 0, v \in C^{+i}, w = 0\}$ and $\{(x_0, y_0) \mid x_0 \in \mathbb{R}, y_0 \geq 0\}$, respectively. Thus $v(y_0) \geq v(\tilde{y}) = v(0)$ and Theorem 2 holds here. However, [2], Theorem 4.1 is not applicable here because G is not C-convex on E.

Lemma 1. [27] Let $x_0 \in K \subseteq X$ and $\text{int} K \neq \emptyset$. If K is convex, then
$$IT_{\text{int} K}(x_0) = \text{intcone}(K - \{x_0\}).$$

The inclusion relation between the generalized second-order adjacent epiderivative and convex cone C and D is established by Wang and Yu in [28], Theorem 5.2. Inspired by [28], Theorem 5.2, we next introduce the equality of the higher-order weak adjacent epiderivative and convex cone C and D to the proof of the strong duality theory.

Lemma 2. Let $(x_0, y_0, z_0) \in \text{gr}(F,G)$ and $z_0 \in -D$. If (x_0, y_0) is a Benson proper efficient solution of (SOP), then for all $x \in \Theta := \text{dom} ED_w^{\flat(m)}(F,G)(x_0, y_0, z_0)$,
$$[ED_w^{\flat(m)}(F,G)(x_0,y_0,z_0)(x) + C \times D + \{(0_Y, z_0)\}] \cap (-((C \setminus \{0_Y\}) \times \text{int} D)) = \emptyset. \tag{7}$$

Proof. We can easily see that (7) is equivalent to
$$[ED_w^{\flat(m)}(F,G)(x_0,y_0,z_0)(x) + C \times D] \cap (-((C \setminus \{0_Y\}) \times (\text{int} D + \{z_0\}))) = \emptyset. \tag{8}$$

Thus we only need to prove that (8) holds. Suppose on the contrary that there exist $x \in \Theta$, $(y,z) \in ED_w^{\flat(m)}(F,G)(x_0,y_0,z_0)(x)$ and $(c_0, d_0) \in C \times D$ such that
$$z + d_0 \in -(\text{int} D + \{z_0\}) \tag{9}$$
and
$$y + c_0 \in -(C \setminus \{0_Y\}). \tag{10}$$

It follows from $(y,z) \in ED_w^{b(m)}(F,G)(x_0,y_0,z_0)(x)$ that $(x,y,z) \in T_{epi(F,G)}^{b(m)}(x_0,y_0,z_0)$. Then for any sequence $\{t_n\}$ with $t_n \to 0^+$, there exists $\{(x_n, y_n, z_n)\} \subseteq epi(F,G)$ such that

$$\left(\frac{x_n - x_0}{t_n}, \frac{(y_n, z_n) - (y_0, z_0)}{t_n^m}\right) \to (x, y, z). \tag{11}$$

From (9) and (11), there exists a sufficiently large natural number N_1 such that

$$\bar{z}_n := \frac{z_n - z_0 + t_n^m d_0}{t_n^m} \in -(intD + \{z_0\}) \subseteq -intcone(D + \{z_0\}) \tag{12}$$
$$\subseteq -IT_{int}D(-z_0), \forall n > N_1,$$

where the last inclusion follows from Lemma 1. According to Definition 13, there exists $\lambda > 0$ such that

$$-z_0 + t_n \mu' \in intD, \forall t_n \in (0, \lambda), \mu' \in B_Y(-\bar{z}_n, \lambda), n > N_1. \tag{13}$$

Since $t_n \to 0^+$, there exists a sufficiently large natural number N_2 with $N_2 \geq N_1$ such that $t_n^m \in (0, \lambda)$. Combining this with (13), one has

$$-z_0 + t_n^m(-\bar{z}_n) \in intD, \forall n > N_2. \tag{14}$$

From (12) and (14), we have

$$-z_0 - (z_n - z_0 + t_n^m d_0) = -z_n - t_n^m d_0 \in intD, \forall n > N_2.$$

It follows from $d_0 \in D$, $t_n^m \to 0^+$ and $intD + D \subseteq intD$ that

$$z_n \in -intD, \forall n > N_2. \tag{15}$$

Noting that $\{(x_n, y_n, z_n)\} \subseteq epi(F, G)$, there exist $x_n \in E$, $\hat{z}_n \in G(x_n)$, $\hat{y}_n \in F(x_n)$ and $(c_n, d_n) \in C \times D$ such that $y_n = \hat{y}_n + c_n$ and $z_n = \hat{z}_n + d_n$. By (15), $\hat{z}_n \in -intD - \{d_n\} \subseteq -intD \subseteq -D, \forall n > N_2$. Therefore

$$x_n \in M, \forall n > N_2. \tag{16}$$

Clearly, we have

$$\frac{y_n - y_0}{t_n^m} + c_0 = \frac{y_n + t_n^m c_0 - y_0}{t_n^m} \in \frac{F(x_n) + C - \{y_0\}}{t_n^m}$$
$$\subseteq \frac{F(M) + C - \{y_0\}}{t_n^m}$$
$$\subseteq clcone(F(M) + C - \{y_0\}).$$

It follows from (11) and (16) that $y + c_0 \in clcone(F(M) + C - \{y_0\})$. Combining this with (10), one has

$$y + c_0 \in clcone(F(M) + C - \{y_0\}) \cap (-(C \setminus \{0_Y\})),$$

which contradicts that (x_0, y_0) is a Benson proper efficient solution of (SOP). Thus (7) holds and the proof is complete. □

According to Theorem 2.3 of [29], we have the following lemma.

Lemma 3. *[29] Let W be a locally convex space, H and Q be cones in W. If H is closed, Q have a compact base and $H \cap Q = \{0_W\}$, then there is a pointed convex cone $\tilde{A}$ such that $Q \setminus \{0_W\} \subseteq int\tilde{A}$ and $\tilde{A} \cap H = \{0_W\}$.*

Theorem 3. *(Strong Duality) Let E be a convex subset of X, $(x_0, y_0, z_0) \in \text{gr}(F, G)$ and $z_0 \in -D$. Suppose that the following conditions are satisfied:*

(i) (F, G) is $C \times D$-convex on E;
(ii) $P(x) := \{(y, z) \in Y \times Z \mid (x, y, z) \in T^{\flat(m)}_{\text{epi}(F,G)}(x_0, y_0, z_0)\}$ fulfills the weak domination property for all $x \in \text{dom}P$;
(iii) C has a compact base;
(iv) (x_0, y_0) be a Benson proper efficient solution of (SOP);
(v) for any $x \in E$, $G(x) \cap (-D) \neq \emptyset$.
Then there exist $\nu \in C^{+i}$ and $\omega \in D^+$ such that $(x_0, y_0, z_0, \nu, \omega)$ is a maximal solution of (DSOP).

Proof. Define
$$\Psi := ED^{\flat(m)}_w(F, G)(x_0, y_0, z_0)(\Theta) + C \times D + \{(0_Y, z_0)\},$$
where $\Theta := \text{dom}ED^{\flat(m)}_w(F, G)(x_0, y_0, z_0)$.

Step 1. We firstly prove that Ψ is a convex set. Indeed, it is sufficient to show the convexity of $\Psi_0 := \Psi - \{(0_Y, z_0)\}$.

Let $(y_i, z_i) \in \Psi_0$ $(i = 1, 2)$. Then there exist $x_i \in \Theta$, $(y'_i, z'_i) \in ED^{\flat(m)}_w(F, G)(x_0, y_0, z_0)(x_i)$ and $(c_i, d_i) \in C \times D$ $(i = 1, 2)$ such that

$$(y_i, z_i) = (y'_i, z'_i) + (c_i, d_i) \quad (i = 1, 2). \tag{17}$$

According to the definition of $ED^{\flat(m)}_w(F, G)(x_0, y_0, z_0)$, one has $(x_i, y'_i, z'_i) \in T^{\flat(m)}_{\text{epi}(F,G)}(x_0, y_0, z_0)$ $(i = 1, 2)$.

Since (F, G) is $C \times D$-convex on E, $\text{epi}(F, G)$ is a convex set. From Proposition 3, $T^{\flat(m)}_{\text{epi}(F,G)}(x_0, y_0, z_0)$ is a convex set. So for any $t \in [0, 1]$,

$$t(x_1, y'_1, z'_1) + (1 - t)(x_2, y'_2, z'_2) \in T^{\flat(m)}_{\text{epi}(F,G)}(x_0, y_0, z_0).$$

By (ii), we have

$$t(y'_1, z'_1) + (1 - t)(y'_2, z'_2) \in ED^{\flat(m)}_w(F, G)(x_0, y_0, z_0)(tx_1 + (1 - t)x_2) + \text{int}(C \times D) \cup \{(0_Y, 0_Z)\}$$
$$\subseteq ED^{\flat(m)}_w(F, G)(x_0, y_0, z_0)(tx_1 + (1 - t)x_2) + C \times D.$$

Combining this with (17), one has

$$t(y_1, z_1) + (1 - t)(y_2, z_2) \in \Psi_0 + C \times D = \Psi_0.$$

Therefore Ψ_0 is a convex set and so $\Psi = \Psi_0 + \{(0_Y, z_0)\}$ is a convex set.

Step 2. We prove that there exist $\nu \in C^{+i}$ and $\omega \in D^+$ such that $(x_0, y_0, z_0, \nu, \omega)$ is a feasible solution of (DSOP).

Define
$$\Phi := \text{clcone}\Psi.$$

Since Ψ is a convex set, Φ is a convex cone. According to Lemma 2, we have

$$\Phi \cap (-((C \setminus \{0_Y\}) \times \text{int}D)) = \emptyset. \tag{18}$$

Hence, we can conclude
$$\Phi \cap (-(C \times \{0_Z\})) = \{(0_Y, 0_Z)\}. \tag{19}$$

In fact, assume that (19) does not hold. Since Φ is a cone, there exists $b \in -C \setminus \{0_Y\}$ such that

$$(b, 0_Z) \in \Phi \cap (-((C \setminus \{0_Y\}) \times \{0_Z\})).$$

Then there exist $x^n \in \Theta$, $(y^n, z^n) \in ED_w^{\flat(m)}(F,G)(x_0, y_0, z_0)(x^n)$, $(c_n, d_n) \in C \times D$ and $\lambda_n \geq 0$ such that

$$b = \lim_{n \to \infty} \lambda_n (y^n + c_n). \tag{20}$$

According to the definition of $ED_w^{\flat(m)}(F,G)(x_0, y_0, z_0)$, for any $t_k \to 0^+$, there exists $(x_k^n, y_k^n, z_k^n) \in \mathrm{epi}(F,G)$ such that

$$\lim_{k \to \infty} \left(\frac{x_k^n - x_0}{t_k}, \frac{y_k^n - y_0 + t_k^m c_n}{t_k^m}, \frac{z_k^n - z_0 + t_k^m d_n}{t_k^m} \right) = (x^n, y^n + c_n, z^n + d_n). \tag{21}$$

This together with condition (v) implies

$$\begin{aligned}
\lambda_n \frac{y_k^n - y_0 + t_k^m c_n}{t_k^m} &\in \lambda_n \frac{F(x_k^n) + C - \{y_0\} + t_k^m c_n}{t_k^m} \\
&\subseteq \mathrm{clcone}[F(E) + C - \{y_0\}] \\
&\subseteq \mathrm{clcone}[F(M) + C - \{y_0\}].
\end{aligned} \tag{22}$$

It follows from (20), (21), (22) and $b \in -(C \setminus \{0_Y\})$ that

$$b \in \mathrm{clcone}(F(M) + C - \{y_0\}) \cap (-(C \setminus \{0_Y\})),$$

which contradicts that (x_0, y_0) is a Benson proper efficient solution of (SOP). Thus (19) holds.

Since C has a compact base, $-(C \times \{0_Z\})$ also has a compact base. Combining this with (19) and Lemma 3, replacing H and Q with Φ and $-(C \times \{0_Z\})$, there exists a pointed convex cone $\tilde{A}$ such that

$$-(C \times \{0_Z\}) \setminus (0_Y, 0_Z) \subseteq \mathrm{int}\tilde{A} \tag{23}$$

and

$$\Phi \cap \tilde{A} = \{(0_Y, 0_Z)\}. \tag{24}$$

Let $\tilde{B} := A \cup \{(0_Y, 0_Z)\}$, where $A := -((C \setminus \{0_Y\}) \times (\mathrm{int}D \cup \{0_Z\})) + \tilde{A}$. Thus $\tilde{B}$ is a convex cone. Next, we further prove that $\tilde{B}$ is a pointed cone. According to Proposition 1, we get $(0_X, 0_Y, 0_Z) \in T_{\mathrm{epi}(F,G)}^{\flat(m)}(x_0, y_0, z_0)$. Combining this with the weak domination property of P, we get

$$(0_Y, 0_Z) \in ED_w^{\flat(m)}(F,G)(x_0, y_0, z_0)(0_X) + C \times D. \tag{25}$$

For $z_0 \in G(x_0) \cap (-D)$ and $(c, d) \in C \times D$, we have

$$\begin{aligned}
(c, d) &= (0_Y, 0_Z) + (c, d - z_0) + (0_Y, z_0) \\
&\in ED_w^{\flat(m)}(F,G)(x_0, y_0, z_0)(0_X) + C \times D + \{(0_Y, z_0)\} \\
&\subseteq \Phi,
\end{aligned}$$

and so

$$C \times D \subseteq \Phi. \tag{26}$$

It follows from (24) and (26) that $(C \times D) \cap \tilde{A} = \{(0_Y, 0_Z)\}$. Hence,

$$((C \setminus \{0_Y\}) \times (\mathrm{int}D \cup \{0_Z\})) \cap \tilde{A} = \emptyset.$$

Combining with the definition of A, one has

$$(0_Y, 0_Z) \notin A. \tag{27}$$

Thus

$$A \cap (-A) = \emptyset. \tag{28}$$

To obtain this result, we suppose on the contrary that there exists $(c,d) \in A \cap (-A)$. Then there exist $(c_i, d_i) \in (C \setminus \{0_Y\}) \times (\text{int} D \cup \{0_Z\})$ $(i = 1, 2)$ and $(c'_i, d'_i) \in \tilde{A}$ $(i = 1, 2)$ such that

$$(c, d) = -(c_1, d_1) + (c'_1, d'_1)$$

and

$$(c, d) = (c_2, d_2) - (c'_2, d'_2).$$

So
$$(-(c_1, d_1) + (c'_1, d'_1)) - ((c_2, d_2) - (c'_2, d'_2))$$
$$= -(c_1 + c_2, d_1 + d_2) + (c'_1 + c'_2, d'_1 + d'_2)$$
$$= (0_Y, 0_Z) \in A,$$

which contradicts (27). Therefore (28) holds. Then $\tilde{B}$ is a pointed convex cone and $(0_Y, 0_Z) \notin \text{int}\tilde{B}$.

Now, we can conclude

$$\Phi \cap \tilde{B} = \{(0_Y, 0_Z)\}. \tag{29}$$

To see the conclusion, we suppose on the contrary that there exists $(y, z) \neq (0_Y, 0_Z)$ such that

$$(y, z) \in \Phi \cap \tilde{B}, \tag{30}$$

because $\tilde{B}$ is a pointed convex cone and Φ is a convex cone. From the definition of $\tilde{B}$, there exist $(y_1, z_1) \in -((C \setminus \{0_Y\}) \times (\text{int}D \cup \{0_Z\}))$ and $(y_2, z_2) \in \tilde{A}$ such that

$$(y, z) = (y_1, z_1) + (y_2, z_2).$$

According to the definition of Φ, there exist $x'_n \in \Theta$, $(y'_n, z'_n) \in ED_w^{p(m)}(F, G)(x_0, y_0, z_0)(x'_n)$, $(c'_n, d'_n) \in C \times D$ and $\lambda'_n \geq 0$ such that

$$(y, z) = \lim_{n \to \infty} \lambda'_n (y'_n + c'_n, z'_n + d'_n + z_0).$$

Since $(y, z) \neq (0_Y, 0_Z)$, without loss of generality, we may assume that $\lambda'_n > 0$. It follows from the definition of Φ that

$$(y, z) - (y_1, z_1) = \lim_{n \to \infty} \lambda'_n (y'_n + c'_n, z'_n + d'_n + z_0) - (y_1, z_1)$$
$$= \lim_{n \to \infty} \lambda'_n (y'_n + c'_n - \frac{y_1}{\lambda'_n}, z'_n + d'_n - \frac{z_1}{\lambda'_n} + z_0)$$
$$\in \Phi,$$

and so

$$(y_2, z_2) = (y, z) - (y_1, z_1) \in \Phi \cap \tilde{A} = \{(0_Y, 0_Z)\}.$$

Thus

$$(y, z) = (y_1, z_1) \in -((C \setminus \{0_Y\}) \times \text{int}D). \tag{31}$$

By (30) and (31), we have

$$(y,z) \in \Phi \cap (-((C \setminus \{0_Y\}) \times \text{int}D)),$$

which contradicts (18).

We claim that
$$-((C \setminus \{0_Y\}) \times (\text{int}D \cup \{0_Z\})) \subseteq \text{int}\tilde{B}. \tag{32}$$

To obtain this conclusion, we replace B and C in [30], Theorem 2.2 with $-((C \setminus \{0_Y\}) \times (\text{int}D \cup \{0_Z\}))$ and $\text{int}\tilde{A}$, respectively, which together with the fact: $(0_Y, 0_Z) \notin \text{int}\tilde{B}$ yields that

$$\text{int}\tilde{B} = -(C \setminus \{0_Y\}) \times (\text{int}D \cup \{0_Z\}) + \text{int}\tilde{A}. \tag{33}$$

Let $c \in C \setminus \{0_Y\}$ and $d \in \text{int}D \cup \{0_Z\}$. Then by (23) and (33), one has

$$-(c,d) = -(\frac{c}{2},d) - (\frac{c}{2},0_Z) \in -((C \setminus \{0_Y\}) \times (\text{int}D \cup \{0_Z\})) + \text{int}\tilde{A} = \text{int}\tilde{B},$$

and so (32) holds.

According to the separation theorem for convex set and (29), there exist $\nu \in Y^*$ and $\omega \in Z^*$ such that

$$\nu(\bar{y}) + \omega(\bar{z}) < 0, \forall (\bar{y}, \bar{z}) \in \text{int}\tilde{B} \tag{34}$$

and

$$\nu(\check{y}) + \omega(\check{z}) \geq 0, \forall (\check{y}, \check{z}) \in \Phi. \tag{35}$$

By (32) and (34), we have

$$\nu(\bar{y}) + \omega(\bar{z}) > 0, \forall (\bar{y}, \bar{z}) \in (C \setminus \{0_Y\}) \times (\text{int}D \cup \{0_Z\}). \tag{36}$$

Taking $\bar{z} = 0_Z$ in (36), one has $\nu(\bar{y}) > 0, \forall \bar{y} \in C \setminus \{0_Y\}$, thus $\nu \in C^{+i}$. For any $\varepsilon > 0$, take $\bar{y} \in (C \setminus \{0_Y\}) \cap B(0_Y, \varepsilon)$ in (36). Then we can observe that $\omega(\bar{z}) \geq 0, \forall \bar{z} \in \text{int}D$, which implies $\omega \in D^+$.

It follows from (35) that

$$\nu(y) + \omega(z) \geq 0, \forall (y,z) \in ED_w^{\flat(m)}(F,G)(x_0,y_0,z_0)(\Theta) + C \times D + \{(0_Y, z_0)\}. \tag{37}$$

Together with (25), we get $\omega(z_0) \geq 0$. It follows from $z_0 \in -D$ and $\omega \in D^+$ that $\omega(z_0) \leq 0$. Thus,

$$\omega(z_0) = 0.$$

Combining this with (37), one has

$$\nu(y') + \omega(z') \geq 0, \forall (y', z') \in ED_w^{\flat(m)}(F,G)(x_0,y_0,z_0)(\Theta) + C \times D,$$

and so

$$\nu(y) + \omega(z) \geq 0, \forall (y,z) \in ED_w^{\flat(m)}(F,G)(x_0,y_0,z_0)(\Theta).$$

Thus $(x_0, y_0, z_0, \nu, \omega)$ is a feasible solution of (DSOP).

Step 3. We prove that $(x_0, y_0, z_0, \nu, \omega)$ is a maximal solution of (DSOP).

Suppose on the contrary that there exists a feasible solution $(\hat{x}, \hat{y}, \hat{z}, \nu', \omega')$ such that $\hat{y} - y_0 \in C \setminus \{0_Y\}$. By $\nu' \in C^{+i}$, we have

$$\nu'(\hat{y}) > \nu'(y_0) \tag{38}$$

Since (x_0, y_0) is a feasible solution of (SOP), it follows from Theorem 2 that $\nu'(y_0) \geq \nu'(\hat{y})$, which contradicts (38). The proof is complete. □

Theorem 4. *(Converse Duality) Let E be a star-shaped set at $x_0 \in E$. Let $y_0 \in F(x_0), z_0 \in G(x_0) \cap (-D)$, $v \in C^{+i}$ and $w \in D^+$ such that (x_0, y_0, z_0, v, w) is a feasible solution of (DSOP). Then (x_0, y_0) is a Benson proper efficient solution of (SOP) if the following conditions are satisfied:*
  (i) (F, G) is decreasing-along-rays at x_0;
  (ii) (F, G) is a generalized $C \times D$-convex at x_0 on E;
  (iii) the set $P_{(F,G)}(x - x_0) := \{y \in Y \mid (x - x_0, y, z) \in T^{\flat(m)}_{epi(F,G)}(x_0, y_0, z_0)\}$ fulfills the weak domination property for all $x \in \text{dom} P_{(F,G)}$.

Proof. It follows from (1), (3) and (4) that

$$v(y) + w(z) \geq 0, \forall (y, z) \in ED^{\flat(m)}_w(F, G)(x_0, y_0, z_0)(x) + C \times D, \tag{39}$$
$$\forall x \in \Theta := \text{dom} ED^{\flat(m)}_w(F, G)(x_0, y_0, z_0).$$

According to Proposition 4, we get

$$(y - y_0, z - z_0) \in ED^{\flat(m)}_w(F, G)(x_0, y_0, z_0)(x - x_0) + C \times D, \tag{40}$$
$$\forall x \in M, y \in F(x), z \in G(x) \cap (-D).$$

By (2), we have $w(z_0) \geq 0$. It follows from $z_0 \in G(x_0) \cap (-D)$ and $w \in D^+$ that $w(z_0) \leqslant 0$, thus $w(z_0) = 0$. Then

$$w(z - z_0) = w(z) - w(z_0) = w(z) \leqslant 0, \forall z \in G(x) \cap (-D), x \in M. \tag{41}$$

It follows from (39), (40) and (41) that

$$v(y - y_0) \geq 0, \forall y \in F(x), x \in M.$$

Further more, we can get

$$v(y + c - y_0) \geq 0, \forall y \in F(x), x \in M,$$

and so

$$v(y) \geq 0, \forall y \in \text{clcone}(F(M) + C - \{y_0\}). \tag{42}$$

Assume that the feasible solution (x_0, y_0) is not a Benson proper efficient solution of (SOP). Then there exists $y' \in -(C \setminus \{0_Y\})$ such that $y' \in \text{clcone}(F(M) + C - \{y_0\})$. This together with (42) implies that

$$v(y') \geq 0. \tag{43}$$

It follows from $v \in C^{+i}$ and $y' \in -(C \setminus \{0_Y\})$ that $v(y') < 0$, which contradicts (43). Thus (x_0, y_0) is a Benson proper efficient of (SOP) and the proof is complete. □

Remark 4. *Example 9 also illustrates that Theorem 4 extends [2], Theorem 4.3 from the cone convexity to generalized cone convexity. Indeed, take $(x_0, y_0, z_0) = (0, 0, 0)$. Then simple calculations show that*

$$T^{\flat(2)}_{epi(F,G)}(0,0,0) = \{(x, y, z) \in X \times Y \times Z \mid x \leq 0, y \geq 0, z \geq 0\} \cup$$
$$\{(x, y, z) \in X \times Y \times Z \mid x > 0, y \geq 0, z \in \mathbb{R}\}$$

and

$$ED^{\flat(2)}_w(F, G)(0,0,0)(x) = \begin{cases} \{(y, z) \in Y \times Z \mid y = 0, z \geq 0\} \cup \{(y, z) \in Y \times Z \mid y \geq 0, z = 0\}, x \leq 0, \\ \{(y, z) \in Y \times Z \mid y = 0, z \in \mathbb{R}\}, x > 0. \end{cases}$$

Then we can choose $\nu = 1$ and $\omega = 0$ such that $(x_0, y_0, z_0, \nu, \omega) = (0, 0, 0, 1, 0)$ is a feasible solution of (DSOP). It is easy to show that the all conditions of Theorem 4 are fulfilled and $(0,0)$ is a Benson proper efficient solution of (SOP). Thus Theorem 4 holds here. However, [2], Theorem 4.3 is not applicable here because G is not C-convex on E.

Author Contributions: The authors made equal contributions to this paper.

Funding: This research was funded by Chongqing Jiaotong University Graduate Education Innovation Foundation Project (No. 2018S0152), Chongqing Natural Science Foundation Project of CQ CSTC(Nos. 2015jcyjA30009, 2015jcyjBX0131, 2017jcyjAX0382), the Program of Chongqing Innovation Team Project in University (No. CXTDX201601022) and the National Natural Science Foundation of China (No. 11571055). Ching-Feng Wen was supported by the Taiwan MOST [grant number 107-2115-M-037-001].

Conflicts of Interest: The author declares no conflict of interest.

References

1. Corley, H.W. Optimality conditions for maximizations of set-valued functions. *J. Optim. Theory Appl.* **1988**, *58*, 1–10. [CrossRef]
2. Chen, C.R.; Li, S.J.; Teo, K.L. Higher order weak epiderivatives and applications to duality and optimality conditions. *Comput. Math. Appl.* **2009**, *57*, 1389–1399. [CrossRef]
3. Li, S.J.; Teo, K.L.; Yang, X.Q. Higher order Mond-Weir duality for set-valued optimization. *J. Comput. Appl. Math.* **2008**, *217*, 339–349. [CrossRef]
4. Wang, Q.L.; Li, S.J.; Chen, C.R. Higher-order generalized adjacent derivative and applications to duality for set-valued optimization. *Taiwan. J. Math.* **2011**, *15*, 1021–1036. [CrossRef]
5. Wen, C.F.; Wu, H.C. Approximate solutions and duality theorems for continuous-time linear fractional problems. *Numer. Funct. Anal. Optim.* **2012**, *33*, 80–129. [CrossRef]
6. Anh, N.L.H. On higher-order mixed duality in set-valued optimization. *Bull. Malays. Math. Sci. Soc.* **2018**, *41*, 723–739. [CrossRef]
7. Li, S.J.; Sun, X.K.; Zhu, S.K. Higher-order optimality conditions for sirict minimality in set-valued optimization. *J. Nonlinear Convex Anal.* **2012**, *13*, 281–291.
8. Anh, N.L.H. Mixed type duality for set-valued optimization problems via higher-order radial epiderivatives. *Numer. Func. Anal. Optim.* **2016**, *37*, 823–838. [CrossRef]
9. Aubin, J.P.; Frankowska, H. *Set-Valued Analysis*; Birkhauser: Boston, MA, USA, 1990.
10. Luc, D.T. *Theory of Vector Optimization*; Springer: Berlin, Germany, 1989.
11. Jahn, J.; Rauh, R. Contingent epiderivatives and set-valued optimization. *Math. Meth. Oper. Res.* **1997**, *46*, 193–211. [CrossRef]
12. Chen, G.Y.; Jahn, J. Optimality conditions for set-valued optimization problems. *Math. Methods Oper. Res.* **1998**, *48*, 187–200. [CrossRef]
13. Jahn, J.; Khan, A.A.; Zeilinger, P. Second-order optimality conditions in set optimization. *J. Optim. Theory Appl.* **2005**, *125*, 331–347. [CrossRef]
14. Li, S.J.; Chen, C.R. Higher order optimality conditions for Henig efficient solutions in set-valued optimization. *J. Math. Anal. Appl.* **2006**, *323*, 1184–1200. [CrossRef]
15. Wang, Q.L.; Li, S.J.; Teo, K.L. Higher order optimality conditions for weakly efficient solutions in nonconvex set-valued optimization. *Optim. Lett.* **2010**, *4*, 425–437. [CrossRef]
16. Anh, N.L.H.; Khanh, P.Q. Higher-order optimality conditions in set-valued optimization using radial sets and radial derivatives. *J. Glob. Optim.* **2013**, *56*, 519–536. [CrossRef]
17. Studniarski, M. Necessary and sufficient conditions for isolated local minima of nonsmooth functions. *SIAM J. Control Optim.* **1986**, *24*, 1044–1049. [CrossRef]
18. Anh, N.L.H. Higher-order optimality conditions in set-valued optimization using Studniarski derivatives and applications to duality. *Positivity* **2014**, *18*, 449–473. [CrossRef]
19. Anh, N.L.H. Higher-order optimality conditions for strict and weak efficient solutions in set-valued optimization. *Positivity* **2016**, *20*, 499–514. [CrossRef]
20. Anh, N.L.H. Higher-order generalized Studniarski epiderivative and its applications in set-valued optimization. *Positivity* **2018**, *22*, 1371–1385. [CrossRef]

21. Lalitha, C.S.; Arora, R. Weak Clarke epiderivative in set-valued optimization. *J. Math. Anal. Appl.* **2008**, *342*, 704–714. [CrossRef]
22. Jahn, J. *Vector Optimization: Theory, Applications, and Extensions*; Springer: Berlin, Germany, 2004.
23. Bobylev, N.A. The Helly theorem for star-shaped sets. *J. Math. Sci.* **2001**, *105*, 1819–1825. [CrossRef]
24. Chinaie, M.; Zafarani, J. Image Space Analysis and Scalarization of Multivalued optimization. *J. Optim. Theory Appl.* **2009**, *142*, 451–467. [CrossRef]
25. Long, X.J.; Peng, J.W.; Wong, M.M. Generalized radial epiderivatives and nonconvex set-valued optimization problems. *Appl. Anal.* **2012**, *91*, 1891–1900. [CrossRef]
26. Jimenez, B.; Novo, V. Second order necessary conitions in set cnstrained differentiable vector optimization. *Math. Methods Oper. Res.* **2013**, *58*, 299–317. [CrossRef]
27. Gong, X.H.; Dong, H.B.; Wang, S.Y. Optimality conditions for proper efficient solutions of vector set-valued optimization. *J. Math. Anal. Appl.* **2003**, *284*, 332–350. [CrossRef]
28. Wang, Q.L.; Yu, G.L. Second-order optimality conditions for set-valued optimization problems under benson proper efficiency. *Abstr. Appl. Anal.* **2011**. [CrossRef]
29. Dauer, J.P.; Saleh, O.A. A characterization of proper minimal points as a solution of sunlinear optimization problems. *J. Math. Anal. Appl.* **1993**, *178*, 227–246. [CrossRef]
30. Zowe, J. A remark on a regularity assumption for the mathematical propramming pronlem in Banach spaces. *J. Optim. Theoy Appl.* **1978**, *25*, 375–381. [CrossRef]

© 2019 by the authors. Licensee MDPI, Basel, Switzerland. This article is an open access article distributed under the terms and conditions of the Creative Commons Attribution (CC BY) license (http://creativecommons.org/licenses/by/4.0/).

Article

An Inequality Approach to Approximate Solutions of Set Optimization Problems in Real Linear Spaces

Elisabeth Köbis [1,*,†], **Markus A. Köbis** [2,†] **and Xiaolong Qin** [3,†]

1. Institute of Mathematics, Faculty of Natural Sciences II, Martin-Luther-University Halle-Wittenberg, 06120 Halle, Germany
2. Department of Mathematics and Computer Science, Institute of Mathematics, Free University Berlin, 14195 Berlin, Germany; mkoebis@zedat.fu-berlin.de
3. General Education Center, National Yunlin University of Science and Technology, Douliou 64002, Taiwan; qinxl@yuntech.edu.tw
* Correspondence: elisabeth.koebis@mathematik.uni-halle.de
† These authors contributed equally to this work.

Received: 29 October 2019; Accepted: 10 January 2020; Published: 20 January 2020

Abstract: This paper explores new notions of approximate minimality in set optimization using a set approach. We propose characterizations of several approximate minimal elements of families of sets in real linear spaces by means of general functionals, which can be unified in an inequality approach. As particular cases, we investigate the use of the prominent Tammer–Weidner nonlinear scalarizing functionals, without assuming any topology, in our context. We also derive numerical methods to obtain approximate minimal elements of families of finitely many sets by means of our obtained results.

Keywords: set optimization; set relations; nonlinear scalarizing functional; algebraic interior; vector closure

MSC: 90C29; 90C26

---

1. Introduction

Set optimization has become an important research area and has gained tremendous interest within the optimization community due to its wide and important applications; see, e.g., [1–4]. There exist various research fields that directly lead to problems which can most satisfactorily be modeled and solved in the unified framework provided by set optimization. For example, duality in vector optimization, gap functions for vector variational inequalities, fuzzy optimization, as well as many problems in image processing, viability theory, economics etc. all lead to optimization problems that can be modeled as set-valued optimization problems. For an introduction to set optimization and its applications, we refer to [5].

For example, it is well known that uncertain optimization problems can be modeled by means of set optimization. Uncertainty here means that some parameters are not known. Instead, possibly only an estimated value or a set of possible values can be determined. As inaccurate data can have severe impacts on the model and therefore on the computed solution, it is important to take such uncertainty into account when modeling an optimization problem. If uncertainty is included in the optimization model, one is left with not only one objective function value, but possibly a whole set of values. This leads to a set-valued optimization problem, where the objective map is set-valued.

Recently, it has been shown that certain concepts of robustness for dealing with uncertainties in vector optimization can be described using approaches from set-valued optimization (see [2,3] and a practical application in the context of layout optimization of photovoltaic powerplants in [6]).

The concept of interval arithmetic for computations with strict error bounds [7] is also a special case of dealing with set-valued mappings.

To obtain minimal solutions of a set-valued optimization problem, one must analyze whether one set dominates another set in a certain sense, i.e., by means of a given set relation. As it turns out, however, (depending on the chosen set relation), this intuitive and natural mathematical modeling framework often reaches its limitations and leads to very large or—even worse—empty solution sets. This is especially important throughout the design and implementation process of numerical algorithms for set optimization problems: The criteria involved in the definition of the set relations are usually based on set inclusions which for continuous problems are very sensitive to numerical inaccuracies or even just round-off errors.

A simple way to remedy this is to use *approximate solution* concepts: Here, the strict set inclusions are in a way relaxed by extending (enlarging/translating) the quantities that are to be compared such that one obtains more robust results for the involved inclusion tests.

The goal of this paper lies in the characterization of several well-known set relations by means of a very broad, manageable and easy-to-compute functional in the context of approximate solutions to set optimization problems using the set approach. In contrast to recent results in this area (for example see [8–11]), we assume that the spaces in which the sets are compared are not endowed with a particular topology. Therefore, our results generalize those found in the literature by dismissing topological properties. Please note that the references [10,11] present results on scalarizing functionals, but the functional acts on a real linear topological space and no relation to approximate solutions is presented there. Moreover, in [8,9], the oriented distance functional (which implicitly requires a topology) is used to derive characterizations of set relations. To the best of our knowledge, our approach of combining algebraic tools with approximate minimality notions in set optimization is original. That way, our results are not only valid in a broader mathematical setting but also provide some further insight into the purely algebraic tools and theoretical requirements necessary to acquire our findings. This is not only mathematically interesting, but deepens the theoretical understanding of approximate minimality in set optimization. It is furthermore in line with the recent increased interest in studying optimality conditions and separation concepts in spaces without a particular topology underneath it, see [12–21] and the references therein.

2. Preliminaries

Throughout this work, let Y be a real linear space. Following the nomenclature of [22], for a nonempty set $F \subseteq Y$, we denote by

$$\text{core } F := \{y \in Y \mid \forall v \in Y \ \exists \lambda > 0 \text{ s.t. } y + [0, \lambda] v \subseteq F\}$$

the algebraic interior of F and for any given $k \in Y$, let

$$\text{vcl}_k F := \{y \in Y \mid \forall \lambda > 0 \ \exists \lambda' \in [0, \lambda] \text{ s.t. } y + \lambda' k \in F\}.$$

We say that F is k-vectorially closed if $\text{vcl}_k F = F$. Obviously, it holds $F \subseteq \text{vcl}_k F$ for all $k \in Y$.

We denote by $\mathcal{P}(Y) := \{A \subseteq Y \mid A \text{ is nonempty}\}$ the power set of Y without the empty set. For two elements A, B of $\mathcal{P}(Y)$, we denote the sum of sets by

$$A + B := \{a + b \mid a \in A, b \in B\}.$$

The set $F \subseteq Y$ is a cone if for all $f \in F$ and $\lambda \geq 0$, $\lambda f \in F$ holds true. The cone F is convex if $F + F \subseteq F$.

Now let $\emptyset \neq C \subseteq Y$ and $k \in Y \setminus \{0\}$. We recall the functional $z^{C,k} \colon Y \to \mathbb{R} \cup \{+\infty\} \cup \{-\infty\} =: \overline{\mathbb{R}}$ from Gerstewitz [23] (which has very recently been extended to the space Y without assuming any topology, see [24] and the references therein)

$$z^{C,k}(y) := \begin{cases} +\infty & \text{if } y \notin \mathbb{R}k - C, \\ \inf\{t \in \mathbb{R} \mid y \in tk - C\} & \text{otherwise}. \end{cases} \quad (1)$$

The functional $z^{C,k}$ was originally introduced as scalarizing functional in vector optimization. Please note that the construction of $z^{C,k}$ was mentioned by Krasnosel'skiĭ [25] (see Rubinov [26]) in the context of operator theory. Figure 1 visualizes the functional $z^{C,k}$, where $C = \mathbb{R}^2_+$ has been taken as the natural ordering cone in $\mathbb{R}^2$ and $k \in \mathrm{core}\, C$. We can see that the set $-C$ is moved along the line $\mathbb{R} \cdot k$ up until y belongs to $tk - C$. The functional $z^{C,k}$ assigns the smallest value t such that the property $y \in tk - C$ is fulfilled.

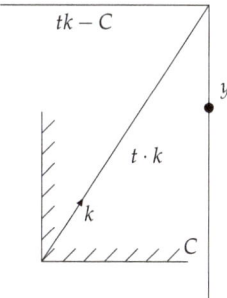

Figure 1. Illustration of the functional $z^{C,k}(y) := \inf\{t \in \mathbb{R} | y \in tk - C\}$.

The functional $z^{C,k}$ plays an important role as nonlinear separation functional for not necessarily convex sets. Applications of $z^{C,k}$ include coherent risk measures in financial mathematics (see, for instance, [27]) and uncertain programming (see [2,3]). Several important properties of $z^{C,k}$ (in the case that Y is endowed with a topology) were studied in [28,29]. Now let us recall the definition of E-monotonicity of a functional.

Definition 1. *Let $E \in \mathcal{P}(Y)$. A functional $z \colon Y \to \overline{\mathbb{R}}$ is called E-**monotone** if*

$$y_1, y_2 \in Y \colon y_1 \in y_2 - E \Rightarrow z(y_1) \leq z(y_2).$$

Below we provide some properties of the functional $z^{C,k}$ introduced in (1).

Proposition 1 ([22]). *Let C and E be nonempty subsets of Y, and let $k \in Y \setminus \{0\}$. Then the following properties hold.*

(a) $\forall y \in Y \colon z^{C,k}(y) \leq 0 \Longleftrightarrow y \in (-\infty, 0]k - \mathrm{vcl}_k\, C$.
(b) $\forall y \in Y \colon z^{C,k}(y) < 0 \Longleftrightarrow y \in (-\infty, 0)k - \mathrm{vcl}_k\, C$.
(c) $z^{C,k}$ *is E-monotone if and only if* $E + C \subset [0, +\infty)k + \mathrm{vcl}_k\, C$.
(d) $\forall y \in Y, \forall r \in \mathbb{R} \colon z^{C,k}(y + rk) = z^{C,k}(y) + r$.

The set relations to be defined below rely on set inclusions where the set C is attached pointwise to the considered sets $A, B \in \mathcal{P}(Y)$. The following corollary relates $A + C$ and $A - C$ respectively by means of the functional $z^{C,k}$ in the case that C is a convex cone.

Corollary 1 ([14], Corollary 2.3). *Let $C \subseteq Y$ be a convex cone, $A \in \mathcal{P}(Y)$ and $k \in Y \setminus \{0\}$. Then it holds*

$$\sup_{a \in A} z^{C,k}(a) = \sup_{y \in A-C} z^{C,k}(y) \text{ and } \inf_{a \in A} z^{C,k}(a) = \inf_{y \in A+C} z^{C,k}(y).$$

A well-known set relation is the upper set less order relation introduced by Kuroiwa [30,31]. We recall a generalized version of this relation here, where the underlying set C is not necessarily a convex cone and thus the resulting relation is not necessarily an order.

Definition 2 (Upper Set Less Relation, [32]). *Let $C \subseteq Y$. The **upper set less relation** $\preceq_C^u$ is defined for two sets $A, B \in \mathcal{P}(Y)$ by*

$$A \preceq_C^u B :\iff A \subseteq B - C.$$

The following theorem shows a first connection between the upper set less relation and the nonlinear scalarizing functional $z^{C,k}$.

Theorem 1 ([14], Theorem 3.2). *Let $C \subseteq Y$ be a convex cone, $A, B \in \mathcal{P}(Y)$ and $k \in Y \setminus \{0\}$. Then*

$$A \preceq_C^u B \implies \sup_{a \in A} z^{C,k}(a) \leq \sup_{b \in B} z^{C,k}(b).$$

The converse implication in Theorem 1 is not generally fulfilled, even if the underlying sets are convex, see ([33], Example 3.2). However, we have the following result.

Theorem 2 ([14], Theorem 3.3). *Let $C \subseteq Y$. For two sets $A, B \in \mathcal{P}(Y)$ and $k \in Y \setminus \{0\}$, it holds*

$$A \preceq_C^u B \implies \sup_{a \in A} \inf_{b \in B} z^{C,k}(a-b) \leq 0.$$

Assume on the other hand that there exists a $k_0 \in Y \setminus \{0\}$ such that $\inf_{b \in B} z^{C,k_0}(a-b)$ is attained for all $a \in A$, C is k_0-vectorially closed and $[0, +\infty)k_0 + C \subseteq C$. Then

$$\sup_{a \in A} \inf_{b \in B} z^{C,k_0}(a-b) \leq 0 \implies A \preceq_C^u B.$$

Remark 1. (1) Please note that for any $A, B \in \mathcal{P}(Y)$, the set relation $A \preceq_C^u B$ by Theorem 2 also implies $\sup_{k \in Y \setminus \{0\}} \sup_{a \in A} \inf_{b \in B} z^{C,k}(a-b) \leq 0$.
(2) Let $A, B \in \mathcal{P}(Y)$ and $C \subseteq Y$. If there exists an element $k_0 \in C \setminus \{0\}$ such that $\inf_{b \in B} z^{C,k_0}(a-b)$ is attained for all $a \in A$, C is k_0-vectorially closed and $[0, +\infty)k_0 + C = C$, then it follows from Theorem 2 that

$$A \preceq_C^u B \iff \sup_{a \in A} \inf_{b \in B} z^{C,k_0}(a-b) \leq 0$$

$$\iff \sup_{k \in Y \setminus \{0\}} \sup_{a \in A} \inf_{b \in B} z^{C,k}(a-b) \leq 0.$$

In the second part of Theorem 2, we need the assumption that there exists a $k_0 \in Y \setminus \{0\}$ such that $\inf_{b \in B} z^{C,k_0}(a-b)$ is attained for all $a \in A$. Sufficient conditions for such an attainment property, i.e., assertions concerning the existence of solutions of the corresponding optimization problems (extremal principles) are given in the literature. The well-known Theorem of Weierstrass says that a lower semi-continuous function on a nonempty weakly compact set in a reflexive Banach space has a minimum. An extension of the Theorem of Weierstrass is given by Zeidler ([34], Proposition 9.13): A proper lower semi-continuous and quasi-convex function on a nonempty closed bounded convex subset of a reflexive Banach space has a minimum. Since the functional z^{C,k_0} is studied here in the context of real linear spaces that are not endowed with a particular topology, we cannot rely on continuity assumptions. Therefore, we propose the following theorem without any attainment property.

Theorem 3 ([14], Theorem 3.6). *Let $C \subseteq Y$, $A, B \in \mathcal{P}(Y)$ and $k_0 \in Y \setminus \{0\}$ such that $(-\infty, 0)k_0 -$ $\mathrm{vcl}_{k_0} C \subseteq -C$ and $\mathrm{vcl}_{-k_0}(B - C) \subseteq B - C$. Then*

$$\sup_{a \in A} \inf_{b \in B} z^{C,k_0}(a - b) \leq 0 \implies A \preceq_C^u B.$$

We also consider the following set relation, which compares sets based on their lower bounds (compare [30,31] for the according definition for orders).

Definition 3 (Lower Set Less Relation, [32]). *Let $C \subseteq Y$. The **lower set less relation** $\preceq_C^l$ is defined for two sets $A, B \in \mathcal{P}(Y)$ by*

$$A \preceq_C^l B :\iff B \subseteq A + C.$$

Because $A \preceq_C^u B$ is equivalent to $-B \preceq_C^l -A$, we obtain the following corollaries from Theorems 1, 2 and 3.

Corollary 2 ([14], Corollary 3.9). *Let $C \subseteq Y$ be a convex cone, $A, B \in \mathcal{P}(Y)$ and $k \in Y \setminus \{0\}$. Then*

$$A \preceq_C^l B \implies \inf_{a \in A} z^{C,k}(a) \leq \inf_{b \in B} z^{C,k}(b).$$

Corollary 3 ([14], Corollary 3.10). *Let $C \subseteq Y$. For two sets $A, B \in \mathcal{P}(Y)$ and $k \in Y \setminus \{0\}$, it holds*

$$A \preceq_C^l B \implies \sup_{b \in B} \inf_{a \in A} z^{C,k}(a - b) \leq 0.$$

Assume on the other hand that there exists a $k_0 \in Y \setminus \{0\}$ such that $\inf_{a \in A} z^{C,k_0}(a - b)$ is attained for all $b \in B$, C is k_0-vectorially closed and $[0, +\infty)k_0 + C \subseteq C$. Then

$$\sup_{b \in B} \inf_{a \in A} z^{C,k_0}(a - b) \leq 0 \implies A \preceq_C^l B.$$

Corollary 4 ([14], Corollary 3.11). *Let $C \subseteq Y$, $A, B \in \mathcal{P}(Y)$ and $k_0 \in Y \setminus \{0\}$ such that $(-\infty, 0)k_0 -$ $\mathrm{vcl}_{k_0} C \subseteq -C$ and $\mathrm{vcl}_{-k_0}(A - C) \subseteq -A - C$. Then*

$$\sup_{b \in B} \inf_{a \in A} z^{C,k_0}(a - b) \leq 0 \implies A \preceq_C^l B.$$

We also study the so-called *set less relation* (see [35,36] for the case where the underlying set C is a convex cone).

Definition 4 (Set Less Relation, [32]). *Let $C \subseteq Y$. The **set less relation** $\preceq_C^s$ is defined for two sets $A, B \in \mathcal{P}(Y)$ by*

$$A \preceq_C^s B :\iff A \preceq_C^u B \text{ and } A \preceq_C^l B.$$

We immediately obtain the following results.

Corollary 5 ([14], Corollary 3.13). *Let $C \subseteq Y$ be a convex cone, $A, B \in \mathcal{P}(Y)$ and $k \in Y \setminus \{0\}$. Then*

$$A \preceq_C^s B \implies \sup_{a \in A} z^{C,k}(a) \leq \sup_{b \in B} z^{C,k}(b) \text{ and } \inf_{a \in A} z^{C,k}(a) \leq \inf_{b \in B} z^{C,k}(b).$$

Corollary 6 ([14], Corollary 3.14). *Let $C \subseteq Y$. For two sets $A, B \in \mathcal{P}(Y)$ and $k \in Y \setminus \{0\}$, it holds*

$$A \preceq_C^s B \implies \sup_{a \in A} \inf_{b \in B} z^{C,k}(a-b) \leq 0 \text{ and } \sup_{b \in B} \inf_{a \in A} z^{C,k}(a-b) \leq 0.$$

Assume on the other hand that there exists a $k_0 \in Y \setminus \{0\}$ such that $\inf_{b \in B} z^{C,k_0}(a-b)$ is attained for all $a \in A$, and there exists $k_1 \in Y \setminus \{0\}$ such that $\inf_{a \in A} z^{C,k_1}(a-b)$ is attained for all $b \in B$, C is both k_0- and k_1-vectorially closed, $[0, +\infty)k_0 + C \subseteq C$ and $[0, +\infty)k_1 + C \subseteq C$. Then

$$\sup_{a \in A} \inf_{b \in B} z^{C,k_0}(a-b) \leq 0 \text{ and } \sup_{b \in B} \inf_{a \in A} z^{C,k_1}(a-b) \leq 0 \implies A \preceq_C^s B.$$

Corollary 7 ([14], Corollary 3.15). *Let $C \subseteq Y$, $A, B \in \mathcal{P}(Y)$ and $k_0, k_1 \in Y \setminus \{0\}$ such that $(-\infty, 0)k_0 - \text{vcl}_{k_0} C \subseteq -C$, $(-\infty, 0)k_1 - \text{vcl}_{k_1} C \subseteq -C$, $\text{vcl}_{-k_0}(B - C) \subseteq B - C$ and $\text{vcl}_{-k_1}(A - C) \subseteq A - C$. Then*

$$\sup_{a \in A} \inf_{b \in B} z^{C,k_0}(a-b) \leq 0 \text{ and } \sup_{b \in B} \inf_{a \in A} z^{C,k_1}(a-b) \leq 0 \implies A \preceq_C^s B.$$

3. Approximate Minimal Elements of Set Optimization Problems

The following definition describes minimality in the setting of a family of sets (see ([5], Definition 2.6.19) for the corresponding definition for preorders).

Definition 5 (Minimal Elements). *Let $\mathcal{A}$ be a family of elements of $\mathcal{P}(Y)$. $\overline{A} \in \mathcal{A}$ is called a **minimal element** of $\mathcal{A}$ w.r.t. $\preceq$ if*

$$A \preceq \overline{A}, A \in \mathcal{A} \implies \overline{A} \preceq A.$$

The set of all minimal elements of $\mathcal{A}$ w.r.t. $\preceq$ will be denoted by $\mathcal{A}_{\min}$.

Please note that if the elements of $\mathcal{A}$ are single-valued and $A \preceq \overline{A} :\iff A \in \overline{A} - C$ with $C \subseteq Y$ being a convex cone, then Definition 5 reduces to the standard notion of minimality in vector optimization (compare, for example, ([15], Definition 4.1)). From vector optimization, it is well known that usually, the existence of minimal elements can only be guaranteed under additional assumptions (for an existence result of minimal elements in set optimization, see, for example, [37]). Since the set $\mathcal{A}_{\min}$ may be empty, it is common practice to use a weaker notion of minimality, so-called approximate minimality. For this reason, we extend three notions of approximate minimality that were originally introduced in [38]. In [38], the following definitions are given for $\preceq = \preceq_C^l$ (see Definition 3). In order to stay as general as possible, we define approximate minimality using set relations that are not required to possess any ordering structure.

Definition 6. *Let $\mathcal{A}$ be a family of elements of $\mathcal{P}(Y)$, $H \in \mathcal{P}(Y)$, $H \neq Y$ and $\preceq$ be a binary relation on $\mathcal{A}$.*

(a) *$\overline{A} \in \mathcal{A}$ is called an H^1–approximate minimal element of $\mathcal{A}$ w.r.t. $\preceq$ if*

$$A \preceq \overline{A}, A \in \mathcal{A} \implies \overline{A} \preceq A + H.$$

(b) *$\overline{A} \in \mathcal{A}$ is called an H^2–approximate minimal element of $\mathcal{A}$ w.r.t. $\preceq$ if*

$$A + H \preceq \overline{A}, A \in \mathcal{A} \implies \overline{A} \preceq A + H.$$

(c) *$\overline{A} \in \mathcal{A}$ is called an H^3–approximate minimal element of $\mathcal{A}$ w.r.t. $\preceq$ if $A + H \not\preceq \overline{A}$, for all $A \in \mathcal{A} \setminus \overline{A}$.*

The set of all H^i–approximate minimal elements of $\mathcal{A}$ w.r.t. $\preceq$ ($i = 1, 2, 3$) will be denoted by $\mathcal{A}_{H^i}$.

Please note that Definition 6 (a) is a natural formulation for approximate minimality, while Definition 6 (b) is derived from the standard notion of approximate efficiency for vector-valued maps (see ([38], Remark 2.5)). Definition 6 (c) represents an approximate version of the well-known nondomination concept of vector optimization.

Here we consider a set-valued optimization problem in the following setting: Let $S \subseteq \mathbb{R}^n$, a set-valued mapping $F \colon S \rightrightarrows Y$ and a set relation $\preceq$ be given. We are looking for **approximate minimal elements** w.r.t. the order relation $\preceq$ in the sense of Definition 6 of the problem

$$\min_{x \in S} F(x). \tag{2}$$

We say that $\bar{x} \in S$ is an H^i-**approximate minimal solution** ($i = 1, 2, 3$) of (2) w.r.t. $\preceq$ if $F(\bar{x})$ is an H^i-approximate minimal element of the family of sets $F(x)$, $x \in S$ w.r.t. $\preceq$. The family of sets $F(x)$, $x \in S$, is denoted by $\mathcal{A}$.

Now we will present characterizations of approximate minimal solutions of (2) w.r.t. $\preceq$. In what follows, we will use the following notation. For some $\bar{x} \in S$, let us denote

$$[F(\bar{x})]^{H^1}_{\preceq} := \{x \in S \mid F(x) \preceq F(\bar{x}),\ F(\bar{x}) \preceq F(x) + H\}$$

and

$$[F(\bar{x})]^{H^2}_{\preceq} := \{x \in S \mid F(x) + H \preceq F(\bar{x}),\ F(\bar{x}) \preceq F(x) + H\}.$$

The following proposition will be useful in the theorem below.

Proposition 2. *$\bar{x} \in S$ is an H^1-approximate minimal solution of the problem (2) w.r.t. $\preceq$ if and only if for any $x \in S \setminus [F(\bar{x})]^{H^1}_{\preceq}$, we have $F(x) \not\preceq F(\bar{x})$.*

Proof. First note that $x \in S \setminus [F(\bar{x})]^{H^1}_{\preceq}$ means that $x \in S$ such that $F(x) \not\preceq F(\bar{x})$ or $F(\bar{x}) \not\preceq F(x) + H$. Let $\bar{x} \in S$ be an H^1-approximate minimal solution of the problem (2) w.r.t. $\preceq$. Then we must consider two cases:
Case 1: For $x \in S$ and $F(x) \not\preceq F(\bar{x})$, there is nothing left to show.
Case 2: For $x \in S$ and $F(\bar{x}) \not\preceq F(x) + H$, we obtain $F(x) \not\preceq F(\bar{x})$ due to $\bar{x}$'s H^1-approximate minimality, as desired.

Conversely, assume that for all $x \in S \setminus [F(\bar{x})]^{H^1}_{\preceq}$, $F(x) \not\preceq F(\bar{x})$ holds true. Suppose, by contradiction, that $\bar{x}$ is not an H^1-approximate minimal solution of the problem (2) w.r.t. $\preceq$. This implies the existence of some $x \in S$ with the properties $F(x) \preceq F(\bar{x})$ and $F(\bar{x}) \not\preceq F(x)$, in contradiction to the assumption. □

Now we consider a functional $g^{H^1} \colon S \times S \to \mathbb{R} \cup \{\pm\infty\}$ with the property

$$\forall\, x, \bar{x} \in S: \quad g^{H^1}(x, \bar{x}) \leq 0 \iff F(x) \preceq F(\bar{x}).$$

Then we have the following characterization for H^1-approximate minimal solution of the problem (2) w.r.t. $\preceq$.

Theorem 4. *$\bar{x} \in S$ is an H^1-approximate minimal solution of the problem (2) w.r.t. $\preceq$ if and only if the following system (in the unknown x)*

$$g^{H^1}(x, \bar{x}) \leq 0,\ x \in S \setminus [F(\bar{x})]^{H^1}_{\preceq},$$

is impossible.

Proof. First note that due to Proposition 2, $\bar{x} \in S$ is an H^1-approximate minimal solution of the problem (2) w.r.t. $\preceq$ if and only if for $x \in S \setminus [F(\bar{x})]^{H^1}_{\preceq}$, we have $F(x) \not\preceq F(\bar{x})$. Furthermore, we have

$$g^{H^1}(x,\bar{x}) \leq 0,\ x \in S \setminus [F(\bar{x})]^{H^1}_{\preceq} \text{ is impossible}$$
$$\iff \nexists x \in S \setminus [F(\bar{x})]^{H^1}_{\preceq} : g^{H^1}(x,\bar{x}) \leq 0$$
$$\iff \forall x \in S \setminus [F(\bar{x})]^{H^1}_{\preceq} : g^{H^1}(x,\bar{x}) > 0$$
$$\iff \forall x \in S \setminus [F(\bar{x})]^{H^1}_{\preceq} : F(x) \not\preceq F(\bar{x}).$$

□

In a similar manner as Proposition 2 and Theorem 4, one can verify the following results. For this, we assume that we are given a functional $g^{H^2}: S \times S \to \mathbb{R} \cup \{\pm\infty\}$ with the property

$$\forall x, \bar{x} \in S: \quad g^{H^2}(x,\bar{x}) \leq 0 \iff F(x) + H \preceq F(\bar{x}).$$

Proposition 3. $\bar{x} \in S$ is an H^2-approximate minimal solution of the problem (2) w.r.t. $\preceq$ if and only if for any $x \in S \setminus [F(\bar{x})]^{H^2}_{\preceq}$, we have $F(x) + H \not\preceq F(\bar{x})$.

Theorem 5. $\bar{x} \in S$ is an H^2-approximate minimal solution of the problem (2) w.r.t. $\preceq$ if and only if the following system (in the unknown x)

$$g^{H^2}(x,\bar{x}) \leq 0,\ x \in S \setminus [F(\bar{x})]^{H^2}_{\preceq},$$

is impossible.

Let us now consider problem (2) with the set relation $\preceq = \preceq^u_C$. Motivated by Theorem 3 and Corollary 4 above, we consider the functionals $g^{H^i}_u : S \times S \to \mathbb{R} \cup \{\pm\infty\}$ ($i = 1, 2$) defined by

$$g^{H^1}_u(x,\bar{x}) := \sup_{y \in F(x)} \inf_{\bar{y} \in F(\bar{x})} z^{C,k}(y - \bar{y}),$$

$$g^{H^2}_u(x,\bar{x}) := \sup_{y \in F(x)+H} \inf_{\bar{y} \in F(\bar{x})} z^{C,k}(y - \bar{y}).$$

Assumption 1. *For $C \subseteq Y$, $k \in Y \setminus \{0\}$, and $\bar{x} \in S$ we assume that*

(a-H^1) *C is k-vectorially closed, $[0,+\infty)k + C \subseteq C$, and for all $x \in S \setminus [F(\bar{x})]^{H^1}_{\preceq^u_C}$ and $y \in F(x)$, the infimum $\inf_{\bar{y} \in F(\bar{x})} z^{C,k}(y - \bar{y})$ is attained;*

(a-H^2) *C is k-vectorially closed, $[0,+\infty)k + C \subseteq C$, and for all $x \in S \setminus [F(\bar{x})]^{H^2}_{\preceq^u_C}$ and $y \in F(x) + H$, the infimum $\inf_{\bar{y} \in F(\bar{x})} z^{C,k}(y - \bar{y})$ is attained;*

(b) *$(-\infty,0)k - \mathrm{vcl}_k\, C \subseteq -C$ and $\mathrm{vcl}_{-k}(F(\bar{x}) - C) \subseteq F(\bar{x}) - C$.*

We next present a sufficient and necessary condition for H^1-approximate minimal solutions of the problem (2) w.r.t. the relation $\preceq^u_C$.

Corollary 8. *Let Assumption 1 (a-H^i) or (b) be satisfied. Then $\bar{x} \in S$ is an H^i-approximate minimal solution ($i = 1, 2$) of the problem (2) w.r.t. $\preceq^u_C$ if and only if the following system (in the unknown x)*

$$g^{H^i}_u(x,\bar{x}) \leq 0,\ x \in S \setminus [F(\bar{x})]^{H^i}_{\preceq^u_C},$$

is impossible.

Proof. The proof follows by Theorems 2, 3, 4 and 5. □

Furthermore, let us consider problem (2) with $\preceq = \preceq_C^l$. We define the functions $g_l^{H^i}: S \times S \to \mathbb{R} \cup \{\pm\infty\}$ for $i = 1, 2$ by

$$g_l^{H^1}(x, \bar{x}) := \sup_{\bar{y} \in F(\bar{x})} \inf_{y \in F(x)} z^{C,k}(y - \bar{y})$$

$$g_l^{H^2}(x, \bar{x}) := \sup_{\bar{y} \in F(\bar{x})} \inf_{y \in F(x) + H} z^{C,k}(y - \bar{y}).$$

Assumption 2. *For $C \subseteq Y, k \in Y \setminus \{0\}$, and $\bar{x} \in S$ we assume that*

(a-H^1) *C is k-vectorially closed, $[0, +\infty)k + C \subseteq C$, and for all $x \in S \setminus [F(\bar{x})]_{\preceq_C^l}^{H^1}$ and $\bar{y} \in F(\bar{x})$, the infimum $\inf_{y \in F(x)} z^{C,k}(y - \bar{y})$ is attained;*

(a-H^2) *C is k-vectorially closed, $[0, +\infty)k + C \subseteq C$, and for all $x \in S \setminus [F(\bar{x})]_{\preceq_C^l}^{H^2}$ and $\bar{y} \in F(\bar{x})$, the infimum $\inf_{y \in F(x) + H} z^{C,k}(y - \bar{y})$ is attained;*

(b) *$(-\infty, 0)k - \text{vcl}_k C \subseteq -C$ and for all $x \in S$: $\text{vcl}_{-k}(-F(x) - C) = -F(x) - C$.*

In the following, we present a sufficient and necessary condition for H^i-approximate minimal solutions of the problem (2) w.r.t. $\preceq_C^l$.

Corollary 9. *Let Assumption 2 (a-H^i) or (b) be satisfied. Then $\bar{x}$ is an H^i-approximate minimal solution ($i = 1, 2$) of the problem (2) w.r.t. $\preceq_C^l$ if and only if the following system (in the unknown x)*

$$g_l^{H^1}(x, \bar{x}) \leq 0, \ x \in S \setminus [F(\bar{x})]_{\preceq_C^l}^{H^i},$$

is impossible.

Proof. The proof follows by Corollaries 3 and 4 as well as Theorems 4 and 5. □

Finally, we have the following result for H^i-approximate minimal solutions of the problem (2) w.r.t. $\preceq_C^s$.

Corollary 10. *Let $i \in \{1, 2\}$ and suppose that Assumptions 1 (a-H^i) and 2 (a-H^i) or Assumptions 1 (b) and 2 (b) are satisfied for the same $k \in Y \setminus \{0\}$. Then $\bar{x}$ is an H^i-approximate minimal solution of the problem (2) w.r.t. $\preceq_C^s$ if and only if the following system (in the unknown x):*

$$g_u^{H^i}(x, \bar{x}) \leq 0 \text{ and } g_l^{H^i}(x, \bar{x}) \leq 0, \ x \in S \setminus \left([F(\bar{x})]_{\preceq_C^l}^{H^i} \cup [F(\bar{x})]_{\preceq_C^l}^{H^i}\right),$$

is impossible.

4. Numerical Procedure for Computing H^i-Approximate Minimal Elements of a Family of Finitely Many Elements

Finding H^i-approximate minimal elements of a family of finitely many elements of $\mathcal{P}(Y)$ is very important. A first approach to deriving and implementing numerical methods for obtaining H^i-approximate minimal elements has been presented in [38] for the lower set less relation $\preceq_C^l$. The assumption that the given family is finitely valued is oftentimes not a restriction, as many continuous set optimization problem can be appropriately discretized, see the discussion in [39] and the theoretical investigations for linear programs [40] as well as the numerical studies in [41]. In this section, we propose numerical methods for obtaining approximate minimal elements as proposed in Definition 5 for general set relations under suitable assumptions.

Please note that the following algorithms can be found in [38] for the specific case that the set relation is equal to $\preceq_C^l$. We present them here for general set relations $\preceq$. The following algorithm is an extension of the so-called *Graef-Younes method* [42,43] and it is useful for sorting out elements which do not belong to the set of H^i–approximate minimal elements.

Algorithm 1: (Method for sorting out elements of a family of finitely many sets which are not H^1- (H^2-, H^3-, respectively) approximate minimal elements).

Input: $\mathcal{A} := \{A_1, \ldots, A_m\}$, set relation $\preceq$, $H \in \mathcal{P}(Y)$
% initialization
$\mathcal{T} := \{A_1\}$,
% iteration loop
for $j = 2 : 1 : m$ do
   if $\left(A \preceq A_j, A \in \mathcal{T} \implies A_j \preceq A + H\right)$
     $\left((A + H \preceq A_j, A \in \mathcal{T} \implies A_j \preceq A + H), \text{respectively}\right)$,
     $\left(A + H \not\preceq \overline{A}, A \in \mathcal{T}, \text{respectively}\right)$, then
     $\mathcal{T} := \mathcal{T} \cup \{A_j\}$
   end if
end for
Output: $\mathcal{T}$

Remark 2. 1.   *Please note that the if-condition in Algorithm 1 is usually not implemented straightforwardly but instead an additional loop over the elements of the set $\mathcal{T}$ is performed. We nevertheless use the above notation of this step to be consistent with the literature on algorithms of Graef-Younes type.*
2. *Note also that the if-condition describes approximate minimality in the set $\mathcal{T}$. Therefore, Definition 6 does not have to be applied to the whole set $\mathcal{A}$, but to a smaller set $\mathcal{T}$, which can drastically reduce the numerical effort. In this way, non-approximate minimal elements can be eliminated from the set $\mathcal{A}$, as the following theorem shows.*

Theorem 6. 1.   *Algorithm 1 is well-defined.*
2. *Algorithm 1 generates a nonempty set $\mathcal{T} \subseteq \mathcal{A}$.*
3. *Every H^1- (H^2-, H^3-, respectively) approximate minimal element of $\mathcal{A}$ w. r. t. $\preceq$ also belongs to the set $\mathcal{T}$ generated by Algorithm 1.*

Proof. The statements 1 and 2 are easily checked (We loop over a finite number of elements, all the necessary comparisons are well-defined and after the first step, the set $\mathcal{T}$ already consists of an element.) and therefore, their proofs are omitted. Now let A_j be an H^1- (H^2-, H^3-, respectively) approximate minimal element of $\mathcal{A}$. Then we have

$$A \preceq A_j, A \in \mathcal{A} \implies A_j \preceq A + H$$
$$(A + H \preceq A_j, A \in \mathcal{A} \implies A_j \preceq A + H, \text{respectively}),$$
$$(A + H \not\preceq \overline{A}, A \in \mathcal{A}, \text{respectively}).$$

Because of $\mathcal{T} \subseteq \mathcal{A}$, by the above implications we directly obtain

$$A \preceq A_j, A \in \mathcal{T} \implies A_j \preceq A + H$$
$$(A + H \preceq A_j, A \in \mathcal{T} \implies A_j \preceq A + H, \text{respectively}),$$
$$(A + H \not\preceq A_j, A \in \mathcal{T}, \text{respectively}).$$

which verifies that the if-condition in Algorithm 1 is satisfied and A_j is added to $\mathcal{T}$. □

After the application of Algorithm 1 we have only created a smaller set $\mathcal{T}$ containing all the approximate minimal elements of the original family of sets. To filter out solely the approximate minimal elements, another step is required which we handle in the following algorithm:

Algorithm 2: (Method for finding H^1- (H^2-, H^3-, respectively) approximate minimal elements of a family $\mathcal{A}$ of finitely many sets).

Input: $\mathcal{A}^* := \{A_1, \ldots, A_m\}$, set relation $\preceq$, $H \in \mathcal{P}(Y)$
% initialization
$\mathcal{T} := \{A_1\}$
% forward iteration loop
for $j = 2 : 1 : m$ do
    if $\left(A \preceq A_j, A \in \mathcal{T} \implies A_j \preceq A + H\right)$
    $\left(\left(A + H \preceq A_j, A \in \mathcal{T} \implies A_j \preceq A + H\right),\right.$
    $\left.\left(A + H \not\preceq A_j, A \in \mathcal{T}, \text{respectively}\right)\right)$, then
        $\mathcal{T} := \mathcal{T} \cup \{A_j\}$
    end if
end for
$\{A_1, \ldots, A_p\} := \mathcal{T}$
$\mathcal{U} := \{A_p\}$
% backward iteration loop
for $j = p - 1 : -1 : 1$ do
    if $\left(A \preceq A_j, A \in \mathcal{U} \implies A_j \preceq A + H\right)$
    $\left(\left(A + H \preceq A_j, A \in \mathcal{U} \implies A_j \preceq A + H\right),\right.$
    $\left.\left(A + H \not\preceq A_j, A \in \mathcal{U}, \text{respectively}\right)\right)$, then
        $\mathcal{U} := \mathcal{U} \cup \{A_j\}$
    end if
end for
Output: $\mathcal{U}$
$\{A_1, \ldots, A_q\} := \mathcal{U}$
$\mathcal{V} := \emptyset$
% final comparison
for $j = 1 : 1 : q$ do
    if $\left(A \preceq A_j, A \in \mathcal{A} \setminus \mathcal{U} \implies A_j \preceq A + H\right)$
    $\left(\left(A + H \preceq A_j, A \in \mathcal{A} \setminus \mathcal{U} \implies A_j \preceq A + H\right), \text{respectively}\right),$
    $\left(A + H \not\preceq A_j, A \in \mathcal{A} \setminus \mathcal{U}, \text{respectively}\right)$, then
        $\mathcal{V} := \mathcal{V} \cup \{A_j\}$
    end if
end for
Output: $\mathcal{V}$

Remark 3. 1. Again, for determining whether the implications in the definition of minimality are fulfilled, one must loop over the elements of the sets of $\mathcal{T}, \mathcal{U}$ and $\mathcal{A} \setminus \mathcal{U}$, resp.
2. Please note that we formulated Algorithm 2 to have two outputs $\mathcal{U}$ and $\mathcal{V}$. For practical purposes it would suffice to use $\mathcal{V}$ which in fact contains all the approximate minimal elements and no more. However, the theoretical investigations below show that the set $\mathcal{U}$ is in its own right interesting to be examined further.

We start the investigation of the above algorithms for the (arguably simplest) case of H^3-approximate minimality. The following result shows that every element of the set $\mathcal{U}$ is an H^3-approximate minimal element of $\mathcal{U}$ w.r.t. $\preceq$ (but not necessarily an H^3-approximate minimal element of the set $\mathcal{A}$).

Lemma 1. *Every element of $\mathcal{U}$ generated by Algorithm 2 after the backward iteration is also an H^3-approximate minimal element of $\mathcal{U}$ w.r.t. $\preceq$.*

Proof. Let $A_j \in \mathcal{U} = \{A_1, \ldots, A_q\}$. By the forward iteration, we obtain

$$\forall i < j \, (i \geq 1): \; A_i + H \npreceq A_j.$$

The backward iteration yields

$$\forall i > j \, (i \leq q): \; A_i + H \npreceq A_j.$$

This means that

$$\forall i \neq j \, (1 \leq i \leq q): \; A_i + H \npreceq A_j,$$

which is equivalent to

$$\forall A_i \in \mathcal{U} \setminus \{A_j\}: \; A_i + H \npreceq A_j.$$

This is the definition of an H^3-approximate minimal element of $\mathcal{U}$ w.r.t. $\preceq$. □

Theorem 7. *Algorithm 2 generates exactly all H^3-approximate minimal elements of $\mathcal{A}$ w.r.t. $\preceq$ within the set $\mathcal{V}$.*

Proof. Let A_j be an arbitrary element in $\mathcal{V}$. Then $A_j \in \mathcal{U}$, as $\mathcal{V} \subseteq \mathcal{U}$, and due to the third if-statement in Algorithm 2

$$A + H \npreceq A_j, \; A \in \mathcal{A} \setminus \mathcal{U}. \tag{3}$$

Suppose that A_j is not H^3-approximate minimal in $\mathcal{A}$. Then there exists some $A \in \mathcal{A} \setminus A_j$ such that

$$A + H \preceq A_j. \tag{4}$$

If $A \notin \mathcal{U}$, then this is a contradiction to (3). If $A \in \mathcal{U}$, then due to the H^3-approximate minimality of A_j in $\mathcal{U}$ (see Lemma 1), we obtain $A + H \npreceq A_j$, a contradiction to (4).

Conversely, let A_j be H^3-approximate minimal in $\mathcal{A}$. This means, by definition that

$$A + H \npreceq A_j, \; A \in \mathcal{A} \setminus A_j.$$

Now let us assume, by contradiction, that $A_j \notin \mathcal{V}$. Then, there exists some $A \in \mathcal{A} \setminus \mathcal{U}$ with $A + H \preceq A_j$, a contradiction. □

To obtain similar results as in Lemma 1 and Theorem 7 for H^1- (H^2-, respectively) approximate minimal elements of $\mathcal{U}$ w.r.t. $\preceq$, we need the following assumption.

Assumption 3. *Suppose that one of the following conditions holds:*

1. *The set relation $\preceq$ is irreflexive.*
2. *The set relation $\preceq$ is reflexive and for every $A \in \mathcal{A}$, $A \preceq A + H$.*

Assumption 4. *Suppose that for all $A \in \mathcal{A}$, it we have $A + H \not\preceq A$ or $A \preceq A + H$.*

Below we give some examples of set relations that fulfill the above assumptions.

Example 1. 1. Consider the *certainly less* relation, which is defined as (see ([32], Definition 3.12))

$$A \preceq_C^{cert} B \iff \forall a \in A, \forall b \in B : a \in b - C,$$

where $C \in \mathcal{P}(Y)$. Then $\preceq_C^{cert}$ is irreflexive if C is pointed, i.e., $C \cap (-C) = \emptyset$ (hence, $0 \notin C$).
2. Let us recall the *possibly less* relation, given as (compare [32,37,44])

$$A \preceq_C^p B \iff \exists a \in A, \exists b \in B : a \in b - C,$$

where $C \in \mathcal{P}(Y)$ such that $0 \in C$. Then $\preceq_C^p$ is reflexive. If C is a convex cone with $H \subseteq C$, then $A \preceq_C^p A + H$ for all $A \in \mathcal{A}$.
3. If C is a convex cone with $0 \in C$ and $H \subseteq -C$, then $A \preceq_C^u A + H$ holds true for all $A \in \mathcal{A}$.

Lemma 2. *Let Assumption 3 (Assumption 4, respectively) be fulfilled. Then every element of $\mathcal{U}$ generated by Algorithm 2 is also an H^1- (H^2-, respectively) approximate minimal element of $\mathcal{U}$ w.r.t. $\preceq$.*

Proof. Let $A_j \in \mathcal{U} = \{A_1, \ldots, A_q\}$. By the forward iteration, we obtain

$$\forall i < j \, (i \geq 1) : A_i \preceq A_j \implies A_j \preceq A_i + H, \tag{5}$$

$$\left(\forall i < j \, (i \geq 1) : A_i + H \preceq A_j \implies A_j \preceq A_i + H, \text{ respectively} \right). \tag{6}$$

The backward iteration yields (5) ((6), respectively) for every $i > j$ ($i \leq q$). Together, this means that

$$\forall i \neq j : A_i \preceq A_j \implies A_j \preceq A_i + H, \tag{7}$$

$$\left(\forall i \neq j : A_i + H \preceq A_j \implies A_j \preceq A_i + H, \text{ respectively} \right). \tag{8}$$

Since the set relation is, due to Assumption 3 either irreflexive or reflexive and for every $A \in \mathcal{A}$, $A \preceq A + H$, (7) is equivalent to the implication given in Definition 6 (a), and hence, A_j is an H^1–approximate minimal element of $\mathcal{U}$ w.r.t. $\preceq$. Similarly, according to Assumption 4, it holds for all $A \in \mathcal{A}$ $A + H \not\preceq A$ or $A \preceq A + H$. With this in mind, the implication (8) coincides with Definition 6 (b), and hence, $A_j \in \mathcal{U}_{H^2}$. □

Theorem 8. *Let Assumption 3 (Assumption 4, respectively) be fulfilled. Then Algorithm 2 generates exactly all H^1- (H^2-, respectively) approximate minimal elements of $\mathcal{A}$ w.r.t. $\preceq$.*

Proof. Let A_j be an arbitrary element in $\mathcal{V}$. Then $A_j \in \mathcal{U}$, as $\mathcal{V} \subseteq \mathcal{U}$, and due to the third if-statement in Algorithm 2

$$A \preceq A_j, \, A \in \mathcal{A} \setminus \mathcal{U} \implies A_j \preceq A + H, \tag{9}$$

$$(A + H \preceq A_j, \, A \in \mathcal{A} \setminus \mathcal{U} \implies A_j \preceq A + H, \text{ respectively}). \tag{10}$$

Suppose that A_j is not H^1- (H^2-, respectively) approximate minimal in $\mathcal{A}$. Then there exists some $A \in \mathcal{A}$ such that

$$A \preceq A_j \text{ and } A_j \not\preceq A + H, \tag{11}$$

$$(A + H \preceq A_j \text{ and } A_j \not\preceq A + H, \text{ respectively}) \tag{12}$$

If $A \notin \mathcal{U}$, then this is a contradiction to (9) ((10), respectively). If $A \in \mathcal{U}$, then $A_j \preceq A + H$, as A_j is H^1- (H^2-, respectively) approximate minimal in $\mathcal{U}$ according to Lemma 2. But this contradicts the implication (11) ((12), respectively).

Conversely, let A_j be an H^1- (H^2-, respectively) approximate minimal element in the set $\mathcal{A}$, i.e.,

$$\begin{aligned} A \preceq A_j, A \in \mathcal{A} &\implies A_j \preceq A + H, \\ (A + H \preceq A_j, A \in \mathcal{A} &\implies A_j \preceq A + H, \text{ respectively}). \end{aligned} \tag{13}$$

Now let us assume, by contradiction, that $A_j \notin \mathcal{V}$. Then, there exists some $A \in \mathcal{A} \setminus \mathcal{U}$ with $A \preceq A_j$ ($A + H \preceq A_j$, respectively), but $A_j \npreceq A + H$, a contradiction to (13). □

To illustrate the algorithms, we will apply the forward and backward iteration for a rather academic example in $\mathbb{R}^2$. Note, however, that its (even computerized) application is not limited to these finite-dimensional structures as the algorithms are based on elementary finite iteration loops. So, once a way has been established to numerically assert the relation $A \preceq B$ for two sets A and B out of a certain family of sets, the algorithms can directly be applied. For the case of polyhedral sets, such a comparison principle has, for example, been established in [45] and similar computational approaches were developed in [46].

Example 2. *For this example, let $C := \mathbb{R}^2_+$, $\preceq := \preceq^{cert}_C$ and $H = \{(1,1)^T\}$. As the family of sets $\mathcal{A}$, we have randomly computed 1000 sets, for easy comparison each set is a ball of radius one in $\mathbb{R}^2$. We are interested in the H^2-approximate minimal elements of the set $\mathcal{A}$ and make use of Algorithm 2 to obtain those. Notice that Assumption 4 is trivially fulfilled. Out of the 1000 sets, a total number of 177 are H^2-approximate minimal w.r.t. to $\preceq$. Algorithm 2 generates at first 189 sets in $\mathcal{T}$; then, 177 sets are collected within the set $\mathcal{U}$ and $\mathcal{V}$. We used the same data as in Example 4.7 and 4.14 from [32], and according to our earlier results, a total number of 93 elements are minimal. In Figure 2, the sets within $\mathcal{T}$ are the lightly and darkly filled circles, while the H^2-approximate minimal elements of the set $\mathcal{A}$ (that is, the sets in $\mathcal{U}$ and $\mathcal{V}$) are the darkly filled circles. For comparison, Algorithm 2 is also used on the same family of sets with $H = \{(0,0)^T\}$ (see ([32], Example 4.7 and 4.14)), with 103 sets within $\mathcal{T}$ and 93 sets within $\mathcal{U}$ and $\mathcal{V}$, see Figure 3. Let us note that this example is chosen to illustrate the efficiency of Algorithm 2 as it is to be expected for problems with relatively homogeneous distribution of set size and structure, see the according discussion in the vector-valued case [15,43].*

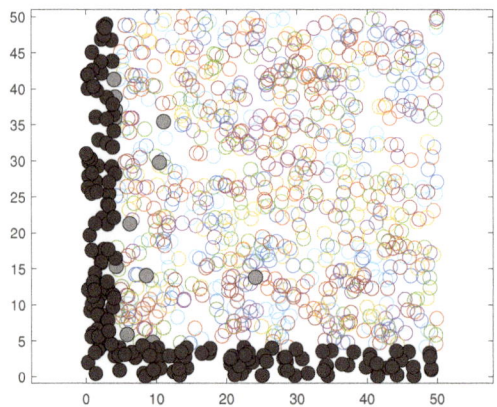

Figure 2. A randomly generated family of sets. The lightly and darkly filled circles belong to the set $\mathcal{T}$ generated by Algorithm 2, while the H^2-approximate minimal elements of the set $\mathcal{A}$ are exactly the darkly filled circles (see Example 2).

Of course, the notion of approximate minimality makes sense when minimal elements do not exist (in the vector-valued case, this can happen when the set of feasible elements in the objective space is open). In the future, we will study continuity notions of set-valued mappings that appear in set optimization problems and investigate existence results.

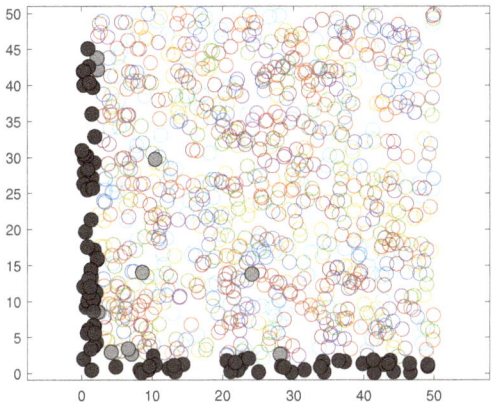

Figure 3. The randomly generated family of sets from Example 2 with $H = \{(0,0)^T\}$, i.e., we do not consider approximate minimal elements here, but look for the minimal elements of the family of sets $\mathcal{A}$. The lightly and darkly filled circles belong to the set $\mathcal{T}$ generated by Algorithm 2, while the minimal elements of the set $\mathcal{A}$ are the darkly filled circles.

5. Conclusions

This paper investigates different kinds of approximate minimal solutions of set optimization problems. In particular, we present an inequality approach to characterize these approximate minimal solutions by means of a prominent scalarizing functional. To be as general as possible, our analysis is developed in real linear spaces without assuming any topology on the spaces and therefore bases only on algebraic relations and set inclusions between all the involved quantities. It would be interesting to study whether different scalarizing functionals may be used for a similar analysis as the separation functionals of Tammer–Weidner type have recently been embedded into a larger class of functionals [47]. We have proposed effective algorithms that select approximate minimal elements out of a family of finitely many sets. As a next step, it will be necessary to test our algorithms on practical examples.

Author Contributions: Conceptualization, E.K., M.A.K. and X.Q.; Methodology, E.K., M.A.K. and X.Q.; Software, E.K. and M.A.K.; Investigation, E.K., M.A.K. and X.Q.; Writing–original draft preparation, E.K. and M.A.K.; writing–review and editing, E.K., M.A.K. and X.Q.; visualization, E.K. and M.A.K. All authors have read and agreed to the published version of the manuscript.

Funding: This research received no external funding.

Conflicts of Interest: The authors declare no conflict of interest.

References

1. Günther, C.; Köbis, E.; Popovici, N. Computing minimal elements of finite families of sets w.r.t. preorder relations in set optimization. *J. Appl. Numer. Optim.* **2019**, *1*, 131–144.
2. Klamroth, K.; Köbis, E.; Schöbel, A.; Tammer, C. A unified approach for different concepts of robustness and stochastic programming via nonlinear scalarizing functionals. *Optimization* **2013**, *62*, 649–671. [CrossRef]

3. Klamroth, K.; Köbis, E.; Schöbel, A.; Tammer, C. A unified approach to uncertain optimization. *Eur. J. Oper. Res.* **2017**, *260*, 403–420. [CrossRef]
4. Li, J.; Tammer, C. Set optimization problems on ordered sets. *Appl. Set-Valued Anal. Optim.* **2019**, *1*, 77–94.
5. Khan, A.; Tammer, C.; Zălinescu, C. *Set-Valued Optimization—An Introduction with Applications*; Springer: Berlin/Heidelberg, Germany, 2015.
6. Bischoff, M.; Jahn, J.; Köbis, E. Hard uncertainties in multiobjective layout optimization of photovoltaic power plants. *Optimization* **2017**, *66*, 361–380. [CrossRef]
7. Moore, R.E. *Interval Analysis*; Prentice & Hall: New York, NY, USA, 1966.
8. Ansari, Q.H.; Köbis, E.; Sharma, P.K. Characterizations of Set Relations with respect to Variable Domination Structures via Oriented Distance Function. *Optimization* **2018**, *69*, 1389–1407. [CrossRef]
9. Chen, J.; Ansari, Q.H.; Yao, J.-C. Characterization of set order relations and constrained set optimization problems via oriented distance function. *Optimization* **2017**, *66*, 1741–1754. [CrossRef]
10. Köbis, E.; Le, T.T.; Tammer, C.; Yao, J.-C. A New Scalarizing Functional in Set Optimization with respect to Variable Domination Structures. *Appl. Anal. Optim.* **2017**, *1*, 311–326.
11. Köbis, E.; Tammer, C. Characterization of Set Relations by Means of a Nonlinear Scalarization Functional. In *Modelling, Computation and Optimization in Information Systems and Management Sciences*; Le Thi, H., Pham Dinh, T., Nguyen, N., Eds.; Volume 359 of Advances in Intelligent Systems and Computing; Springer: Cham, Switzerland, 2015; pp. 491–503.
12. Adán, M.; Novo, V. Proper efficiency in vector optimization on real linear spaces. *J. Optim. Theor. Appl.* **2004**, *121*, 515–540. [CrossRef]
13. Adán, M.; Novo, V. Duality and saddle-points for convex-like vector optimization problems on real linear spaces. *TOP* **2005**, *13*, 343–357. [CrossRef]
14. Hebestreit, N.; Köbis, E. Representation of Set Relations in Real Linear Spaces. *J. Nonlinear Convex Anal.* **2018**, *19*, 287–296.
15. Jahn, J. *Vector optimization—Theory, Applications, and Extensions*; Springer: Berlin/Heidelberg, Germany, 2011.
16. Kiyani, E.; Soleimanin-Damaneh, M. Approximate proper efficiency on real linear vector spaces. *Pac. J. Optim.* **2014**, *10*, 715–734.
17. Hernández, E.; Jiménez, B.; Novo, V. *Benson-Type Proper Efficiency in Set-Valued Optimization on Real Linear Spaces*; Lecture Notes in Econom. and Math. Systems 563; Springer: Berlin, Germany, 2006; pp. 45–59.
18. Hernández, E.; Jiménez, B.; Novo, V. Weak and proper efficiency in set-valued optimization on real linear spaces. *J. convex Anal.* **2007**, *14*, 275–296.
19. Qui, J.H.; He, F. A general vectorial Ekeland's variational principle with a P-distance. *Acta Math. Sin.* **2013**, *29*, 1655–1678.
20. Zhou, Z.A.; Peng, J.W. Scalarization of set-valued optimization problems with generalized cone subconvexlikeness in real ordered linear spaces. *J. Optim. Theory Appl.* **2012**, *154*, 830–841. [CrossRef]
21. Zhou, Z.A.; Yang, X.-M. Scalarization of ϵ-super efficient solutions of set-valued optimization problems in real ordered linear spaces. *J. Optim. Theory Appl.* **2014**, *162*, 680–693. [CrossRef]
22. Gutiérrez, C.; Novo, V.; Ródenas-Pedregosa, J.L.; Tanaka, T. Nonconvex Separation Functional in Linear Spaces with Applications to Vector Equilibria. *SIAM J. Optim.* **2016**, *26*, 2677–2695. [CrossRef]
23. Gerstewitz, C. Nichtkonvexe Dualitität in der Vektoroptimierung. *Wiss. Zeitschr. -Leuna-Mersebg.* **1983**, *25*, 357–364.
24. Gutiérrez, C.; Jiménez, B.; Miglierina, E.; Molho, E. Scalarization in set optimization with solid and nonsolid ordering cones. *J. Global Optim.* **2015**, *61*, 525–552. [CrossRef]
25. Krasnosel'skiĭ, M.A. *Positive Solutions of Operator Equations*; Translated from Russian by Richard E. Flaherty; Leo, F., Boron, P., Eds.; Noordhoff Ltd.: Groningen, The Netherlands, 1964.
26. Rubinov, A.M. Sublinear operators and their applications. *Russ. Math. Surv.* **1977**, *32*, 115–175. [CrossRef]
27. Heyde, F. Coherent risk measures and vector optimization. In *Multicriteria Decision Making and Fuzzy Systems*; Küfer, K.-H., Ed.; Theory, Methods and Applications; Shaker Verlag Aachen: Düren, Germany, 2006; pp. 3–12.
28. Gerth, C.; Weidner, P. Nonconvex separation theorems and some applications in vector optimization. *J. Optim. Theory Appl.* **1990**, *67*, 297–320. [CrossRef]
29. Göpfert, A.; Tammer, C.; Riahi, H.; Zălinescu, C. *Variational Methods in Partially Ordered Spaces*; Springer: Berlin, Germany, 2003.

30. Kuroiwa, D. The natural criteria in set-valued optimization, Sūrikaisekikenkyūsho Kōkyūroku. *Res. Nonlinear Anal. Convex Anal.* **1997**, *1031*, 85–90.
31. Kuroiwa, D. Some duality theorems of set-valued optimization with natural criteria. In Proceedings of the International Conference on Nonlinear Analysis and Convex Analysis, Niigata, Japan, 28–31 July 1998; Tanaka, T., Ed.; World Scientific River Edge: Singapore, 1999; pp. 221–228.
32. Köbis, E.; Kuroiwa, D.; Tammer, C. Generalized set order relations and their numerical treatment. *Appl. Anal. Optim.* **2017**, *1*, 45–65.
33. Köbis, E.; Köbis, M.A. Treatment of set order relations by means of a nonlinear scalarization functional: A full characterization. *Optimization* **2016**, *65*, 1805–1827. [CrossRef]
34. Zeidler, E. *Nonlinear Functional Analysis and its Applications. Part I: Fixed-Point Theorems*; Springer: New York, NY, USA, 1986.
35. Nishnianidze, Z.G. Fixed points of monotone multivalued operators. *Soobshch. Akad. Nauk Gruzin. SSR* **1984**, *114*, 489–491.
36. Young, R.C. The algebra of many-valued quantities. *Math. Ann.* **1931**, *104*, 260–290. [CrossRef]
37. Jahn, J.; Ha, T.X.D. New order relations in set optimization. *J. Optim. Theory Appl.* **2011**, *148*, 209–236. [CrossRef]
38. Gutiérrez, C.; Huerga, L.; Köbis, E.; Tammer, C. Approximate Solutions of Set-Valued Optimization Problems Using Set-Criteria. *Appl. Anal. Optim.* **2017**, *1*, 477–500.
39. Ben-Tal, A.; El Ghaoui, L.; Nemirovski, A. *Robust Optimization*; Princeton University Press: Princeton, NJ, USA, 2009.
40. Fischetti, M.; Monaci, M. Cutting plane versus compact formulations for uncertain (integer) linear programs. *Math. Prog. Comp.* **2012**, *4*, 239–273. [CrossRef]
41. Bertsimas, D.; Dunning, I.; Lubin, M. Reformulation versus cutting-planes for robust optimization. *Comp. Manag. Sci.* **2015**, *13*, 195–217. [CrossRef]
42. Jahn, J.; Rathje, U. Graef-Younes Method with Backward Iteration. In *Multicriteria Decision Making and Fuzzy Systems—Theory, Methods and Applications*; Küfer, K.H., Rommelfanger, H., Tammer, C., Winkler, K., Eds.; Shaker Verlag Aachen: Düren, Germany, 2006; pp. 75–81.
43. Younes, Y.M. Studies on Discrete Vector Optimization. Ph.D. Thesis, University of Demiatta, Demiatta, Egypt, 1993.
44. Chiriaev, D.; Walster, G.W. *Interval Arithmetic Specification*; Technical Report; International Committee for Information Technology Standards (INCITS): Washington, DC, USA, 1998.
45. Jahn, J. Vectorization in set optimization. *J. Optim. Theory Appl.* **2015**, *167*, 783–795. [CrossRef]
46. Schrage, C.; Löhne, A. An algorithm to solve polyhedral convex set optimization problems. *Optimization* **2013**, *62*, 131–141. [CrossRef]
47. Quintana, E.; Bouza, G.; Tammer, C. A Unified Characterization of Nonlinear Scalarizing Functionals in Optimization. *Vietnam. J. Math.* **2019**, *47*, 683–713.

© 2020 by the authors. Licensee MDPI, Basel, Switzerland. This article is an open access article distributed under the terms and conditions of the Creative Commons Attribution (CC BY) license (http://creativecommons.org/licenses/by/4.0/).

MDPI
St. Alban-Anlage 66
4052 Basel
Switzerland
Tel. +41 61 683 77 34
Fax +41 61 302 89 18
www.mdpi.com

Mathematics Editorial Office
E-mail: mathematics@mdpi.com
www.mdpi.com/journal/mathematics

Lightning Source UK Ltd.
Milton Keynes UK
UKHW052155130922
408735UK00006B/117